CANCER CELL SIGNALING

TARGETING SIGNALING PATHWAYS TOWARD THERAPEUTIC APPROACHES TO CANCER

CANCER CELL SIGNALING

TARGETING SIGNALING PATHWAYS TOWARD THERAPEUTIC APPROACHES TO CANCER

Edited by
Kasirajan Ayyanathan, PhD

Apple Academic Press

TORONTO   NEW JERSEY

Apple Academic Press Inc. | Apple Academic Press Inc.
3333 Mistwell Crescent | 9 Spinnaker Way
Oakville, ON L6L 0A2 | Waretown, NJ 08758
Canada | USA

©2015 by Apple Academic Press, Inc.

First issued in paperback 2021

Exclusive worldwide distribution by CRC Press, a member of Taylor & Francis Group

No claim to original U.S. Government works

ISBN 13: 978-1-77463-083-9 (pbk)
ISBN 13: 978-1-77188-067-1 (hbk)

Library of Congress Control Number: 2014939254

Library and Archives Canada Cataloguing in Publication

Cancer cell signaling: targeting signaling pathways toward therapeutic approaches to cancer/edited by Kasirajan Ayyanathan, PhD.

Includes bibliographical references and index.
ISBN 978-1-77188-067-1 (bound)
1. Cancer cells. 2. Cellular signal transduction. 3. Cancer--Treatment. I. Ayyanathan, Kasirajan, author, editor

RC269.7.C38 2014   616.99'407   C2014-902897-0

Apple Academic Press also publishes its books in a variety of electronic formats. Some content that appears in print may not be available in electronic format. For information about Apple Academic Press products, visit our website at **www.appleacademicpress.com** and the CRC Press website at **www.crcpress.com**

ABOUT THE EDITOR

KASIRAJAN AYYANATHAN, PhD

Kasirajan Ayyanathan received his PhD degree from the Department of Biochemistry, Indian Institute of Science, one of the premier research institutions in India. Subsequently, at Temple University School of Medicine, USA, he conducted post-doctoral research on the signal transduction by purinergic receptors, a class of G-Protein Coupled Receptors (GPCR), in erythroleukemia cancer cells. Next, he was trained as a staff scientist at the Wistar Institute, USA, for almost ten years and studied transcription regulation, chromatin, and epigenetic regulatory mechanisms in cancer before becoming an Associate Professor at Florida Atlantic University (FAU). He is a Research Associate Professor at the Center for Molecular Biology and Biotechnology. Presently, he is a visiting scholar at the Wistar Institute. He is the recipient of the Chern memorial award and the Howard Temin career research award.

Dr. Ayyanathan is well trained in molecular biology, cell biology, biochemistry with main focus on studying transcription factors and gene regulation. He has contributed to several projects such as: generation of conditional transcriptional repressors that are directed against the endogenous oncogenes to inhibit malignant growth, establishment of stable cell lines that express chromatin integrated transcriptional repressors and reporter genes in order to study the epigenetic mechanisms of KRAB repression, and identification of novel SNAG repression domain interacting proteins in order to understand their roles in transcriptional repression and oncogenesis. Dr. Ayyanathan has published several research articles in peer-reviewed articles in these subject areas.

CONTENTS

ACKNOWLEDGMENT AND
HOW TO CITE

The editor and publisher thank each of the authors who contributed to this book, whether by granting their permission individually or by releasing their research as open source articles or under a license that permits free use, provided that attribution is made. The chapters in this book were previously published in various places in various formats. To cite the work contained in this book and to view the individual permissions, please refer to the citation at the beginning of each chapter. Each chapter was read individually and carefully selected by the editor. The result is a book that provides a nuanced study of the topic of cancer cell signaling.

LIST OF CONTRIBUTORS

Carlos L. Arteaga
Division of Hematology-Oncology, Department of Medicine, School of Medicine, Vanderbilt University, 2220 Pierce Avenue, 777 PRB, Nashville, TN 37232-6307, USA, Department of Cancer Biology, Vanderbilt University, 2220 Pierce Avenue, 777 PRB, Nashville, TN 37232-6307, USA, Breast Cancer Research Program, Vanderbilt-Ingram Cancer Center, 2220 Pierce Avenue, 777 PRB, Nashville, TN 37232-6307, USA, and Vanderbilt-Ingram Cancer Center, Vanderbilt University, 2220 Pierce Avenue, 777 PRB, Nashville, TN 37232-6307, USA

Martin Augsten
Institute of Biochemistry II, University Hospital Jena, Jena, Germany and Department of Oncology-Pathology, Karolinska Institutet, 171 76 Stockholm, Sweden

Daniela Aust
Institute of Pathology, University Hospital Dresden, Fetscherstr, 74, 01307 Dresden, Germany

Kasirajan Ayyanathan
Center for Molecular Biology and Biotechnology, Charles E. Schmidt College of Science, Florida Atlantic University, Jupiter, Florida, United States of America and Department of Biological Sciences, Charles E. Schmidt College of Science, Florida Atlantic University, Boca Raton, Florida, United States of America

Meraj Aziz
Clinical Translational Research Division, Translational Genomics Research Institute, Scottsdale, AZ 85259, USA

Sahana Suresh Babu
Centre for Blood Research, Department of Medicine, University of British Columbia, 4306-2350 Health Sciences Mall, V6T 1Z3, BC Vancouver, Canada

Michael T. Barrett
Clinical Translational Research Division, Translational Genomics Research Institute, Scottsdale, AZ 85259, USA

Wei Bao
Department of Obstetrics and Gynecology, International Peace Maternity & Child Health Hospital, Shanghai Jiao Tong University School of Medicine, Hengshan Road, Shanghai, China

Xinnan Bao
Department of Orthopaedics, Changzhou No. 1 People's Hospital, Changzhou, Jiangsu 213003, PR China

Anika Böttcher
Institute of Biochemistry II, University Hospital Jena, Jena, Germany and German Research Center for Environmental Health, Neuherberg, Germany

Fernando Calvo
Tumour Cell Biology Laboratory, Cancer Research UK London Research Institute, London, UK and Tumour Microenvironment Team Division of Cancer Biology, The Institute of Cancer Research, London, UK

Siprachanh Chanthaphaychith
Division of Hematology-Oncology, Department of Medicine, School of Medicine, Vanderbilt University, 2220 Pierce Avenue, 777 PRB, Nashville, TN 37232-6307, USA

Siyuan Chen
Division of Oncology, Department of Pediatric Surgery, West China Hospital of Sichuan University, Chengdu 610041, China and Pediatric Intensive Care Unit, West China Hospital of Sichuan University, Chengdu 610041, China

Zhong Chen
Tumor Biology Section, Head and Neck Surgery Branch, National Institute on Deafness and Other Communication Disorders, National Institutes of Health, Bethesda, Maryland 20892-1419, USA

Edward M. Conway
Centre for Blood Research, Department of Medicine, University of British Columbia, 4306-2350 Health Sciences Mall, V6T 1Z3, BC Vancouver, Canada

Hongwei Cui
Clinical Medicine Research Center of The Affiliated Hospital, Inner Mongolia Medical University, No 1 Tongdao North Street, Huimin District, Hohhot, Inner Mongolia 010050, China

Kimberly Brown Dahlman
Department of Cancer Biology, Vanderbilt University, 2220 Pierce Avenue, 777 PRB, Nashville, TN 37232-6307, USA and Vanderbilt-Ingram Cancer Center, Vanderbilt University, 2220 Pierce Avenue, 777 PRB, Nashville, TN 37232-6307, USA

Ken Dawson-Scully
Center for Molecular Biology and Biotechnology, Charles E. Schmidt College of Science, Florida Atlantic University, Jupiter, Florida, United States of America and Department of Biological Sciences, Charles E. Schmidt College of Science, Florida Atlantic University, Boca Raton, Florida, United States of America

Lianghua Ding
Department of Orthopaedics, Changzhou No. 1 People's Hospital, Changzhou, Jiangsu 213003, PR China

Chao Dong
Clinical Medicine Research Center of The Affiliated Hospital, Inner Mongolia Medical University, No 1 Tongdao North Street, Huimin District, Hohhot, Inner Mongolia 010050, China

Marcos R.H. Estecio
Department of Leukemia, the University of Texas MD Anderson Cancer Center, Houston, TX, USA

Lisa Evers
Clinical Translational Research Division, Translational Genomics Research Institute, Scottsdale, AZ 85259, USA

Karlheinz Friedrich
Institute of Biochemistry II, University Hospital Jena, Jena, Germany

Da-xin Gong
Department of Urology, the First Affiliated Hospital of China Medical University, Shenyang, Liaoning 110001, China

Wenjun Guo
Ruth L and Davis S Gottesman Institute for Stem Cell Biology and Regenerative Medicine, Department of Cell Biology, Albert Einstein College of Medicine, Bronx, NY, USA

Christy R. Hagan
Department of Medicine (Hematology, Oncology, and Transplantation) and the Department of Pharmacology, University of Minnesota, Masonic Cancer Center, 420 Delaware St SE, MMC 806, Minneapolis, MN 55455, USA

Shuanghua He
Department of Orthopaedics, Changzhou No. 1 People's Hospital, Changzhou, Jiangsu 213003, PR China

Xiaoying He
Department of Obstetrics and Gynecology, International Peace Maternity & Child Health Hospital, Shanghai Jiao Tong University School of Medicine, Hengshan Road, Shanghai, China

Tara Holley
Clinical Translational Research Division, Translational Genomics Research Institute, Scottsdale, AZ 85259, USA

Jean-Pierre J. Issa
Department of Leukemia, the University of Texas MD Anderson Cancer Center, Houston, TX, USA and Fels Institute for Cancer Research and Molecular Biology, Temple University, Philadelphia, PA, USA

Jaroslav Jelinek
Department of Leukemia, the University of Texas MD Anderson Cancer Center, Houston, TX, USA

Yi Ji
Division of Oncology, Department of Pediatric Surgery, West China Hospital of Sichuan University, Chengdu 610041, China

Won-Kyoung Kang
Department of Surgery, Seoul St. Mary's Hospital, The Catholic University of Korea, Seoul, Korea

Michelle Kassner
Cancer and Cell Biology Division, Translational Genomics Research Institute, Phoenix, AZ 85004, USA

Shailaja Kesaraju
Center for Molecular Biology and Biotechnology, Charles E. Schmidt College of Science, Florida Atlantic University, Jupiter, Florida, United States of America

Hee-Na Kim
Department of Pathology, Seoul Clinical Laboratory Clinic, Seoul, Korea

Won Ki Kim
Colorectal Cancer Branch, Division of Translational and Clinical Research I, Research Institute, National Cancer Center, Gyeonggi, 410-769, Republic of Korea and Laboratory of Immunology, Department of Biological Science, Sungkyunkwan University, Suwon, 440-746, Republic of Korea

Thomas Knösel
Institute of Pathology, Ludwig-Maximilians-University (LMU), Thalkirchnerstr. 36, 80337 Munich, Germany

Chui-ze Kong
Department of Urology, the First Affiliated Hospital of China Medical University, Shenyang, Liaoning 110001, China

Carol A. Lange
Department of Medicine (Hematology, Oncology, and Transplantation) and the Department of Pharmacology, University of Minnesota, Masonic Cancer Center, 420 Delaware St SE, MMC 806, Minneapolis, MN 55455, USA

Choong-Eun Lee
Laboratory of Immunology, Department of Biological Science, Sungkyunkwan University, Suwon, 440-746, Republic of Korea

Myung Ah Lee
Division of Medical Oncology, Department of Internal Medicine, Cancer Research Institute, College of Medicine, The Catholic University of Korea, Seoul St. Mary's Hospital, 222 Banpo-daero, Seocho-gu, 137-701 Seoul, Korea

Victor Lei
Centre for Blood Research, Department of Medicine, University of British Columbia, 4306-2350 Health Sciences Mall, V6T 1Z3, BC Vancouver, Canada

Elizabeth Lenkiewicz
Clinical Translational Research Division, Translational Genomics Research Institute, Scottsdale, AZ 85259, USA

Kai Li
Division of Oncology, Department of Pediatric Surgery, Children's Hospital of Fudan University, Shanghai 201102, China

Li Li
Laboratory of Pathology, West China Hospital of Sichuan University, Chengdu 610041, China

Shoudan Liang
Department of Bioinformatics, the University of Texas MD Anderson Cancer Center, Houston, TX, USA

Yun Liao
Department of Obstetrics and Gynecology, International Peace Maternity & Child Health Hospital, Shanghai Jiao Tong University School of Medicine, Hengshan Road, Shanghai, China

Jiao Liu
Department of Urology, the First Affiliated Hospital of China Medical University, Shenyang, Liaoning 110001, China

Hai Long
Department of Leukemia, the University of Texas MD Anderson Cancer Center, Houston, TX, USA

Yue Lu
Department of Leukemia, the University of Texas MD Anderson Cancer Center, Houston, TX, USA

Gabriel G. Malouf
Department of Leukemia, the University of Texas MD Anderson Cancer Center, Houston, TX, USA and Department of Medical Oncology, Groupe Hospitalier Pitié - Salpêtrière, Assistance Publique Hopitaux de Paris, Faculty of Medicine Pierre et Marie Curie, Institut Universitaire de Cancérologie, Paris, France

Sendurai A. Mani
Department of Translational Molecular Pathology, The University of Texas MD Anderson Cancer Center, Houston, TX, USA and Metastasis Research Center, The University of Texas MD Anderson Cancer Center, Houston, TX, USA

Alice M. O'Byrne
Centre for Blood Research, Department of Medicine, University of British Columbia, 4306-2350 Health Sciences Mall, V6T 1Z3, BC Vancouver, Canada

Seong-Taek Oh
Department of Surgery, Seoul St. Mary's Hospital, The Catholic University of Korea, Seoul, Korea

Shoghag Panjarian
Fels Institute for Cancer Research and Molecular Biology, Temple University, Philadelphia, PA, USA

Jin-Hee Park
Division of Medical Oncology, Department of Internal Medicine, Cancer Research Institute, College of Medicine, The Catholic University of Korea, Seoul St. Mary's Hospital, 222 Banpo-daero, Seocho-gu, 137-701 Seoul, Korea

Pedro A. Perez-Mancera
CRUK Cambridge Institute, University of Cambridge, Li Ka Shing Centre, Robinson Way, Cambridge CB2 0RE, UK

Christian Pilarsky
Department of Surgery, University Hospital Dresden, Fetscherstr, 74, 01307 Dresden, Germany

Meiting Qiu
Department of Obstetrics and Gynecology, International Peace Maternity & Child Health Hospital, Shanghai Jiao Tong University School of Medicine, Hengshan Road, Shanghai, China

Priyanka Ramachandran
Department of Translational Molecular Pathology, The University of Texas MD Anderson Cancer Center, Houston, TX, USA

Ramesh K. Ramanathan
Clinical Translational Research Division, Translational Genomics Research Institute, Scottsdale, AZ 85259, USA and Virginia G. Piper Cancer Center, Scottsdale Healthcare, Scottsdale, AZ 85258, USA

Noel J-M Raynal
Department of Leukemia, the University of Texas MD Anderson Cancer Center, Houston, TX, USA

Cosima Riemenschnitter
Institute of Surgical Pathology, University Hospital Zurich, Zurich, Switzerland

Brent N. Rexer
Division of Hematology-Oncology, Department of Medicine, School of Medicine, Vanderbilt University, 2220 Pierce Avenue, 777 PRB, Nashville, TN 37232-6307, USA, Department of Cancer Biology, Vanderbilt University, 2220 Pierce Avenue, 777 PRB, Nashville, TN 37232-6307, USA, Breast Cancer Research Program, Vanderbilt-Ingram Cancer Center, 2220 Pierce Avenue, 777 PRB, Nashville, TN 37232-6307, USA, and Vanderbilt-Ingram Cancer Center, Vanderbilt University, 2220 Pierce Avenue, 777 PRB, Nashville, TN 37232-6307, USA

Si Young Rhyu
Division of Medical Oncology, Department of Internal Medicine, Cancer Research Institute, College of Medicine, The Catholic University of Korea, Seoul St. Mary's Hospital, 222 Banpo-daero, Seocho-gu, 137-701 Seoul, Korea

Tapasree Roysarkar
Department of Translational Molecular Pathology, The University of Texas MD Anderson Cancer Center, Houston, TX, USA and Metastasis Research Center, The University of Texas MD Anderson Cancer Center, Houston, TX, USA

Ignacio Rubio
Center for Sepsis Control and Care, University Hospital Jena, Jena, Germany and Institute of Molecular Cell Biology, University Hospital Jena, Jena, Germany

Petra Rümmele
Institute of Pathology, University of Regensburg, Franz-Josef-Strauss-Allee 11, 93053 Regensburg, Germany

ji-Yoon Ryu
Laboratory of Immunology, Department of Biological Science, Sungkyunkwan University, Suwon, 440-746, Republic of Korea

Rainer Spanbroek
Institute of Vascular Medicine, University Hospital Jena, Jena, Germany

Liya Su
Clinical Medicine Research Center of The Affiliated Hospital, Inner Mongolia Medical University, No 1 Tongdao North Street, Huimin District, Hohhot, Inner Mongolia 010050, China

Xiulan Su
Clinical Medicine Research Center of The Affiliated Hospital, Inner Mongolia Medical University, No 1 Tongdao North Street, Huimin District, Hohhot, Inner Mongolia 010050, China

Xiaoliang Sun
Department of Orthopaedics, Changzhou No. 1 People's Hospital, Changzhou, Jiangsu 213003, PR China

Tomomitsu Tahara
Department of Leukemia, the University of Texas MD Anderson Cancer Center, Houston, TX, USA

Nanyun Tang
Cancer and Cell Biology Division, Translational Genomics Research Institute, Phoenix, AZ 85004, USA

Joseph H. Taube
Department of Translational Molecular Pathology, The University of Texas MD Anderson Cancer Center, Houston, TX, USA and Metastasis Research Center, The University of Texas MD Anderson Cancer Center, Houston, TX, USA

Ivett Teleki
1st Department of Pathology & Experimental Cancer Research, Semmelweis University, Budapest, Hungary

Agata Tinnirello
Department of Translational Molecular Pathology, The University of Texas MD Anderson Cancer Center, Houston, TX, USA

Verena Tischler
Institute of Surgical Pathology, University Hospital Zurich, Zurich, Switzerland

Yanet Valdez
Centre for Blood Research, Department of Medicine, University of British Columbia, 4306-2350 Health Sciences Mall, V6T 1Z3, BC Vancouver, Canada

Zsuzsanna Varga
Institute of Surgical Pathology, University Hospital Zurich, Zurich, Switzerland and Institute of Surgical Pathology, University Hospital Zurich, Schmelzbergstrasse 12, CH-8091 Zurich, Switzerland

Daniel D. Von Hoff
Clinical Translational Research Division, Translational Genomics Research Institute, Scottsdale, AZ 85259, USA and Virginia G. Piper Cancer Center, Scottsdale Healthcare, Scottsdale, AZ 85258, USA

Xiaoping Wan
Department of Obstetrics and Gynecology, Shanghai First People's Hospital, Shanghai Jiao Tong University School of Medicine, Xinsongjiang Road, Shanghai, China

Jingyun Wang
Department of Obstetrics and Gynecology, International Peace Maternity & Child Health Hospital, Shanghai Jiao Tong University School of Medicine, Hengshan Road, Shanghai, China

Neng Wang
Department of Orthopaedics, Changzhou No. 1 People's Hospital, Changzhou, Jiangsu 213003, PR China

Xuemei Wang
PET-CT Center of The Affiliated Hospital, Inner Mongolia Medical University, No 1 Tongdao North Street, Huimin District, Hohhot, Inner Mongolia 010051, China

Herbert Weissbach
Center for Molecular Biology and Biotechnology, Charles E. Schmidt College of Science, Florida Atlantic University, Jupiter, Florida, United States of America

Bo Xiang
Division of Oncology, Department of Pediatric Surgery, West China Hospital of Sichuan University, Chengdu 610041, China

Andrea Xu
Centre for Blood Research, Department of Medicine, University of British Columbia, 4306-2350 Health Sciences Mall, V6T 1Z3, BC Vancouver, Canada

Chang Xu
Division of Oncology, Department of Pediatric Surgery, West China Hospital of Sichuan University, Chengdu 610041, China

Jumpei Yamazaki
Department of Leukemia, the University of Texas MD Anderson Cancer Center, Houston, TX, USA

Tingting Yang
Department of Obstetrics and Gynecology, Shanghai First People's Hospital, Shanghai Jiao Tong University School of Medicine, Xinsongjiang Road, Shanghai, China

Holly Yin
Cancer and Cell Biology Division, Translational Genomics Research Institute, Phoenix, AZ 85004, USA

Byong Chul Yoo
Colorectal Cancer Branch, Division of Translational and Clinical Research I, Research Institute, National Cancer Center, Gyeonggi, 410-769, Republic of Korea

Ailiang Zhang
Department of Orthopaedics, Changzhou No. 1 People's Hospital, Changzhou, Jiangsu 213003, PR China

Jialing Zhang
Clinical Medicine Research Center of The Affiliated Hospital, Inner Mongolia Medical University, No 1 Tongdao North Street, Huimin District, Hohhot, Inner Mongolia 010050, China

Xiu-Ying Zhang
Department of Bioinformatics, the University of Texas MD Anderson Cancer Center, Houston, TX, USA

Zhe Zhang
Department of Urology, the First Affiliated Hospital of China Medical University, Shenyang, Liaoning 110001, China

Yu-yan Zhu
Department of Urology, the First Affiliated Hospital of China Medical University, Shenyang, Liaoning 110001, China

INTRODUCTION

CD248 is a cell surface glycoprotein, highly expressed by stromal cells and fibroblasts of tumors and inflammatory lesions, but virtually undetectable in healthy adult tissues. CD248 promotes tumorigenesis, while lack of CD248 in mice confers resistance to tumor growth. Mechanisms by which CD248 is downregulated are poorly understood, hindering the development of anti-cancer therapies. In Chapter 1, Babu and colleagues sought to characterize the molecular mechanisms by which CD248 is downregulated by surveying its expression in different cells in response to cytokines and growth factors. Only transforming growth factor (TGFβ) suppressed CD248 protein and mRNA levels in cultured fibroblasts and vascular smooth muscle cells in a concentration- and time-dependent manner. TGFβ transcriptionally downregulated CD248 by signaling through canonical Smad2/3-dependent pathways, but not via mitogen activated protein kinases p38 or ERK1/2. Notably, cancer associated fibroblasts (CAF) and cancer cells were resistant to TGFβ mediated suppression of CD248. The findings indicate that decoupling of CD248 regulation by TGFβ may contribute to its tumor-promoting properties, and underline the importance of exploring the TGFβ-CD248 signaling pathway as a potential therapeutic target for early prevention of cancer and proliferative disorders.

Ras is a membrane-associated small G-protein that funnels growth and differentiation signals into downstream signal transduction pathways by cycling between an inactive, GDP-bound and an active, GTP-bound state. Aberrant Ras activity as a result of oncogenic mutations causes de novo-cell transformation and promotes tumor growth and progression. In Chapter 2, Augsten and colleagues describe a novel strategy to block deregulated Ras activity by means of oligomerized cognate protein modules derived from the Ras-binding domain of c-Raf (RBD), which the authors named MSOR for multivalent scavengers of oncogenic Ras. The introduction of well-characterized mutations into RBD was used to adjust the affinity and hence the blocking potency of MSOR towards activated Ras. MSOR inhibited several oncogenic Ras-stimulated processes including downstream activation of Erk1/2, induction of matrix-degrading enzymes, cell motility and invasiveness in a graded fashion depending on the oligomerization grade and the nature of the individual RBD-modules. The amenability to accurate experimental regulation was further improved by engineering an inducible MSOR-expression system to render the reversal of oncogenic Ras effects controllable. MSOR represent a new tool for the experimental and possibly therapeutic selective blockade of oncogenic Ras signals.

Despite multiple advances in the treatment of HER2+ breast cancers, resistance develops even to combinations of HER2 targeting agents. Inhibition of PI3K pathway signaling is critical for the efficacy of HER2 inhibitors. Activating mutations in PIK3CA can overlap with HER2 amplification and have been shown to confer resistance to HER2 inhibitors in preclinical studies. In Chapter 3, Rexer and colleagues profiled lapatinib-resistant cells for mutations in the PI3K pathway with the SNaPshot assay. Hotspot PIK3CA mutations were retrovirally transduced into HER2-amplified cells. The impact ofPIK3CA mutations on the effect of HER2 and PI3K inhibitors was assayed by immunoblot, proliferation and apoptosis assays. Uncoupling of PI3K signaling from HER2 was investigated by ELISA for phosphoproteins in the HER2-PI3K signaling cascade. The combination of HER2 inhibitors with PI3K inhibition was studied in HER2-amplified xenograft models with wild-type or mutantPIK3CA. The authors describe the acquisition of a hotspot PIK3CA mutation in cells selected for resistance to the HER2 tyrosine kinase inhibitor lapatinib. We also show that the gain of function conferred by these PIK3CA mutations partially uncouples PI3K signaling from the HER2 receptor upstream. Drug resistance conferred by this uncoupling was overcome by blockade of PI3K with the pan-p110 inhibitor BKM120. In mice bearing HER2-amplified wild-type PIK3CA xenografts, dual HER2 targeting with trastuzumab and lapatinib resulted in tumor regression. The addition of a PI3K inhibitor further improved tumor regression and decreased tumor relapse after discontinuation of treatment. In a PIK3CA-mutant HER2+ xenograft, PI3K inhibition with BKM120 in combination with lapatinib and trastuzumab was required to achieve tumor regression. These results suggest that the combination of PI3K inhibition with dual HER2 blockade is necessary to circumvent the resistance to HER2 inhibitors conferred by

PIK3CA mutation and also provides benefit to HER2+ tumors with wild-type PIK3CA tumors.

The wnt/β-catenin signaling pathway is known to affect in cancer oncogenesis and progression by interacting with the tumor microenvironment. However, the roles of wnt3a and wnt5a in colorectal cancer (CRC) have not been thoroughly studied. In Chapter 4, Lee and colleagues investigated the expression of wnt protein and the concordance rate in primary tumor and metastatic sites in CRC. To determine the relationship of wnt proteins with invasion related protein, the authors also analyzed the association between wnt protein expression and the expression of matrix metalloproteinase-9 (MMP-9) and vascular endothelial growth factor receptor-2 (VEGFR-2). Tumor tissue was obtained from eighty-three paraffin- embedded blocks which were using resected tissue from both the primary tumor and metastatic sites for each patient. The authors performed immunohistochemical staining for wnt3a, wnt5a, β-catenin, MMP-9 and VEGFR-2. Wnt3a, wnt5a, β-catenin, and MMP-9 expression was high; the proteins were found in over 50% of the primary tumors, but the prevalence was lower in tissue from metastatic sites. The concordance rates between the primary tumor and metastatic site were 76.2% for wnt5a and 79.4% for wnt3a and β-catenin, but VEGFR-2 was expressed in 67.4% of the metastatic sites even when not found in the primary tumor. Wnt3a expression in primary tumors was significantly associated with lymph node involvement ($p = 0.038$) and MMP-9 expression in the primary tumor ($p = 0.0387$), mesenchyme adjacent to tumor ($p = 0.022$) and metastatic site ($p = 0.004$). There was no other relationship in the expression of these proteins. Vascular invasion in primary tumor tissue may be a potential prognostic marker for liver metastasis, but no significant association was observed among the wnt protein, MMP-9, and VEGFR-2 for peritoneal seeding. In survival analysis, β-catenin expression was significantly correlated with overall survival ($p = 0.05$). Wnt3a and wnt5a expression had a concordance rate higher than 60% with a high concordance rate between the primary tumor and metastatic site. Wnt3a expression is associated with the expression of MMP-9 in primary tumor tissue adjacent mesenchymal tissue, and at the metastatic site. As a prognostic marker, only β-catenin expression showed significant relation with survival outcome.

Wnt5a is classified as a non-transforming Wnt family member and plays complicated roles in oncogenesis and cancer metastasis. However, Wnt5a signaling in osteosarcoma progression remains poorly defined. In Chapter 5, Zhang and colleagues found that Wnt5a stimulated the migration of human osteosarcoma cells (MG-63), with the maximal effect at 100 ng/ml, via enhancing phosphorylation of phosphatidylinositol-3 kinase (PI3K)/Akt. PI3K and Akt showed visible signs of basal phosphorylation and elevated phosphorylation at 15 min after stimulation with Wnt5a. Pharmaceutical inhibition of PI3K with LY294002 significantly blocked the Wnt5a-induced activation of Akt (p-Ser473) and decreased Wnt5a-induced cell migration. Akt siRNA remarkably inhibited Wnt5a-induced cell migration. Additionally, Wnt5a does not alter the total expression and phosphorylation of β-catenin in MG-63 cells. Taken together, the authors demonstrated for the first time that Wnt5a promoted osteosarcoma cell migration via the PI3K/Akt signaling pathway. These findings could provide a rationale for designing new therapy targeting osteosarcoma metastasis.

Increasing evidence suggests that forkhead box A1 (FOXA1) is frequently dysregulated in many types of human cancers. However, the exact function and mechanism of FOXA1 in human endometrial cancer (EC) remains unclear. FOXA1 expression, androgen receptor (AR) expression, and the relationships of these two markers with clinicopathological factors were determined by immunohistochemistry analysis in Chapter 6, by Qiu and colleagues. FOXA1 and AR were up-regulated by transient transfection with plasmids, and were down-regulated by transfection with siRNA or short hairpin RNA (shRNA). The effects of FOXA1 depletion and FOXA1 overexpression on AR-mediated transcription as well as Notch pathway and their impact on EC cell proliferation were examined by qRT-PCR, western blotting, co-immunoprecipitation, ChIP-PCR, MTT, colony-formation, and xenograft tumor–formation assays. The authors found that the expression of FOXA1 and AR in ECs was significantly higher than that in a typical hyperplasia and normal tissues. FOXA1 expression was significantly correlated with AR expression in clinical tissues. High FOXA1 levels positively correlated with pathological grade and depth of myometrial invasion in EC. High AR levels also positively correlated with pathological grade in EC. Moreover, the expression of XBP1, MYC, ZBTB16, and UHRF1, which are downstream targets of AR, was promoted by FOXA1 up-regulation or inhibited by FOXA1 down-regulation. Co-immunoprecipitation showed that FOXA1 interacted with AR in EC cells. ChIP-PCR assays showed that FOXA1 and AR could directly bind to the promoter and enhancer regions upstream of MYC. Mechanistic investigation revealed that over-expression of Notch1 and Hes1 proteins by FOXA1 could be reversed by AR depletion. In addition, the authors showed that down-regulation of AR attenuated FOXA1-up-regulated cell proliferation. However, AR didn't influence the promotion effect of FOXA1 on cell migration and invasion. In vivo xenograft model, FOXA1 knockdown reduced the rate of tumor growth. These results suggest that FOXA1 promotes cell proliferation by AR and activates Notch pathway. It indicated that FOXA1 and AR may serve as potential gene therapy in EC.

Netrin-1 and its receptor UNC5B play important roles in angiogenesis, embryonic development, cancer and inflammation. However, their expression patttern and biological roles in bladder cancer have not been well characterized. Chapter 7, by Liu and colleagues, aims to investigating the clinical significance of PKC α, netrin-1 and UNC5B in bladder cancer as well as their association with malignant biological behavior of cancer cells. Netrin-1 and UNC5B expression was examined in 120 bladder cancer specimens using immunohistochemistry and in 40 fresh cancer tissues by western blot. Immunofluorescence was performed in cancer cell lines. PKC α agonist PMA and PKC siRNA was employed in bladder cancer cells. CCK-8, wound healing assays and flow cytometry analysis were used to examine cell proliferation, migration and cell cycle, respectively. Netrin-1 expression was positively correlated with histological grade, T stage, metastasis and poor prognosis in bladder cancer tissues. Immunofluorescence showed elevated netrin-1 and decreased UNC5B expression in bladder cancer cells compared with normal bladder cell line. Furthermore, cell proliferation, migration and cell cycle progression were promoted with PMA treatment while inhibited by calphostin C. In addition, PMA treatment could induce while calphostin C reduce netrin-1 expression in bladder cancer cells. The present study identified netrin-1/UNC5B, which could be regulated by PKC signaling, was important mediators of bladder cancer progression.

Infantile hemangioma (IH), which is the most common tumor in infants, is a benign vascular neoplasm resulting from the abnormal proliferation of endothelial cells and pericytes. For nearly a century, researchers have noted that IH exhibits diverse and often dramatic clinical behaviors. On the one hand, most lesions pose no threat or potential for complication and resolve spontaneously without concern in most children with IH. On the other hand, approximately 10% of IHs are destructive, disfiguring and even vision- or life-threatening. Recent studies have provided some insight into the pathogenesis of these vascular tumors, leading to a better understanding of the biological features of IH and, in particular, indicating that during hemangioma neovascularization, two main pathogenic mechanisms prevail, angiogenesis and vasculogenesis. Both mechanisms have been linked to alterations in several important cellular signaling pathways. These pathways are of interest from a therapeutic perspective because targeting them may help to reverse, delay or prevent hemangioma neovascularization. In Chapter 8, Ji and colleagues explore some of the major pathways implicated in IH, including the VEGF/VEGFR, Notch, β-adrenergic, Tie2/angiopoietins, PI3K/AKT/mTOR, HIF-α-mediated and PDGF/PDGF-R-β pathways. The authors focus on the role of these pathways in the pathogenesis of IH, how they are altered and the consequences of these abnormalities. In addition, they review the latest preclinical and clinical data on the rationally designed targeted agents that are now being directed against some of these pathways.

Pancreatic ductal adenocarcinoma (PDA) is a highly lethal cancer characterized by complex aberrant genomes. A fundamental goal of current studies is to identify those somatic events arising in the variable landscape of PDA genomes that can be exploited for improved clinical outcomes. In Chapter 9, Evers and colleagues used DNA content flow sorting to identify and purify tumor nuclei of PDA samples from 50 patients. The genome of each sorted sample was profiled by oligonucleotide comparative genomic hybridization and targeted resequencing of STAG2. Transposon insertions within STAG2 in aKRASG12D-driven genetically engineered mouse model of PDA were screened by RT-PCR. The authors then used a tissue microarray to survey STAG2 protein expression levels in 344 human PDA tumor samples and adjacent tissues. Univariate Kaplan Meier analysis and multivariate Cox Regression analysis were used to assess the association of STAG2 expression relative to overall survival and response to adjuvant therapy. Finally, RNAi-based assays with PDA cell lines were used to assess the potential therapeutic consequence of STAG2 expression in response to 18 therapeutic agents. STAG2 is targeted by somatic aberrations in a subset (4%) of human PDAs. Transposon-mediated disruption of STAG2 in a KRASG12D genetically engineered mouse model promotes the development of PDA and its progression to metastatic disease. There was a statistically significant loss of STAG2 protein expression in human tumor tissue (Wilcoxon-Rank test) with complete absence of STAG2 staining observed in 15 (4.3%) patients. In univariate Kaplan Meier analysis nearly complete STAG2 positive staining (>95% of nuclei positive) was associated with a median survival benefit of 6.41 months (P=0.031). The survival benefit of adjuvant chemotherapy was only seen in patients with a STAG2 staining of less than 95% (median survival benefit 7.65 months; P=0.028). Multivariate Cox Regression analysis showed that STAG2 is an independent prognostic factor for survival in pancreatic cancer patients. Finally, we show that RNAi-mediated knockdown of STAG2 selectively sensitizes human PDA cell lines to platinum-based therapy.

The ovarian steroid hormone, progesterone, and its nuclear receptor, the progesterone receptor, are implicated in the progression of breast cancer. Clinical trial data on the effects of hormone replacement therapy underscore the importance of understanding how progestins influence breast cancer growth. The progesterone receptor regulation of distinct target genes is mediated by complex interactions between the progesterone receptor and other regulatory factors that determine the context-dependent transcriptional action of the progesterone receptor. These interactions often lead to post-translational modifications to the progesterone receptor that can dramatically alter receptor function, both in the normal mammary gland and in breast cancer. Chapter 10, by Hagan and Lange, highlights the molecular components that regulate progesterone receptor transcriptional action and describes how a better understanding of the complex interactions between the progesterone receptor and other regulatory factors may be critical to enhancing the clinical efficacy of anti-progestins for use in the treatment of breast cancer.

Epithelial-mesenchymal transition (EMT) is known to impart metastasis and stemness characteristics in breast cancer. In Chapter 11, to characterize the epigenetic reprogramming following Twist1-induced EMT, Malouf and colleagues characterized the epigenetic and transcriptome landscapes using whole-genome transcriptome analysis by RNA-seq, DNA methylation by digital restriction enzyme analysis of methylation (DREAM) and histone modifications by CHIP-seq of H3K4me3 and H3K-27me3 in immortalized human mammary epithelial cells relative to cells induced to undergo EMT by Twist1. EMT is accompanied by focal hyper-methylation and widespread global DNA hypomethylation, predominantly within transcriptionally repressed gene bodies. At the chromatin level, the number of gene promoters marked by H3K4me3 increases by more than one fifth; H3K27me3 undergoes dynamic genomic redistribution charac-terized by loss at half of gene promoters and overall reduction of peak size by almost half. This is paralleled by increased phosphorylation of EZH2 at serine 21. Among genes with highly altered mRNA expression, 23.1% switch between H3K4me3 and H3K27me3 marks, and those point to the master EMT targets and regulators CDH1, PDGFRα and ESRP1. Strik-ingly, Twist1 increases the number of bivalent genes by more than two fold. Inhibition of the H3K27 methyltransferases EZH2 and EZH1, which form part of the Polycomb repressive complex 2 (PRC2), blocks EMT and stemness properties. The authors' findings demonstrate that the EMT pro-gram requires epigenetic remodeling by the Polycomb and Trithorax com-plexes leading to increased cellular plasticity. This suggests that inhibiting epigenetic remodeling and thus decrease plasticity will prevent EMT, and the associated breast cancer metastasis.

Expression of transcription-factors as Slug and Sox9 was recently described to determine mammary stem-cell state. Sox10 was previously shown to be present also in breast cancer. Protein overexpression of Slug, Sox9 and Sox10 were associated with poor overall survival and with tri-ple-negative phenotype in breast cancer. In Chapter 12, Riemenschnitter and colleagues tested the stability of Slug, Sox9 and Sox10 expression during chemotherapy and addressed their prognostic role of in neoadju-vant treated primary breast-cancer and their correlation to pathological-re-sponse and overall survival. The authors analyzed immunohistochemical expression of Slug, Sox9 and Sox10 in tissue microarrays of 96 breast can-cers prior to and after neoadjuvant chemotherapy. Expression was evalu-ated in invasive tumor cells and in tumor stroma and scored as 0, 1+, 2+ 3+. Expression-profile prior to and after chemotherapy was correlated to overall survival (Kaplan Meier) and with established clinico-pathological parameter. Sox9, Sox10 and Slug were expressed in 82–96% of the tumor cells prior to chemotherapy. Slug was expressed in 97% of the cases in tumor stroma before therapy. Change in expression-profile after chemotherapy occurred only in Slug expression in tumor-cells (decreased from 82 to 51%, p=0.0001, Fisher's exact test). The other markers showed no significant change after chemotherapy. Stromal Sox9 expression (0 to 2+) correlated to better overall survival after chemotherapy (p=0.004) and reached almost statistical significance prior to chemotherapy (p=0.065). There was no correlation between Sox9 and hormone-receptor expression. In multivariate-analysis, the stromal Sox9 expression after chemotherapy proved to be an independent and better prognostic marker than hormone-receptor status. Other clinico-pathological parameter (as HER2-status or pathological-stage) showed no correlation to the analyzed markers. Strong stromal Sox9 expression in breast cancer after chemotherapy was found to bear negative prognostic information and was associated with shortened overall survival. Slug expression was significantly changed (reduced) in samples after neoadjuvant chemotherapy.

A great challenge of cancer chemotherapy is to eliminate cancer cells and concurrently maintain the quality of life (QOL) for cancer patients. In an older article, Su and colleagues identified a novel anti-cancer bioactive peptide (ACBP), a peptide induced in goat spleen or liver following immunization with human gastric cancer protein extract. ACBP alone exhibited anti-tumor activity without measurable side effects. Thus, in Chapter 13, the authors hypothesize that ACBP and combined chemotherapy could improve the efficacy of treatment and lead to a better QOL. In this study, ACBP was isolated and purified from immunized goat liver, and designated as ACBP-L. The anti-tumor activity was investigated in a previously untested human gastric cancer MGC-803 cell line and tumor model. ACBP-L inhibited cell proliferation in vitro in a dose and time dependent manner, titrated by MTT assay. The effect of ACBP-L on cell morphology was observed through light and scanning electron microscopy. In vivo ACBP-L alone significantly inhibited MGC-803 tumor growth in a xenograft nude mouse model without measurable side effects. Treatment with the full dosage of Cisplatin alone (5 mg/kg every 5 days) strongly suppressed tumor growth. However, the QOL in these mice had been significantly affected when measured by food intakes and body weight. The combinatory regiment of ACBP-L with a fewer doses of Cisplatin (5 mg/kg every 10 days) resulted in a similar anti-tumor activity with improved QOL. 18F-FDG PET/CT scan was used to examine the biological activity in tumors of live animals and indicated the consistent treatment effects. The tumor tissues were harvested after treatment, and ACBP-L and Cisplatin treatment suppressed Bcl-2, and induced Bax, Caspase 3, and Caspase 8 molecules as detected by RT-PCR and immunohistochemistry. The combinatory regiment induced stronger Bax and Caspase 8 protein expression. The authors' current finding in this gastric cancer xenograft animal model demonstrated that ACBP-L could lower Cisplatin dose to achieve a similar anti-tumor efficacy as the higher dose of Cisplatin alone, through enhanced modulation of apoptotic molecules. This newly developed combination regiment improved QOL in tumor bearing hosts, which could lead to clinical investigation for the new strategy of combination therapy.

Sulindac is an FDA-approved non-steroidal anti-inflammatory drug with documented anticancer activities. Ayyanathan and colleagues' recent studies showed that sulindac selectively enhanced the killing of cancer cells exposed to oxidizing agents via production of reactive oxygen species (ROS) resulting in mitochondrial dysfunction. This effect of sulindac and oxidative stress on cancer cells could be related to the defect in respiration in cancer cells, first described by Warburg 50 years ago, known as the Warburg effect. In Chapter 14, the authors postulated that sulindac might enhance the selective killing of cancer cells when combined with any compound that alters mitochondrial respiration. To test this hypothesis they have used dichloroacetate (DCA), which is known to shift pyruvate metabolism away from lactic acid formation to respiration. One might expect that DCA, since it stimulates aerobic metabolism, could stress mitochondrial respiration in cancer cells, which would result in enhanced killing in the presence of sulindac. In this study, the authors have shown that the combination of sulindac and DCA enhances the selective killing of A549 and SCC25 cancer cells under the conditions used. As predicted, the mechanism of killing involves ROS production, mitochondrial dysfunction, JNK signaling and death by apoptosis. The results suggest that the sulindac-DCA drug combination may provide an effective cancer therapy.

CHAPTER 1

TGFβ-MEDIATED SUPPRESSION OF CD248 IN NON-CANCER CELLS VIA CANONICAL SMAD-DEPENDENT SIGNALING PATHWAYS IS UNCOUPLED IN CANCER CELLS

SAHANA SURESH BABU, YANET VALDEZ, ANDREA XU, ALICE M. O'BYRNE, FERNANDO CALVO, VICTOR LEI, AND EDWARD M. CONWAY

1.1 BACKGROUND

CD248, also referred to as endosialin and tumor endothelial marker (TEM-1) [1] (reviewed in [2]), is a member of a family of type I transmembrane glycoproteins containing C-type lectin-like domains, that includes thrombomodulin [3] and CD93 [4]. Although the mechanisms are not fully elucidated, these molecules all modulate innate immunity, cell proliferation and vascular homeostasis and are potential therapeutic targets for several diseases, including cancer, inflammatory disorders and thrombosis.

This chapter was originally published under the Creative Commons Attribution License. Babu SS, Valdez Y, Xu A, O'Byrne AM, Calvo F, Lei V, and Conway EM. TGFβ-Mediated Suppression of CD248 in Non-Cancer Cells via Canonical Smad-Dependent Signaling Pathways is Uncoupled in Cancer Cells. BMC Cancer 14,113 (2014). doi:10.1186/1471-2407-14-113.

CD248 is expressed by cells of mesenchymal origin, including murine embryonic fibroblasts (MEF), vascular smooth muscle cells, pericytes, myofibroblasts, stromal cells and osteoblasts [5-12]. During embryonic development, CD248 is prominently and widely expressed in the fetus (reviewed in [2]). However, after birth, CD248 protein levels are dramatically downregulated [7,13-15], resulting in only minimal expression in the healthy adult, except in the endometrium, ovary, renal glomerulus and osteoblasts [11,16-18].

While largely absent in normal tissues, CD248 is markedly upregulated in almost all cancers. Highest expression is found in neuroblastomas and in subsets of carcinomas, such as breast and colon cancers, and in addition, in glioblastomas and mesenchymal tumors, such as fibrosarcomas and synovial sarcomas [8,14,15,17,19,20], where it is mostly detected in perivascular and tumor stromal cells, but also in the tumor cells themselves [21,22]. CD248 is also expressed in placenta and during wound healing and in wounds such as ulcers. It is also prominently expressed in synovial fibroblasts during inflammatory arthritis [10]. In some tumors and in chronic kidney disease, CD248 expression directly correlates with worse disease and/or a poor prognosis [9,23,24]. The contributory role of CD248 to these pathologies was confirmed in gene inactivation studies. Mice lacking CD248 are generally healthy, except for an increase in bone mass [11,25] and incomplete post-natal thymus development [26]. However, in several models, they are protected against tumor growth, tumor invasiveness and metastasis [25,27] and they are less sensitive to anti-collagen antibody induced arthritis [10].

While the mechanisms by which CD248 promotes tumorigenesis and inflammation are not clearly defined, the preceding observations have stimulated interest in exploring CD248 as a therapeutic target, primarily by using anti-CD248 antibodies directed against its ectodomain [19,20,28,29]. Likely due to limited knowledge of CD248 regulatory pathways, other approaches to interfere with or suppress CD248 have not been reported. CD248 is upregulated in vitro by high cell density, serum starvation, by the oncogene v-mos[5] and by hypoxia [30]. We previously showed that fibroblast expression of CD248 is suppressed by contact with endothelial cells [27]. Otherwise, factors which down-regulate CD248 have not heretofore been reported, yet such insights might reveal novel sites for therapeutic intervention.

In this study, we evaluated the effects of several cytokines on the expression of CD248. We show that TGFβ specifically and dramatically downregulates CD248 expression in normal cells of mesenchymal origin and that this is mediated via canonical Smad-dependent intracellular signaling pathways. Notably, cancer cells and cancer associated fibroblasts are resistant to TGFβ mediated suppression of CD248. The findings suggest that CD248 not only promotes tumorigenesis, but may be a marker of the transition of TGFβ from a tumor suppressor to a tumor promoter. Delineating the pathways that couple TGFβ and CD248 may uncover novel therapeutic strategies.

1.2 METHODS

1.2.1 REAGENTS

Rabbit anti-human CD248 antibodies (Cat no #18160-1AP) were from ProteinTech (Chicago, USA); goat anti-human actin antibodies (#sc-1616) from Santa Cruz (USA); rabbit anti-SMAD1,5-Phospho (Cat no #9516), rabbit anti-Smad2-Phospho (#3101), rabbit anti-ERK1/2-phospho (#9101S), rabbit anti-p38-phospho (#9211), rabbit anti-SMAD2/3 (#5678) and rabbit anti-SMAD3 (#9513) were from Cell Signaling (USA). Murine anti-rabbit α-smooth muscle actin monoclonal antibodies (#A5228) were from Sigma-Aldrich (Canada). Secondary antibodies included goat anti-rabbit IRDye® 800 (LIC-926-32211). Goat anti-rabbit IRDye® 680 (LIC-926-68071) or donkey anti-goat IRDye® 680 antibodies (LIC-926-68024) and anti-rabbit Alexa green-488 were from Licor (Nebraska, USA).

Basic fibroblast growth factor (bFGF), recombinant human transforming growth factor β-1 (TGFβ) (240-B/CF), recombinant human bone morphogenic protein (BMP-2) (355-BM-010/CF), recombinant human/mouse/Rat Activin A, CF (338-AC-010/CF), recombinant rat platelet derived growth factor-BB (PDGF) (250-BB-050), recombinant human vascular endothelial growth factor (VEGF), and recombinant mouse interleukin-6 (IL-6) (406-ML/CF), recombinant mouse tumor necrosis factor-α (TNF-α) (410-MT/CF) and recombinant mouse interferon-γ (IFN-γ) (485-MI/CF) were purchased from R&D Systems (Minneapolis, USA). Phorbol 12-Myristate 13-Acetate (PMA) (P1585) and α-amanitin were from Sigma-Aldrich (Oakville, Canada). The inhibitors SB431542 (for ALK5), SB202190 (for p38) and U0126 (for ERK1/2) were from Tocris Biosciences, Canada.

1.2.2 MICE

Transgenic mice lacking CD248 (CD248$^{KO/KO}$) were previously generated and genotyped as described [10]. Mice were maintained on a C57Bl6 genetic background and corresponding sibling-derived wild-type mice (CD-248$^{WT/WT}$) were used as controls.

1.2.3 CELL CULTURE

Murine embryonic fibroblasts (MEF) were isolated from CD248$^{WT/WT}$ or CD248$^{KO/KO}$ mice as previously described [10]. Cells were cultured in DMEM (Invitrogen, Canada) with 10% fetal calf serum (FCS) and 1% Penicillin/Streptomycin (Invitrogen, Karlsruhe, Germany) and used at passages 2-5. Upon reaching confluence, cells were incubated for 14 hrs in low serum media (1% FCS) and then treated as indicated in the Results with TGFβ (0.1-12 ng/ml), BMP-2 (50-100 ng/ml), PDGF (50 ng/ml), VEGF (20 ng/ml), bFGF (10 ng/ml), IL-6 10 ng/ml), PMA (60 ng/ml), SB43152 (1 μM), and/or α-amanitin (20 μg/ml), for different time periods as noted. Using previously reported methods [31,32], vascular smooth muscle cells (SMC) were isolated from the aortae of CD248$^{WT/WT}$ or CD-248$^{KO/KO}$ pups, cultured in SMC growth media (Promocell, Heidelberg, Germany) with 15% FCS and 1% Penicillin/Streptomycin (Invitrogen) and used at passages 2-5. Wehi-231 and A20 (mouse B-lymphoma) cell lines (gift of Dr. Linda Matsuuchi, University of British Columbia) were cultured in RPMI media with 10% fetal calf serum (FCS), 1% Penicillin/ Streptomycin and 0.1% mercaptoethanol. Normal fibroblasts (NF) derived from normal mouse mammary glands, and cancer associated fibroblasts (CAF) from mammary carcinoma in mice containing the MMTV-PyMT transgene [33] were provided by Dr. Erik Saha (Cancer Research London UK Research Institute, London, UK), and cultured in DMEM with 10% FCS, 1% Penicillin/Streptomycin and 1% insulin-transferrin-selenium.

1.2.4 PROTEIN ELECTROPHORESIS AND WESTERN BLOTTING

Cells were scraped from culture dishes, suspended in PBS, pelleted by centrifugation and lysed with 50 µl RIPA buffer (30 mM Tris–HCl, 15 mM NaCl, 1% Igepal, 0.5% deoxycholate, 2 mM EDTA, 0.1% SDS). Centrifugation-cleared lysates were quantified for protein content. Equal quantities of cell lysates (25 µg) were separated by SDS-PAGE under reducing or non-reducing conditions as noted, using 8% and 12% low-bisacrylamide gels (acrylamide to bis-acrylamide = 118:1). In pilot studies, these gels provided highest resolution of the bands of interest [34]. Proteins were transferred to a nitrocellulose membrane and after incubating with blocking buffer (1:1 PBS:Odyssey buffer) (Licor, Nebraska, U.S.A.), they were probed with rabbit anti-CD248 antibodies 140 µg/ml, goat anti-actin antibodies, rabbit anti-Smad1-Phospho, anti-Smad2-Phospho, anti-Smad2-Total or anti-Smad3 antibodies in blocking buffer overnight. After washing and incubation of the filter with the appropriate secondary antibodies (100 ng/ml IRDye® 800 goat anti-rabbit or IRDye® Donkey anti-goat–Licor, Nebraska, USA) in blocking buffer for 1 hr at room temperature, detection was accomplished using a Licor Odyssey® imaging system (Licor, Nebraska, USA) and intensity of bands of interest were quantified relative to actin using Licor software (Licor, Nebraska, U.S). All studies were performed a minimum of 3 times, and representative Western blots are shown.

1.2.5 IMMUNOFLUORESCENCE ANALYSIS

Preconfluent cells were grown on cover slips and fixed at room temperature with acetone (100%) for 2 minutes, followed by a 30 minute incubation with blocking buffer (1% BSA in PBS). Cells were then incubated with anti-CD248 rabbit antibodies 40 µg/ml, for 1 hr followed by extensive washes and incubation with Alexa green 488 anti-rabbit antibody (5 mg/ml) for 1 hr. The cells were washed and fixed with antifade containing DAPI (Invitrogen, Canada) for subsequent imaging with a confocal microscopic (Nikon C2 model, Nikon, Canada).

1.2.6 DETERMINATION OF STABILITY OF CD248 MRNA

α-Amanitin, an inhibitor of RNA-polymerase II, was used to quantify the half-life of CD248 mRNA using previously reported methods [35]. Briefly, 90% confluent MEF were incubated with DMEM with 1% fetal calf serum (FCS) overnight, after which the media was refreshed, and subsequently stimulated with α-Amanitin 20 µg/ml ± TGFβ for the indicated time periods. RNA was isolated for gene expression analysis.

1.2.7 GENE EXPRESSION ANALYSIS

RNA was isolated from the MEF and reverse transcribed to cDNA/mRNA according to the manufacturer's instructions (Qiagen RNeasy kit and QuantiTech reverse transcription kit, Hilden, Germany). Expression of CD248 mRNA was analyzed by RT-PCR and quantified with SYBR green using real time PCR (Applied Biosystems® Real-Time PCR Instrument, Canada). CD248 mRNA levels were reported relative to the expression of the housekeeping gene, Glyceraldehyde 3-Phosphate dehydrogenase (GAPDH). The following amplification primers were used: CD248 forward (5′-GGGCCCCTACCACTCCTCAGT-3′); CD248 reverse (5′-AGGTGGGTGGACAGGGCTCAG-3′); GAPDH forward (5′-GACCACAGTCCATGCCATCACTGC-3′); GAPDH reverse (5′-AT-GACCTTGCCCACAGCCTTGG-3′).

1.2.8 ANIMAL CARE

Experimental animal procedures were approved by the Institutional Animal Care Committee of the University of British Columbia.

1.2.9 STATISTICS

Experiments were performed in triplicate and data were analyzed using Bonferroni post-test to compare replicates (GraphPad Prism software Inc, California, USA). Error bars on figures represent standard errors of the mean (SEM). $P < 0.05$ was considered statistically significant.

1.3 RESULTS

1.3.1 SCREEN FOR CYTOKINES THAT MODULATE EXPRESSION OF CD248

In view of the established links between CD248 and cell proliferation, migration and invasion, we screened a number of growth factors, cytokines and PMA for effects on the expression of CD248 by MEF. These factors and the chosen concentrations were selected based on the fact that all reportedly induce MEF to undergo inflammatory, migratory and/or proliferative changes. We previously determined that these cells express CD248 at readily detectable levels, as assessed by Western blot, where it is often seen as a monomer (~150 kDa) and a dimer (~300 kDa). An incubation time of 48 hrs was chosen based on our previous findings that CD248-dependent release and activation of matrix metalloproteinase (MMP9) induced by TFGβ was observed over that period [10]. As seen in Figure 1A, bFGF, VEGF, PDGF, PMA, IL-6, TNF-α, and IFN-γ had no effects on CD248 expression. However, TGFβ suppressed expression of CD248 in MEF to almost undetectable levels (Figure 1A). The same pattern of response was evident in the murine fibroblast cell line 10 T1/2 (Figure 1B), and in mouse primary aortic smooth muscle cells (SMC) (Figure 1C), suggesting that CD248 specifically responds to TGFβ and that the response is active in diverse cell lines.

1.3.2 TGFβ SUPPRESSES EXPRESSION OF CD248 BY MEF

TGFβ exerts a range of cellular effects by binding to and activating its cognate serine/threonine kinase receptors, TGFβ type I (TGFβRI, ALK-5) and type II (TGFβRII), which in turn mediate intracellular signaling events via canonical Smad-dependent and Smad-independent signaling pathways (e.g. p38 mitogen-activated protein kinase (MAPK) pathway) (for reviews [36-38]). The canonical Smad-dependent pathway results in recruitment and phosphorylation of Smad2 and Smad3 which complex with Smad4 to enter the nucleus and form a transcriptional complex that modulates target gene expression in a context-dependent manner. Diversity in the response to TGFβ signaling is achieved by Smad2/3-independent, "non-canonical" signaling pathways, which may include, among others, activation of combinations of mitogen-activated protein kinases ERK1/2 and p38, PI3K/Akt, cyclo-oxygenase, Ras, RhoA, Abl and Src (for reviews [36-38]). We characterized the pathways by which TGFβ suppresses CD248. MEF were exposed to a range of concentrations of TGFβ (0.1 to 12 ng/ml) for a period of 48 hrs. Western blots of cell lysates showed that TGFβ downregulated the expression of CD248 in a concentration-dependent manner. As expected, TGFβ also induced phosphorylation of Smad2 and Smad3 in a concentration-dependent manner (Figure 2A,B). Confocal microscopy was used to visualize the effects of TGFβ on expression of CD248 by MEF (Figure 2C). At 48 hrs without TGFβ, CD248 was readily detected on the surface of CD248WT/WT MEF, but was entirely absent in TGFβ-treated cells as well as in CD248KO/KO MEF.

We next evaluated the temporal response of CD248 to a fixed concentration of TGFβ (3 ng/ml) (Figure 3A,B) and found that CD248 expression was suppressed in a time-dependent manner to <50% by 6 hrs of exposure to TGFβ. Once again, TGFβ induced phosphorylation of Smad2. Notably, as seen in experiments using CD248KO/KO MEF (lacking CD248) (Figure 3C), CD248 was not required for TGFβ-mediated phosphorylation of Smad2, indicating that CD248 is not a co-receptor for TGFβ signaling.

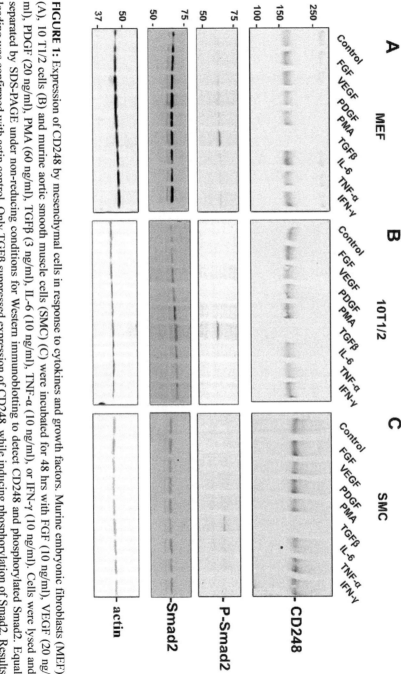

FIGURE 1: Expression of CD248 by mesenchymal cells in response to cytokines and growth factors. Murine embryonic fibroblasts (MEF) (A), 10 T1/2 cells (B) and murine aortic smooth muscle cells (SMC) (C) were incubated for 48 hrs with FGF (10 ng/ml), VEGF (20 ng/ml), PDGF (20 ng/ml), PMA (60 ng/ml), TGFβ (3 ng/ml), IL-6 (10 ng/ml), TNF-α (10 ng/ml), or IFN-γ (10 ng/ml). Cells were lysed and separated by SDS-PAGE under non-reducing conditions for Western immunoblotting to detect CD248 and phosphorylated Smad2. Equal loading was confirmed with actin control. Only TGFβ suppressed expression of CD248, while inducing phosphorylation of Smad2. Results are representative of 3 independent experiments. Molecular weight markers in kDa are shown on the left.

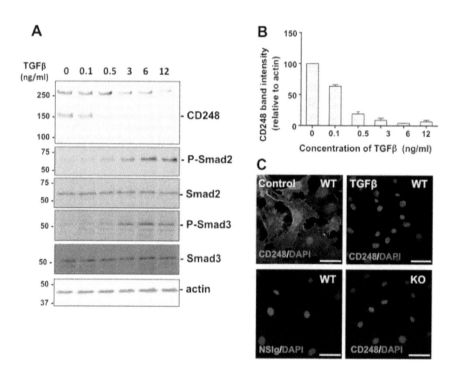

FIGURE 2: Expression of CD248 in response to increasing concentrations of TGFβ. (A) MEF were incubated for 48 hrs with increasing concentrations of TGFβ. Expression of CD248 (seen as monomers (~160 kDa) and dimers) and phosphorylation of Smad2, were detected by Western blot. (B) CD248 expression relative to actin expression was quantified by densitometry (n=3 experiments) and results were normalized to the no-treatment condition. (C) CD248 expression by MEF (wild-type, WT; or lacking CD248, KO) was detected with specific anti-CD248 antibodies after exposure to carrier (Control) or TGFβ for 48 hrs. TGFβ suppresses CD248 in a concentration-dependent manner, with simultaneous increase in phosphorylated Smad2 and ERK1/2. Scale bar=50 μm.

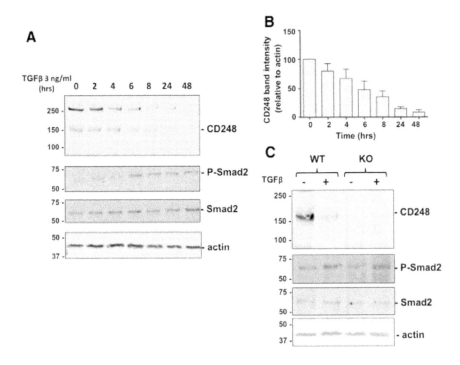

FIGURE 3: Temporal response of CD248 to TGFβ. (A) MEF were incubated for 0-48 hrs with TGFβ 3 ng/ml. Expression of CD248 and phosphorylation of Smad2, were detected by Western blot. (B) CD248 expression relative to actin expression was quantified by densitometry (n=3 experiments) and results were normalized to the no-treatment condition. CD248 expression decreases as Smad2 is phosphorylated. (C) CD248[WT/WT] (WT) or CD248[KO/KO] (KO) MEF were exposed to TGFβ (0 or 3 ng/ml) for 48 hrs and lysates were Western blotted. Representative blots from 3 experiments are shown. Smad2 and ERK1/2 are phosphorylated in response to TGFβ even in cells that lack CD248.

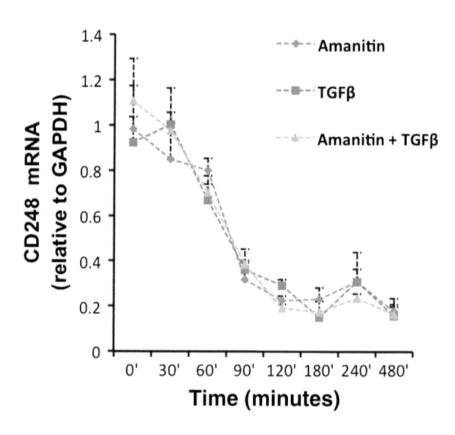

FIGURE 4: Stability of CD248 mRNA is unaffected by TGFβ. MEF were treated with TGFβ 3 ng/ml alone, α-amanatin 20 μg/ml alone, or with a combination of TGFβ and α-amanitin as described in Methods. CD248 mRNA levels, relative to the mRNA levels of the housekeeping gene GAPDH, were quantified at different time intervals by qRT-PCR. Results were normalized from 3 independent experiments, each done in triplicate. The half-life of CD248 mRNA is approximately 75 minutes, which is unaltered by TGFβ.

1.3.3 TGFβ SUPPRESSES CD248 MRNA ACCUMULATION

We evaluated the mechanism by which TGFβ suppresses CD248. CD248 mRNA levels in MEF were quantified by qRT-PCR at different time intervals following exposure of the cells to 3 ng/ml TGFβ. TGFβ suppressed CD248 mRNA levels in a time-dependent manner and by 75 minutes, mRNA accumulation had diminished to ~50% (Figure 4) and was ~20% by 2 hrs.

Using the RNA polymerase II inhibitor, α-amanitin (20 μg/ml), we measured the stability of CD248 mRNA in MEF and assessed whether it is altered by TGFβ. As seen in Figure 4, the time-dependent reduction in CD248 mRNA with α-amanitin alone was almost identical to the pattern seen with TGFβ alone, i.e., the half-life was determined to be approximately 75 minutes. The addition of TGFβ to α-amanitin did not alter the half-life. The findings suggest that TGFβ acts primarily at the level of CD248 transcription and does not alter the stability of CD248 mRNA.

1.3.4 SUPPRESSION OF CD248 BY TGFβ IS MEDIATED BY ALK-5 SIGNALING

In MEF, TGFβ reportedly signals exclusively through complexes involving ALK5 [39]. SB431542 is a selective inhibitor of TGFβ superfamily type I activin receptor-like kinase (ALK) receptors, ALK4, ALK5 and ALK7, which does not affect components of the ERK, JNK, or p38 MAP kinase pathways [40]. We tested whether ALK5 is required for TGFβ-mediated suppression of CD248. MEF were incubated with the inhibitor (1 μM) for 1 hr prior to the addition of 3 ng/ml TGFβ. Expression of CD248 at 48 hrs was assessed by Western blot, immunofluorescence analysis and qRT-PCR (Figure 5A-C). When added alone, neither the inhibitor SB431542 nor its vehicle DMSO, had any effect on CD248 expression. As before, TGFβ dramatically suppressed CD248, while simultaneously inducing phosphorylation of Smad2 (Figure 5A). This effect of TGFβ was entirely abrogated by preincubation of the cells with SB431542. Thus, addition of TGFβ down-regulates CD248 via activation of ALK-5.

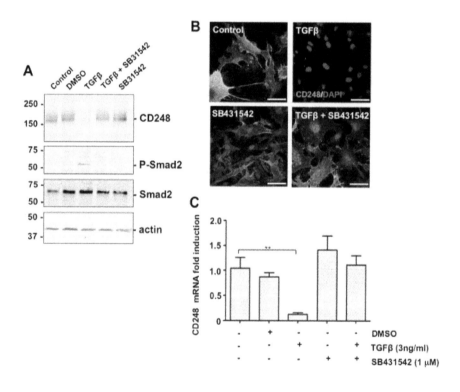

FIGURE 5: TGFβ-induced suppression of CD248 is mediated via canonical signaling pathways. (A, B, C) MEF were incubated for 48 hrs with TGFβ 3 ng/ml and the ALK-inhibitor SB431542 1 μM either singly or in combination. Controls included carriers for SB431542 (DMSO) or for TGFβ (0.1% BSA). (A) Western blots and (B) immunofluorescence were used to detect expression of CD248 (green). (C) CD248 mRNA levels were also quantified (n = 3 experiments, each in triplicate; *p < 0.05). Results indicate that TGFβ-mediated suppression of CD248 protein and mRNA requires integrity of canonical ALK5-Smad2 signaling pathway. Scale bar = 50 μm.

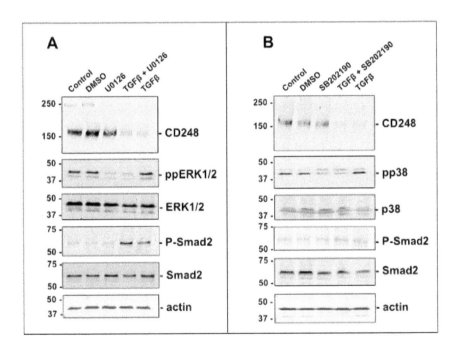

FIGURE 6: TGFβ-mediated suppression of CD248 via ALK5 is specific. (A, B) MEF were incubated with TGFβ (3 ng/ml) for 48 hrs in the presence or absence of the inhibitor of phosphorylated ERK1/2, U0126 10 μM (A) or phosphorylated p38, SB202190 10 μM (B). Representative Western blots from 3 independent experiments are shown and were used to assess the effect on CD248 expression. TGFβ-coupling to either ERK1/2 or to p38 is not involved in its suppressive effects on CD248.

1.3.5 TGFβ-MEDIATED SUPPRESSION OF CD248 IS INDEPENDENT OF ERK1/2 AND P38 SIGNALING

We also tested whether suppression of CD248 expression by TGFβ is mediated via one or more non-canonical Smad2/3-independent pathways. Using U0126, a specific inhibitor of ERK1/2 phosphorylation [41], we showed that TGFβ does not rely on signaling via ERK1/2 to suppress CD248 (Figure 6A). In a similar manner, using the p38 inhibitor, SB202190 [42], we also demonstrated that phosphorylation of p38 is not required for TGFβ to downregulate expression of CD248 (Figure 6B). Thus, in MEF, TGFβ suppresses CD248 expression via signaling pathways that do not require activation of these two Smad2/3-independent pathways.

1.3.6 REGULATION OF CD248 BY BONE MORPHOGENIC PROTEIN 2 (BMP2) AND ACTIVIN

The TGFβ family of cytokines comprises over 35 members, including the prototypic TGFβ isoforms (TGFβ1, β2, β3), bone morphogenic proteins (BMPs), growth and differentiation factors, activins and nodal. These regulate cell survival, proliferation, differentiation, adhesion, migration and death in a cell type-and context-dependent manner. To further assess the specificity of action of TGFβ on CD248 expression, we tested whether BMP2 and activin had similar effects. MEF were treated for 24 and 48 hrs with 50 and 100 ng/ml of activin or BMP2 (Figure 7A). At these concentrations of BMP2, Smad1 was, as expected, phosphorylated, while Smad2 was not [43]. Notably, BMP2 had no effect on CD248 expression, and thus does not participate in its regulation under these conditions. Activin induced phosphorylation of Smad2, which reportedly occurs via ALK-4/7 activation [44] (Figure 7B). In contrast to TGFβ, activin caused only a slight reduction in CD248 expression after 48 hrs of exposure.

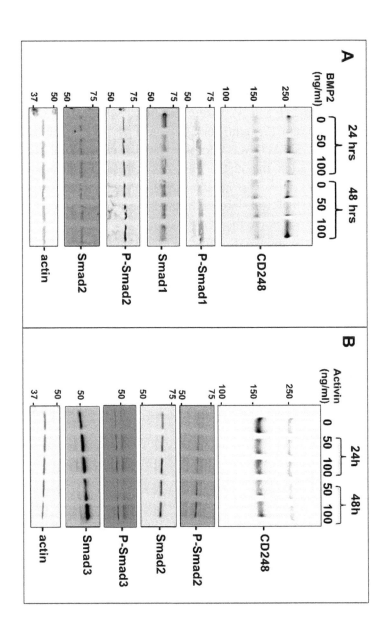

FIGURE 7: Regulation of CD248 by BMP-2 and Activin. MEF were incubated with different concentrations of BMP2 (A) or activin (B) for 24 or 48 hrs. Representative Western blots from 3 independent experiments are shown and were used to assess the effect on CD248 expression.

1.3.7 CANCER CELL LINES ARE RESISTANT TO TGFβ SUPPRESSION OF CD248

Since elevated CD248 is associated with tumorigenesis, we tested whether TGFβ could suppress CD248 in tumor cell lines as effectively as in the healthy non-cancerous cells examined above. Mouse B lymphoma cell lines, Wehi-231 and A20 were incubated with TGFβ at concentrations of 3 ng/ml and 12 ng/ml for 24 hrs and 48 hrs (Figure 8). Under these conditions, SMAD2 was phosphorylated, with minimal effect on Smad3 phosphorylation. In both the Wehi-231 cells (Figure 8A) and the A20 cells (Figure 8B), there was no significant suppression of CD248 expression in response to TGFβ. Indeed, in the latter, there was a slight increase in CD248 in response to the TGFβ.

We also examined the effect of TGFβ on the expression of CD248 by normal and cancer associated fibroblasts (NF and CAF, respectively) that were derived from mouse mammary tissues [33]. Protein levels of CD248 were relatively low in both of these cell lines, making it difficult to assess changes by Western blot. CD248 mRNA levels were therefore quantified by qRT-PCR (Figure 8C). Following exposure of the cells to 3 ng/ml or 12 ng/ml TGFβ for 24 and 48 hrs, CD248 mRNA accumulation was significantly suppressed in the NF, while in contrast, there was no effect on CD248 mRNA levels in the CAF. Overall, the preceding findings indicate that the expression of CD248 in cancer cells is resistant to regulation by TGFβ.

1.4 DISCUSSION

Since the discovery of CD248 [45], clinical and genetic evidence has pointed to it as a promoter of tumor growth and inflammation (reviewed in [2]). Increased expression of CD248 is detected in stromal cells surrounding most tumors, and high levels often correlate with a poor prognosis [20,23]. Means of interfering with the tumorigenic effects of CD248 have eluded investigators due to a lack of knowledge surrounding the regulation of CD248. This has limited opportunities for the design of innovative therapeutic approaches. In this report, we show that expression of CD248 by non-cancerous cells of mesenchymal origin is specifically and dramatically downregulated at a transcriptional and protein level by the pleiotropic cytokine, TGFβ, and that the response is dependent on canonical Smad2/3-dependent signaling. Notably, CD248 expression by cancer cells and cancer associated fibroblasts is not altered by TGFβ. The findings suggest that a TGFβ-based strategy to suppress CD248 may be useful as a therapeutic intervention to prevent early stage, but not later stage, tumorigenesis.

Members of the TGFβ family regulate a wide range of cellular processes (e.g. cell proliferation, differentiation, migration, apoptosis) that are highly context-dependent, i.e., stage of development, stage of disease, cell/tissue type and location, microenvironmental factors, and epigenetic factors. Under normal conditions, TGFβ plays a dominant role as a tumor suppressor at early stages of tumorigenesis, inhibiting cell proliferation and cell migration (reviewed in [46,47]). TGFβ ligands signal via TGFβRI (ALK-5) and TGFβRII. A third accessory type III receptor (TGFβRIII) lacks kinase activity, but facilitates the tumor-suppressor activities of TGFβ. TGFβ binds to TGFβRII which transphosphorylates ALK-5. In canonical signaling, ALK-5 then phosphorylates Smad2 and Smad3, inducing the formation of heteromeric complexes with Smad4, for translocation into the nucleus, interaction with transcription factors, and regulation of promoters of several target genes [48,49]. Disruption of TGFβ signaling has been associated with several cancers and a poor prognosis [47], and mice that lack TGFβ spontaneously develop tumors and inflammation [50].

TGFβ signaling is not, however, restricted to Smads 2 and 3, but can couple to non-canonical (Smad2/3-independent) effectors [48,51-54]. Recent data support the notion that canonical signaling favours tumor suppression, while non-canonical signaling tips the balance, such that TGFβ switches to become a promoter of tumor growth, invasion and metastasis, overriding the tumor-suppressing activities transmitted via Smad2/3. This dichotomous nature is known as the "TGFβ Paradox", a term coined to describe the conversion in function of TGFβ from tumor suppressor to tumor promoter [55-57]. The mechanisms underlying this switch are steadily being delineated, as regulation of the multiple effector molecules that are coupled to TGFβ are identified and characterized (reviewed in [47]). Our findings suggest that CD248 may be one such TGFβ-effector molecule that undergoes a context-dependent change in coupling, and thus may be a potential therapeutic target.

Upon determining that TGFβ suppresses CD248, we first showed that the response is dependent on Smad 2 signaling. This is consistent with the almost undetectable levels of CD248 in normal tissues, its expression presumably held in check at least in part by TGFβ's tumor suppressor properties. The fact that TGFβ induces phosphorylation of Smad2 in MEF that lack CD248, indicates that CD248 is not required for Smad2 phosphorylation. Rather, in the TGFβ-signaling pathway, CD248 is positioned "downstream" of Smad2/3 phosphorylation. We also showed that CD248 is downregulated by TGFβ primarily at a transcriptional level, and without affecting the stability of its mRNA. We have not determined which regions of the CD248 promoter are required for TGFβ-induced suppression. However, intriguingly, the murine promoter of the CD248 gene contains the sequence 5'-TTTGGCGG (position -543 to -536) [5] that overlaps with a consensus E2F transcription factor binding site. This is almost identical to the unique Smad3 DNA binding site in the c-myc promoter that is crucial for TGFβ-induced gene suppression [58]. Detailed mapping of the promoter will provide insights into precisely how CD248 is regulated by TGFβ.

We also examined whether TGFβ coupling to non-canonical effector molecules, ERK1/2 and p38, alters expression of CD248. Neither ERK1/2 nor p38, pathways implicated in TGFβ-induced metastasis, affected CD248 expression. Thus, based on current data, TGFβ-induced suppression of CD248 occurs primarily, if not exclusively, via canonical Smad2/3 signaling.

The specificity of the response of CD248 to TGFβ extends beyond Smad2/3-related signaling. In a survey of growth factors and cytokines, we could not identify other factors that similarly suppress (or conversely, increase) CD248 expression in MEF, 10 T1/2 cells or primary vascular smooth muscle cells. Even BMP2 and activin, members of the TGFβ superfamily and pleiotropic cytokines that also exhibit tumor promoter and suppressor activities, had little effect on CD248 expression. Although our survey was limited in range, concentration and time of exposure, the find-

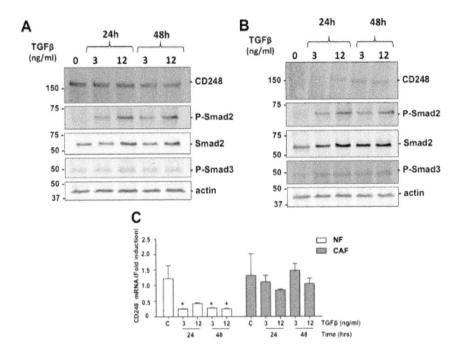

FIGURE 8: Regulation of CD248 in cancer cells. (A, B) Wehi-231 (A) and A20 (B) mouse lymphoma cells were incubated with different concentrations of TGFβ for 24 or 48 hrs and lysates were assessed by Western immunoblot. CD248 levels were minimally affected in spite of phosphorylation of Smad2. Results are representative of 3 independent experiments. (C) Normal fibroblasts (NF) and cancer associated fibroblasts (CAF) from murine mammary tissue were exposed to TGFβ for 24 or 48 hrs and CD248 mRNA levels were quantified and normalized to levels from untreated NF. CD248 mRNA levels in NF were significantly suppressed by TGFβ, whereas there was no effect on CD248 in CAF. *$p < 0.05$, $n = 3$.

ings suggest specificity, and highlight the central role that TGFβ likely plays in regulating expression of CD248 in non-cancerous cells.

Most notably, in two tumor cell lines and in cancer associated fibroblasts, the regulation of expression of CD248 was resistant to TGFβ. Indeed, in these cells, TGFβ neither decreased nor increased CD248, suggesting a decoupling of the regulatory link between TGFβ and CD248. Thus, with the switch from a tumor suppressor to a tumor promoter, TGFβ loses it ability to regulate CD248. Although TGFβ does not appear to directly participate in enhancing CD248 expression during late tumorigenesis, loss of its ability to suppress CD248 may be relevant in tumor progression and metastasis.

1.5 CONCLUSIONS

We have shown that the tumor suppressor properties of TGFβ, observed in early stage cancer, are likely mediated in part via suppression of CD248, the latter which is mediated via canonical Smad-dependent pathways. Upregulation of CD248 might be an early detection marker of tumor growth and metastasis, and may be valuable in monitoring TGFβ-based therapies. The clinical relevance of understanding how CD248 is regulated is highlighted by ongoing Phase 1 and 2 clinical trials in which the anti-CD248 antibody, MORAb-004, is being tested for efficacy in solid tumors and lymphomas (http://www.clinicaltrials.gov). Delineating the molecular mechanism(s) by which TGFβ loses its ability to suppress CD248 will be key for the design of additional therapeutic interventions to prevent and/or reduce CD248-dependent tumor cell proliferation and metastasis.

REFERENCES

1. Christian S, Ahorn H, Koehler A, Eisenhaber F, Rodi HP, Garin-Chesa P, Park JE, Rettig WJ, Lenter MC: Molecular cloning and characterization of endosialin, a C-type lectin- like cell surface receptor of tumor endothelium. J Biol Chem 2001, 276(10):7408-7414.
2. Valdez Y, Maia M, Conway EM: CD248: reviewing its role in health and disease. Curr Drug Targets 2012, 13(3):432-439.

3. Morser J: Thrombomodulin links coagulation to inflammation and immunity. Curr Drug Targets 2012, 13(3):421-431.

4. Greenlee-Wacker MC, Galvan MD, Bohlson SS: CD93: recent advances and implications in disease. Curr Drug Targets 2012, 13(3):411-420.

5. Opavsky R, Haviernik P, Jurkovicova D, Garin MT, Copeland NG, Gilbert DJ, Jenkins NA, Bies J, Garfield S, Pastorekova S, Oue A, Wolff L: Molecular characterization of the mouse Tem1/endosialin gene regulated by cell density in vitro and expressed in normal tissues in vivo. J Biol Chem 2001, 276(42):38795-38807.

6. Brady J, Neal J, Sadakar N, Gasque P: Human endosialin (tumor endothelial marker 1) is abundantly expressed in highly malignant and invasive brain tumors. J Neuropathol Exp Neurol 2004, 63(12):1274-1283.

7. MacFadyen JR, Haworth O, Roberston D, Hardie D, Webster MT, Morris HR, Panico M, Sutton-Smith M, Dell A, van der Geer P, Wienke D, Buckley CD, Isacke CM: Endosialin (TEM1, CD248) is a marker of stromal fibroblasts and is not selectively expressed on tumour endothelium. FEBS Let 2005, 579(12):2569-2575.

8. Christian S, Winkler R, Helfrich I, Boos AM, Besemfelder E, Schadendorf D, Augustin HG: Endosialin (Tem1) is a marker of tumor-associated myofibroblasts and tumor vessel-associated mural cells. Am J Pathol 2008, 172(2):486-494.

9. Simonavicius N, Robertson D, Bax DA, Jones C, Huijbers IJ, Isacke CM: Endosialin (CD248) is a marker of tumor-associated pericytes in high-grade glioma. Mod Pathol 2008, 21(3):308-315.

10. Maia M, de Vriese A, Janssens T, Moons M, van Landuyt K, Tavernier J, Lories RJ, Conway EM: CD248 and its cytoplasmic domain: a therapeutic target for arthritis. Arthritis Rheum 2010, 62(12):3595-3606.

11. Naylor AJ, Azzam E, Smith S, Croft A, Poyser C, Duffield JS, Huso DL, Gay S, Ospelt C, Cooper MS, Isacke C, Goodyear SR, Rogers MJ, Buckley CD: The mesenchymal stem cell marker CD248 (endosialin) is a negative regulator of bone formation in mice. Arthritis Rheum 2012, 64(10):3334-3343.

12. Simonavicius N, Ashenden M, van Weverwijk A, Lax S, Huso DL, Buckley CD, Huijbers IJ, Yarwood H, Isacke CM: Pericytes promote selective vessel regression to regulate vascular patterning. Blood 2012, 120(7):1516-1527.

13. Huber MA, Kraut N, Schweifer N, Dolznig H, Peter RU, Schubert RD, Scharffetter-Kochanek K, Pehamberger H, Garin-Chesa P: Expression of stromal cell markers in distinct compartments of human skin cancers. J Cutan Pathol 2006, 33(2):145-155.

14. Rupp C, Dolznig H, Puri C, Sommergruber W, Kerjaschki D, Rettig WJ, Garin-Chesa P: Mouse endosialin, a C-type lectin-like cell surface receptor: expression during embryonic development and induction in experimental cancer neoangiogenesis. Cancer Immun 2006, 6:10.

15. MacFadyen J, Savage K, Wienke D, Isacke CM: Endosialin is expressed on stromal fibroblasts and CNS pericytes in mouse embryos and is downregulated during development. Gene Expr Patterns 2007, 7(3):363-369.

16. St Croix B, Rago C, Velculescu V, Traverso G, Romans KE, Montgomery E, Lal A, Riggins GJ, Lengauer C, Vogelstein B, Kinzler KW: Genes expressed in human tumor endothelium. Science (New York, NY) 2000, 289(5482):1197-1202.

17. Dolznig H, Schweifer N, Puri C, Kraut N, Rettig WJ, Kerjaschki D, Garin-Chesa P: Characterization of cancer stroma markers: in silico analysis of an mRNA expression database for fibroblast activation protein and endosialin. Cancer Immun 2005, 5:10.

18. Huang HP, Hong CL, Kao CY, Lin SW, Lin SR, Wu HL, Shi GY, You LR, Wu CL, Yu IS: Gene targeting and expression analysis of mouse Tem1/endosialin using a lacZ reporter. Gene Expr Patterns 2011, 11(5-6):316-326.

19. Rouleau C, Curiel M, Weber W, Smale R, Kurtzberg L, Mascarello J, Berger C, Wallar G, Bagley R, Honma N, Hasegawa K, Ishida I, Kataoka S, Thurberg BL, Mehraein K, Horten B, Miller G, Teicher BA: Endosialin protein expression and therapeutic target potential in human solid tumors: sarcoma versus carcinoma. Clin Cancer Res 2008, 14(22):7223-7236.

20. Rouleau C, Smale R, Fu YS, Hui G, Wang F, Hutto E, Fogle R, Jones CM, Krumbholz R, Roth S, Curiel M, Ren Y, Bagley RG, Wallar G, Miller G, Schmid S, Horten B, Teicher BA: Endosialin is expressed in high grade and advanced sarcomas: evidence from clinical specimens and preclinical modeling. Int J Oncol 2011, 39(1):73-89.

21. Carson-Walter EB, Winans BN, Whiteman MC, Liu Y, Jarvela S, Haapasalo H, Tyler BM, Huso DL, Johnson MD, Walter KA: Characterization of TEM1/endosialin in human and murine brain tumors. BMC Cancer 2009, 9:417.

22. Davies G, Cunnick GH, Mansel RE, Mason MD, Jiang WG: Levels of expression of endothelial markers specific to tumour-associated endothelial cells and their correlation with prognosis in patients with breast cancer. Clin Exp Metastasis 2004, 21(1):31-37.

23. Zhang ZY, Zhang H, Adell G, Sun XF: Endosialin expression in relation to clinicopathological and biological variables in rectal cancers with a Swedish clinical trial of preoperative radiotherapy. BMC Cancer 2011, 11:89.

24. Smith SW, Eardley KS, Croft AP, Nwosu J, Howie AJ, Cockwell P, Isacke CM, Buckley CD, Savage CO: CD248+ stromal cells are associated with progressive chronic kidney disease. Kidney Int 2011, 80(2):199-207.

25. Nanda A, Karim B, Peng Z, Liu G, Qiu W, Gan C, Vogelstein B, St Croix B, Kinzler KW, Huso DL: Tumor endothelial marker 1 (Tem1) functions in the growth and progression of abdominal tumors. Proc Natl Acad Sci USA 2006, 103(9):3351-3356.

26. Lax S, Ross EA, White A, Marshall JL, Jenkinson WE, Isacke CM, Huso DL, Cunningham AF, Anderson G, Buckley CD: CD248 expression on mesenchymal stromal cells is required for post-natal and infection-dependent thymus remodelling and regeneration. FEBS Open Bio 2012, 2:187-190.

27. Maia M, DeVriese A, Janssens T, Moons M, Lories RJ, Tavernier J, Conway EM: CD248 facilitates tumor growth via its cytoplasmic domain. BMC Cancer 2011, 11:162.

28. Marty C, Langer-Machova Z, Sigrist S, Schott H, Schwendener RA, Ballmer-Hofer K: Isolation and characterization of a scFv antibody specific for tumor endothelial marker 1 (TEM1), a new reagent for targeted tumor therapy. Cancer Let 2006, 235(2):298-308.

29. Zhao A, Nunez-Cruz S, Li C, Coukos G, Siegel DL, Scholler N: Rapid isolation of high-affinity human antibodies against the tumor vascular marker Endosialin/TEM1, using a paired yeast-display/secretory scFv library platform. J Immunol Methods 2011, 363(2):221-232.

30. Ohradanova A, Gradin K, Barathova M, Zatovicova M, Holotnakova T, Kopacek J, Parkkila S, Poellinger L, Pastorekova S, Pastorek J: Hypoxia upregulates expression of human endosialin gene via hypoxia-inducible factor 2. Bri J Cancer 2008, 99(8):1348-1356.

31. Ray JL, Leach R, Herbert JM, Benson M: Isolation of vascular smooth muscle cells from a single murine aorta. Methods Cell Sci 2001, 23(4):185-188.

32. Suresh Babu S, Wojtowicz A, Freichel M, Birnbaumer L, Hecker M, Cattaruzza M: Mechanism of stretch-induced activation of the mechanotransducer zyxin in vascular cells. Sci Signal 2012, 5(254):ra91.

33. Calvo F, Ege N, Grande-Garcia A, Hooper S, Jenkins RP, Chaudhry SI, Harrington K, Williamson P, Moeendarbary E, Charras G, Sahai E: Mechanotransduction and YAP-dependent matrix remodelling is required for the generation and maintenance of cancer-associated fibroblasts. Nat Cell Biol 2013, 15(6):637-646.

34. Garate M, Campos EI, Bush JA, Xiao H, Li G: Phosphorylation of the tumor suppressor p33(ING1b) at Ser-126 influences its protein stability and proliferation of melanoma cells. FASEB J 2007, 21(13):3705-3716.

35. Conway EM, Rosenberg RD: Tumor necrosis factor suppresses transcription of the thrombomodulin gene in endothelial cells. Mol Cell Biol 1988, 8:5588-5592.

36. Xu P, Liu J, Derynck R: Post-translational regulation of TGF-beta receptor and Smad signaling. FEBS Let 2012, 586(14):1871-1884.

37. Moustakas A, Heldin CH: The regulation of TGFbeta signal transduction. Development 2009, 136(22):3699-3714.

38. Chen G, Deng C, Li YP: TGF-beta and BMP signaling in osteoblast differentiation and bone formation. Int J Biol Sci 2012, 8(2):272-288.

39. Karlsson G, Liu Y, Larsson J, Goumans MJ, Lee JS, Thorgeirsson SS, Ringner M, Karlsson S: Gene expression profiling demonstrates that TGF-beta1 signals exclusively through receptor complexes involving Alk5 and identifies targets of TGF-beta signaling. Physiol Genomics 2005, 21(3):396-403.

40. Inman GJ, Nicolas FJ, Callahan JF, Harling JD, Gaster LM, Reith AD, Laping NJ, Hill CS: SB-431542 is a potent and specific inhibitor of transforming growth factor-beta superfamily type I activin receptor-like kinase (ALK) receptors ALK4, ALK5, and ALK7. Mol Pharmacol 2002, 62(1):65-74.

41. Favata MF, Horiuchi KY, Manos EJ, Daulerio AJ, Stradley DA, Feeser WS, Van Dyk DE, Pitts WJ, Earl RA, Hobbs F, Copeland RA, Magolda RL, Scherle PA, Trzaskos JM: Identification of a novel inhibitor of mitogen-activated protein kinase kinase. J Biol Chem 1998, 273(29):18623-18632.

42. Hippenstiel S, Soeth S, Kellas B, Fuhrmann O, Seybold J, Krull M, Eichel-Streiber C, Goebeler M, Ludwig S, Suttorp N: Rho proteins and the p38-MAPK pathway are important mediators for LPS-induced interleukin-8 expression in human endothelial cells [In Process Citation]. Blood 2000, 95(10):3044-3051.

43. Sun F, Pan Q, Wang J, Liu S, Li Z, Yu Y: Contrary effects of BMP-2 and ATRA on adipogenesis in mouse mesenchymal fibroblasts. Biochem Genetics 2009, 47(11-12):789-801.

44. Tojo M, Hamashima Y, Hanyu A, Kajimoto T, Saitoh M, Miyazono K, Node M, Imamura T: The ALK-5 inhibitor A-83-01 inhibits Smad signaling and epithelial-

to-mesenchymal transition by transforming growth factor-beta. Cancer Sci 2005, 96(11):791-800.

45. Rettig WJ, Garin-Chesa P, Healey JH, Su SL, Jaffe EA, Old LJ: Identification of endosialin, a cell surface glycoprotein of vascular endothelial cells in human cancer. Proc Natl Acad Sci U S A 1992, 89(22):10832-10836.

46. Wendt MK, Tian M, Schiemann WP: Deconstructing the mechanisms and consequences of TGF-beta-induced EMT during cancer progression. Cell Tissue Res 2012, 347(1):85-101.

47. Drabsch Y, ten Dijke P: TGF-beta signalling and its role in cancer progression and metastasis. Cancer Metastasis Rev 2012, 31(3-4):553-568.

48. Prud'homme GJ: Pathobiology of transforming growth factor beta in cancer, fibrosis and immunologic disease, and therapeutic considerations. Lab Invest 2007, 87(11):1077-1091.

49. Ikushima H, Miyazono K: Cellular context-dependent "colors" of transforming growth factor-beta signaling. Cancer Sci 2010, 101(2):306-312.

50. Zhang Y, Wen G, Shao G, Wang C, Lin C, Fang H, Balajee AS, Bhagat G, Hei TK, Zhao Y: TGFBI deficiency predisposes mice to spontaneous tumor development. Cancer Res 2009, 69(1):37-44.

51. Daroqui MC, Vazquez P, Bal de Kier Joffe E, Bakin AV, Puricelli LI: TGF-beta autocrine pathway and MAPK signaling promote cell invasiveness and in vivo mammary adenocarcinoma tumor progression. Oncol Rep 2012, 28(2):567-575.

52. Fleming YM, Ferguson GJ, Spender LC, Larsson J, Karlsson S, Ozanne BW, Grosse R, Inman GJ: TGF-beta-mediated activation of RhoA signalling is required for efficient (V12)HaRas and (V600E)BRAF transformation. Oncogene 2009, 28(7):983-993.

53. Wakefield LM, Roberts AB: TGF-beta signaling: positive and negative effects on tumorigenesis. Curr Opin Genetics Dev 2002, 12(1):22-29.

54. Wendt MK, Smith JA, Schiemann WP: p130Cas is required for mammary tumor growth and transforming growth factor-beta-mediated metastasis through regulation of Smad2/3 activity. J Biol Chem 2009, 284(49):34145-34156.

55. Rahimi RA, Leof EB: TGF-beta signaling: a tale of two responses. J Cell Biochem 2007, 102(3):593-608.

56. Schiemann WP: Targeted TGF-beta chemotherapies: friend or foe in treating human malignancies? Exp Rev Anticancer Ther 2007, 7(5):609-611.

57. Tian M, Schiemann WP: The TGF-beta paradox in human cancer: an update. Future Oncol 2009, 5(2):259-271.

58. Frederick JP, Liberati NT, Waddell DS, Shi Y, Wang XF: Transforming growth factor beta-mediated transcriptional repression of c-myc is dependent on direct binding of Smad3 to a novel repressive Smad binding element. Mol Cell Biol 2004, 24(6):2546-2559.

CHAPTER 2

GRADED INHIBITION OF ONCOGENIC RAS-SIGNALING BY MULTIVALENT RAS-BINDING DOMAINS

MARTIN AUGSTEN, ANIKA BÜTTCHER, RAINER SPANBROEK, IGNACIO RUBIO, AND KARLHEINZ FRIEDRICH

2.1 BACKGROUND

The prototypical Ras isoforms H-Ras, K-Ras and N-Ras (collectively Ras) are membrane-associated small G-proteins that cycle between an active, GTP-bound and an inactive, GDP-bound state. Ras becomes activated, that is GTP-loaded, by guanine nucleotide exchange factors (GEFs) such as Sos or RasGRP, which are themselves engaged and activated downstream of various cell surface receptors via adapter proteins, like Shc and Grb-2 and/or via second messenger lipids like phosphatidic acid or diacylglycerol [1,2]. Inactivation of GTP-loaded Ras occurs through a GTP-hydrolase (GTPase) activity intrinsic to Ras and enhancement of this reaction by GTPase activating proteins (GAPs) [1,3]. Ras function is also controlled

This chapter was originally published under the Creative Commons Attribution License. Augsten M, Böttcher A, Spanbroek R, Rubio I, and Friedrich K. Graded Inhibition of Oncogenic Ras-Signaling by Multivalent Ras-Binding Domains. Cell Communication and Signaling 12,1 (2014). doi:10.1186/1478-811X-12-1.

by a series of obligatory post-translational modifications which include an initial farnesylation step and the reversible attachment of palmitate groups to N-Ras and H-Ras [4]. Although many details of this complex processing remain unknown, it is well established that the correct post-translational processing is required to direct Ras to cellular membranes and specific microdomains within the plasma membrane (PM) [5].

Ras proteins play important roles in receptor-mediated signal trans-duction pathways that control cell proliferation and differentiation and are moreover critically involved in the regulation of cell motility and in-vasiveness [3,6,7]. Ras regulates these processes by feeding signals into various major signaling pathways, prominently the Erk kinase pathway, a cascade of protein kinases which ultimately drives the transcription of key target genes for cell cycle progression and other processes [8]. Ras-dependent activation of the Erk kinase pathway relies on the productive contact of Ras-GTP with members of the Raf family of serine/threonine kinases (collectively Raf), which together with other coincident inputs result in Raf activation [9,10]. Raf binds Ras-GTP via a N-terminally located Ras-binding domain (RBD), roughly 80 amino acid residues in size, that features several orders of magnitude higher affinity for Ras-GTP than Ras-GDP [11,12]. Several amino acid residues in the RBD are critical for the interaction with Ras-GTP and mutation of these sites impairs the high affinity binding of RBD to Ras-GTP [13,14].

Tight regulation of the Ras activation status is critical for cell physi-ology. Mutations that convert Ras into an oncoprotein are found in up to 25% of human tumors [15] (http://www.sanger.ac.uk). Oncogenic mu-tations, including substitutions of glycine 12 and glutamine 61, com-promise the intrinsic and GAP-promoted GTPase activity of Ras. In agreement with a critical role of continuous aberrant Ras-GTP elicited signaling in oncogenesis, defects in GAP function or gain-of-function mutations in GEFs do also result in cell transformation and other patho-logical conditions [1,16-18]. Aberrant activation of the Ras/Raf-pathway contributes to essential aspects of tumor development and progression such as cell cycle deregulation, avoidance of apoptosis, cell motility and drug resistance and are moreover known to be important for tumor maintenance and cancer cell viability at late stages of tumorogenesis [19,20]. Due to its nodal role in cell transformation, Ras was early on identified as an attractive target for pharmaceutical intervention. Soon after the identification and characterization of farnesyl transferase (FTase) as the enzyme responsible for the first in the series of Ras-modifications, FTase inhibitors which efficiently blocked Ras mediated cell transformation in cell culture and animal models were developed [21-23]. However, the results of clinical trials with a large panel of FTase inhibitors were disappointing and discouraged many from pursuing further efforts to target oncogenic Ras. Later, Ras neutralizing antibodies were employed as oncogenic Ras blockers in cell culture experimentation [24-26] and mutant Ras epitopes were exploited for their suitability as antigens in the development of cancer vaccines [27]. Further approaches to target oncogenic Ras rested on antisense oligonucleotides directed to the Ras mRNA [28], and more recently on exploiting structural information and improved in silico approaches to identify and target druggable pockets or moieties that affect Ras nucleotide exchange [29,30], Ras activation [31,32], effector interaction [33,34] or binding to escort proteins critical for subcellular trafficking [35]. Moreover, numerous studies have targeted Ras downstream effector pathways such as Raf kinases, MEK or PI3Ks [36,37]. However, to date, Raf, MEK and PI3K inhibitors have shown little efficacy in the treatment of oncogenic Ras driven tumours, essentially evidencing that we still do not understand all intricacies of Ras signaling in the context of oncogenesis. In sum, in the light of the high prevalence of Ras mutations in human tumors it is sobering that 30 years after its discovery as the first human oncogene no strategy for the direct blockade of oncogenic Ras has reached clinical use.

In the present study we have developed and characterized a novel approach for the blockade of Ras-GTP dependent signaling. We demonstrate that oligovalent, Ras-GTP scavenging probes composed of up to 3 wild-type or mutant RBD modules, behave as "multivalent scavengers of oncogenic ras" (MSOR) that can be applied to inhibit various parameters of Ras-dependent oncogenic cell transformation in an adjustable fashion.

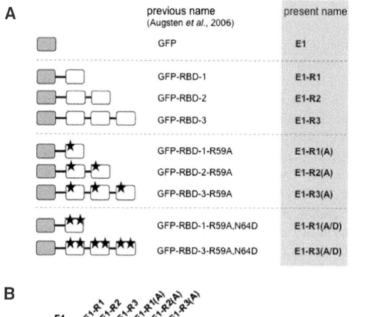

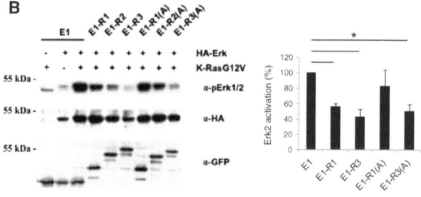

FIGURE 1: MSOR inhibit oncogenic Ras-induced signaling. (A) Schematic presentation of the EGFP-fused RBD mono- and oligomers explored in this study. The different mono- di and trivalent probes (R1, R2, R3) are composed of either wild-type or mutant c-Raf-derived RBDs. The RBD-mutations R59A (*) and R59A/N64D (**) are abbreviated by (A) and (A/D), respectively. Oligovalent probes consisting of two or three RBDs are collectively described as MSOR for multivalent scavengers of oncogenic Ras. (B) The influence of RBD monomers and MSOR on Ras-induced signaling was studied in NIH3T3 cells transiently expressing K-RasG12V, HA-tagged Erk2 and mono-, di- or trivalent EGFP-RBDs (wild type or R59A-mutant). Cell lysates were subjected to western blot analysis detecting phosphorylated and total Erk2 and expression of EGFP-RBD-constructs. Signals from four independent experiments were quantified and expressed as ratio of phosphorylated and total Erk2.

2.2 RESULTS

2.2.1 OLIGOVALENT RAS-BINDING DOMAINS BLOCK ONCOGENIC RAS-INDUCED SIGNALING

We have previously employed modular probes consisting of oligomerized Ras-binding domain (RBD) units as novel Ras-GTP-specific probes. Fused to EGFP, these oligomers are instrumental for the visualization of growth factor-stimulated activation of endogenous Ras in live cells [38-41]. In the course of those studies we noticed that oligomeric RBD-variants sequestered Ras-GTP in vitro in an oligomerization grade-dependent fashion and interfered with Ras-dependent signaling in COS-7 cells [38]. This prompted us to test whether or not RBD-oligomers can be used to block the action of oncogenic Ras. In the present study we use the MSOR nomenclature introduced in ref. [39] which is recapitulated in Figure 1A.

In order to confirm the previously observed inhibitory effect of MSOR on oncogenic Ras-signaling we compared the impact of mono-, di-and trimeric wildtype RBDs (E1-R1, E1-R2, E1-R3, respectively) on oncogenic K-RasG12V induced Erk kinase activation in mouse fibroblasts. NIH3T3 cells were transfected with various combinations of constitutively active, oncogenic K-RasG12V, HA-tagged Erk2 and different RBD-expressing plasmids. As expected, K-RasG12V enhanced activation of the co-transfected Erk2 kinase (as assessed by Erk2 phosphorylation) and this activation was diminished in the presence of mono- and oligovalent wild-type RBD constructs (Figure 1B). Importantly, the blocking efficiency of RBDs increased as the degree of oligomerization rose from single (E1-R1) to triple (E1-R3) with the latter abolishing RasG12V-dependent signaling.

To substantiate this observation and to ascertain the specificity of the blocking effect, we tested RBD-variants containing the R59A mutation which lowers the affinity of RBD for Ras-GTP by about 30fold [14,42]. This type of mutations is commonly used in the context of full-length Raf to disrupt Ras-to-Raf signal propagation in cell biological studies [11]. In line with its inability to interact with Ras-GTP in vitro[38] the RBD-

R59A-monomer E1-R1(A) did not significantly block Ras-K-RasG12V-induced phosphorylation of Erk2 (Figure 1B). However, expression of the same RBD-R59A module as a dimer (E1-R2(A)) or trimer (E1-R3(A)) inhibited RasG12V-induced signaling with gradually increasing strength, albeit always with lower potency than the wild-type MSOR counterparts. Noteworthy, E1-R3(A) expression was lower than that of its monomeric counterpart E1-R1(A), arguing that the gradual increase in blocking strength did not reflect the mere increase in numbers of RBD modules but rather was contingent on the presence of concatenated RBD units. These data recapitulated previous findings from COS-7 cells [38], and illustrated the validity of the oligomerization principle as a means to raise and tune the avidity and affinity of oligovalent binding domains for Ras-GTP.

2.2.2 RBD-OLIGOMERS INHIBIT DIFFERENT PARAMETERS OF RAS-MEDIATED CELLULAR TRANSFORMATION

Oncogenic Ras-signaling stimulates several pro-tumorigenic pathways that regulate cell proliferation, migration and invasion, among other events. Given their ability to inhibit K-RasG12V-signaling, we hypothesized that MSOR might block aspects of oncogenic Ras-driven transformation. First, we tested the ability of E1-R1 and E1-R3 to block K-RasG12V-induced invasion in matrigel. As shown in Figure 2A, both wild-type RBD-variants interfered with the K-RasG12V-induced invasion of COS-7 cells in matrigel-coated trans-well migration chambers. Secondly, we investigated whether MSOR would also affect anchorage-independent growth, another important hallmark of cellular transformation. To this end we chose to study NIH3T3 cells, since these cells retain numerous features of untransformed cells including cell-cell contact inhibition or the requirement for substrate attachment for productive growth and proliferation. However, NIH3T3 cells do not express EGFR, the prototypical receptor tyrosine kinase commonly used to robustly activate Ras [43], but instead express high levels of PDFGR which is a poor Ras activator. To study Ras signaling in these cells we employed an engineered subline termed NIH-TM which responds to stimulation with Nerve Growth Factor (NGF) owing to the stable expression of a TrkA/c-Met hybrid receptor composed of the extracellular part of Trk and the intracellular domain of c-Met [44]. Stimulation of c-Met activates Ras via the canonical Grb-2/Sos pathway and induces proliferation of NIH3T3 cells [45]. Moreover, over-activation of this receptor tyrosine kinase promotes tumor growth and metastasis [46]. Accordingly, NGF-treatment of NIH-TM cells lead to increased colony formation in soft agar and this effect was completely reversed in the presence of E1-R1 or E1-R3 (Figure 2B), consistent with the ability of wild-type RBD-constructs to also block growth factor-stimulated Ras signaling.

Anchorage-independent growth and cell invasion depend on the action of matrix-degrading enzymes. The promoter region of several protease-encoding genes contains a Ras-responsive element (RRE) or an RRE-like enhancer motif [47,48]. Microarray analysis confirmed that oncogenic K-Ras induced the expression of several protease genes of the ADAM's and cathepsin families that act both intra- and extracellularly and are involved in matrix remodeling (Figure 2C, Additional file 1). Importantly, the Ras-stimulated upregulation of these proteases was abrogated by E1-R3 (Figure 2C, Additional file 1). Furthermore, this MSOR-construct decreased RasG12V-dependent activation of the RRE-containing MMP-1 promoter in NIH3T3 cells, as assayed using a luciferase reporter system (Figure 2D). Interestingly, in this case the single RBD unit (E1-R1) was unable to even partially inhibit the effect of K-RasG12V (Figure 2D) or H-RasG12V (Additional file 2), highlighting once more the oligomerization dependent, adjustable blocking potency of MSOR. Moreover, these data suggested that distinct end points of oncogenic Ras signaling exhibit varying sensitivities to the action of RBD polypeptides.

2.2.3 MSOR INTERFERE WITH RAS-DEPENDENT CELL SURVIVAL SIGNALING AND INDUCE APOPTOSIS

So far, the impact of MSOR was studied in the context of oncogenic Ras signaling. However, we noticed previously that expression of high affinity MSOR in the absence of constitutively active Ras has a profound effect on the morphology and viability of various types of cells [38]. Figure 3A shows fluorescence images of COS-7 cells expressing E1-R1, E1-R2 or E1-R3 in the absence of Ras co-transfection. Whereas expression of E1-

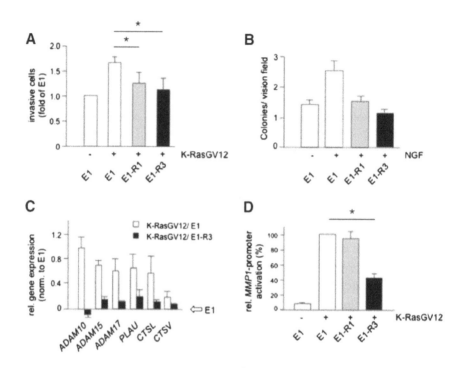

FIGURE 2: MSOR mitigate different parameters associated with cellular transformation. (A) The influence of mono- and trivalent wild-type RBDs on Ras-driven invasion was analyzed after transient transfection of COS-7 cells with expression constructs for K-RasG12V and E1, E1-R1 or E1-R3 and subsequent transmigration of transfected cells through a Matrigel® layer. The figure shows the average of three independent experiments. (B) The impact of mono- and trivalent wild-type RBD probes on c-Met-stimulated anchorage-independent growth was investigated by seeding NIH3T3-TM cells transiently expressing E1, E1-R1 or E1-R3 into soft agar and subsequent culture in the presence or absence of 25 ng/ml NGF. Colony formation was evaluated by counting of colonies in at least ten arbitrarily selected vision fields. The figure shows the average of three independent experiments. (C) Effect of the trivalent wild-type RBD construct on the Ki-RasG12V-induced protease gene expression. COS-7 cells were transiently transfected with a plasmid encoding E1 alone or an expression construct for K-RasG12V along with E1 or E1-R3. The impact of RBD constructs on K-RasG12V-stimulated expression of different proteases was analyzed on a custome oligonucleotide microarray. Signals were assessed densitometrically and normalized to the E1 expression level. See Material and methods for a more detailed description. Data are derived from three independent experiments. (D) Consequences of mono- and trivalent wild-type RBD probes on the K-RasG12V-stimulated induction of the human MMP1-promoter. NIH3T3 cells were transiently transfected with E1, E1-R1

or E1-R3 together with an expression construct encoding K-RasG12V as indicated. Then, a MMP-1-firefly-luciferase reporter plasmid was co-transfected along with a reference renilla luciferase construct and the relative luciferase activity was determined. The figure shows the average of three independent experiments each performed in duplicates.

R1 had no obvious effect on morphology and overall appearance of COS-7 cells, expression of the more avid MSOR variants E1-R2 and E1-R3 induced dramatic changes in cell morphology giving rise to spindle-like and asymmetric shapes, fragmented nuclei, vacuoles and membrane blebbing (Figure 3A).

Since membrane blebbing and other phenotypic changes in cells expressing E1-R3 were reminiscent of apoptotic cells we investigated whether or not MSOR induced apoptosis of cells expressing native wild-type Ras. Annexin V-staining confirmed the increased occurrence of apoptosis among MSOR-transfected COS-7 cells (Figure 3B). These data are compatible with a MSOR-mediated blockade of basal, endogenous Ras-GTP signaling, which reportedly protects cells from apoptosis [49]. This notion was further supported by microarray data showing that E3-R3 upregulated the expression of caspases (Figure 3C, Additional file 1), even so in the presence of co-transfected oncogenic Ras. Importantly, the higher potency of E1-R3 versus E1-R1 in apoptosis induction was not a result of an overall higher total number of RBD units but caused by the presence of the oligovalent polypeptides, because cells expressing up to 5 fold higher levels of E1-R1 did not exhibit the same signs of cellular breakdown (unpublished observation). We concluded from these findings that MSOR impair cell survival by the sustained strong sequestration and blockade of basal Ras-GTP signaling.

2.2.4 ADJUSTED INHIBITION OF RAS-MEDIATED CELLULAR EFFECTS BY INDUCIBLE MSOR EXPRESSION

The cytotoxic effects of E1-R2 and E1-R3 prompted us to develop strategies that allowed tuning the action of MSOR. First, we employed a tetra-

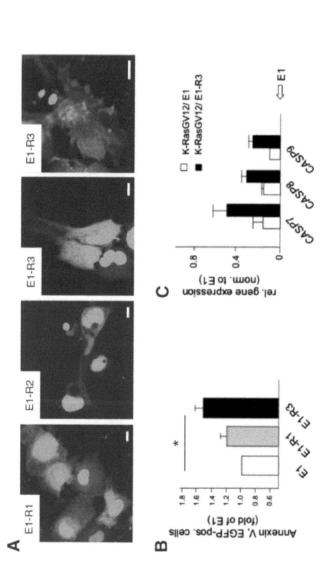

FIGURE 3: Targeting endogenously active Ras by MSOR impacts on cell survival. (A) Confocal images of Cos-7 cells transiently expressing EGFP or mono-, di- or trivalent wild-type RBD probes. Scale bar 10 μm. (B) The fraction of dead cells among E1-, E1-R1- or E1-R3-expressing cells was determined by measuring Annexin V-positive cells using FACS and normalized to the EGFP-expressing condition. Data represent three independent experiments. (C) Caspase gene expression in COS-7 cells, transiently transfected with an expression construct for K-RasG12V along with E1 or E1-R3 was analyzed on a custom oligonucleotide microarray. Signals were assessed densitometrically and normalized to the E1 expression level. See Material and methods for a more detailed description. Data are derived from three independent experiments. See Material and methods for a more detailed description.

cycline controllable system (Tet-off system) to regulate the expression of highly avid MSOR like E1-R3. COS-7 cells were transiently transfected with Tet-off constructs driving the expression of monomeric E1-R1 and trimeric E1-R3. In a non-repressed setting, expression of E1-R1 and E1-R3 was readily detectable (Figure 4A) but did not induce the prominent morphological changes observed under conditions of enhanced expression (Figure 3A). Addition of increasing concentrations of the tetracycline-derivative doxycycline (Dox) to the culture medium inhibited the MSOR expression in a concentration-dependent manner (Figure 4A), thus confirming the proper function of the inducible expression system.

Next, the effect of experimentally induced expression of RBD-constructs on the RasG12V-stimulated Erk2-activation in COS-7 was assessed (Figure 4B). Induction of E1-R3 expression decreased RasG12V-sparked Erk2-phosphorylation while the corresponding monomer was ineffective under the same conditions. This finding contrasts with the blocking action of E1-R1 in transient overexpression experiments (see Figure 1B) and suggested that MSOR-dependent blockade of distinct Ras elicited effects may depend on the expression levels achieved in individual experiments and/or may sometimes require sustained action of the MSOR proteins over a longer period of time.

In agreement with its blocking of Erk2 activation, the wild-type trimer but not the monomer was able to blunt RasG12V-stimulated activation of the *MMP-1*-reporter in NIH3T3 cells (Figure 4C) and EGF-driven invasion of COS-7 cells (Figure 4D). Taken together these data illustrate the efficacy of inducible MSOR to control and tune Ras action.

2.2.5 CONTROLLED INHIBITION OF ONCOGENIC RAS BY ATTENUATED MSOR

Another potential approach for reducing the cytotoxicity of MSOR constructs was the introduction of specific mutations in the RBD that strongly decrease their affinity for Ras-GTP, like the R59A mutation described above. This approach was successfully applied previously, and lead to the development of the double point mutant RBD-R59A/N64D, which in its trimeric form E1-R3(A/D) retained high avidity for Ras-GTP while exhibiting little cytotoxicity [38,39].

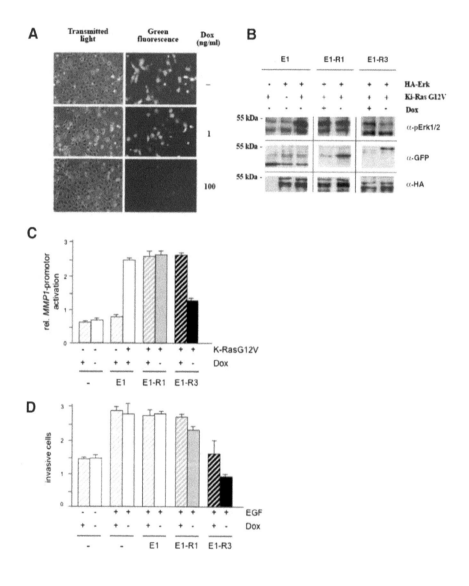

FIGURE 4: Adjusted, low-level expressed MSOR maintain their Ras-blocking activity. (A)*Tet-off* promoter-controlled expression of RBD-monomers and MSOR. COS-7 cells were transfected with pNRTIS 12-derived plasmids encoding E1-R3 and subsequently left untreated or treated with different concentrations of doxycycline (Dox) to modulate MSOR expression levels. (B) The impact of inducible mono- and trivalent wild-type RBD probes on oncogenic Ras-stimulated MAP-kinase signaling. COS-7 cells were transiently transfected with different constructs as indicated. Next, cells were divided into samples

that were grown either in the absence (-) or presence (+) of 100 ng/ml Dox. Cells lysates were analyzed by Western blot detecting phosphorylated and total Erk2 and the expression of EGFP-constructs. One representative experiment out of three independent experiments is shown. (C) Influence of inducible mono- and trivalent wild-type RBD probes on oncogenic Ras-stimulated *MMP-1* promoter activation. NIH3T3 cells were transiently transfected with the *MMP-1*-luciferase reporter, the K-RasG12V-encoding construct and Tet-off-controlled EGFP-constructs E1, E1-R1 or E1-R3. Expression of EGFP-constructs was turned on (-Dox) or off (+100 ng/ml Dox) and reporter gene activity was measured. The figure shows the average result from three independent experiments each performed in triplicate. (D) Influence of inducible mono- and trivalent wild-type RBD probes on EGF-stimulated cell invasion. COS-7 cells were transfected with the Tet-off-regulated EGFP-constructs indicated, and cultured in the absence of Dox. Subsequently, EGFP-expressing cells were collected by preparative fluorescence activated cell sorting and cultured in absence or presence of 100 ng/ml Dox to regulate the expression of E1, E1-R1 and E1-R3. Then cells were collected, seeded onto Matrigel-coated Transwells and subjected to invasion in absence or presence of 50 ng/ml EGF. Results represent the means of two entirely independent experiments.

In line with those features, over-expression of E1-R3(A/D) or its monomeric counterpart E1-R1(A/D) in COS-7 cells did not induce morphological changes or apoptosis (Figure 5A) as observed with the wild-type MSOR E1-R3 (Figure 3A). Similarly to E1-R1(A), the E1-R1(A/D) monomer did not impact on oncogenic K-Ras-driven signal transduction (Figure 5B). However, the trivalent double point mutant E1-R3(A/D) clearly diminished the RasG12V-induced Erk2-activation in both COS-7 and NIH3T3 cells (Figure 5B). Moreover, E1-R3(A/D) did also abrogate aspects of cellular transformation such as MMP1-activation (Figure 5C) and cell invasion (Figure 5D). Collectively, these findings illustrated that even low-affinity, biologically inert modules like the double point mutant RBD-R59A/N64D can be converted into robust scavengers of oncogenic Ras by increasing their avidity for Ras-GTP via oligomerization.

2.3 DISCUSSION

This study describes a novel application for the RBD of c-Raf as a building block of multivalent probes for the adjustable and graded inhibition of oncogenic Ras signaling. The data presented herein illustrate that MSOR are able to specifically target and block various events downstream of aber-

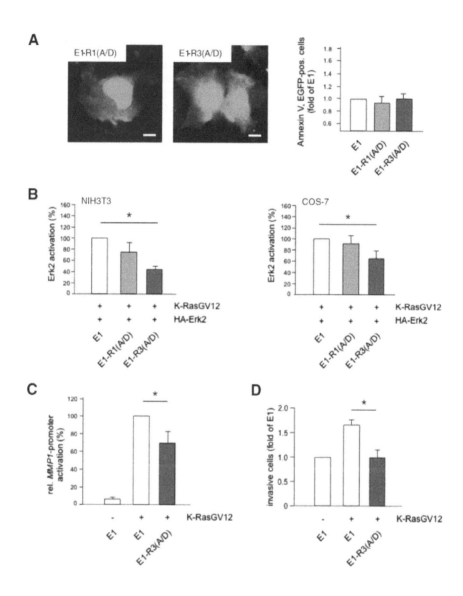

FIGURE 5: A non-toxic MSOR-variant efficiently blocks Ras-induced signaling and transformation. (A) Confocal images of COS-7 cells transiently expressing the mono- and trivalent low affinity RBD-R59A/N64D probes (left panel). Scale bar 10 μm. Cell death analysis of COS-7 cells transiently expressing E1, E1-R1(A/D) or E1-R3(A/D) using Annexin V-staining (right panel). The average of three independent experiments is shown. (B) Quantitative assessment of Western blot analysis of oncogenic Ras-induced

Erk2 activation. NIH3T3 (left panel) and COS-7 (right panel) cells transiently expressing K-RasG12V, HA-tagged Erk2 and E1, E1-R1(A/D) or E1-R3(A/D) were subjected to western blot analysis detecting phosphorylated and total Erk2. Signals were quantified and expressed as ratio of phosphorylated and total Erk2. The ratios derived from the different conditions were normalized to EGFP, and the average of three independent experiments is shown in the figure. (C) Impact of low-affinity RBD constructs on the K-RasG12V-stimulated induction of the human *MMP1*-promoter. NIH3T3 cells were transiently transfected with the different RBD-constructs indicated. Then, a *MMP-1*-firefly-luciferase reporter plasmid was co-transfected along with a reference renilla luciferase construct and the relative luciferase activity was determined after 24 h. The figure shows the result of three independent experiments each performed in duplicate. (D) Influence of the low-affinity RBD probes on the oncogenic Ras-stimulated invasiveness of COS-7 in a Matrigel-coated Transwell assay. The data depicted show the average from three independent experiments.

rant Ras-signaling including Erk-activation (Figures 1B and 5B), induction of matrix-remodeling enzymes (Figures 2C, 2D and 5C), Ras-stimulated matrix invasion (Figures 2A and 5D) and growth factor-induced contact-independent growth (Figure 2B). Moreover, it is worth emphasizing that MSOR not only counteracted the action of oncogenic Ras itself but also abrogated several parameters of cellular transformation sparked by cell surface growth factor receptors that signal via Ras (Figures 2B and 4D), suggesting a potentially broader application of MSOR in pro-tumorigenic settings that involve aberrant Ras-signaling. Importantly, the binding properties of MSOR are amenable to manipulation at three different levels: 1st, by varying their oligomerization grade and thus the avidity towards Ras-GTP [38], 2nd, by introducing point mutations in single RBD modules, affecting the affinity of individual RBDs to Ras-GTP and 3rd by regulating their protein expression levels. Several observations reported here strongly indicate that different combinations of the three parameters enumerated above will generate MSOR with distinct binding and inhibitory properties. For example, the wild-type RBD-monomer R1 effectively blocked different aspects of enhanced Ras-signaling (Figures 1B, 2A and 2B) when over-expressed to high levels in cells but it was ineffective at low expression levels in most cell types studied (Figure 4). In contrast, the trivalent protein R3 exhibited strong inhibitory effect in the same settings irrespective of its expression levels, suggesting that a higher avidity for Ras-GTP effectively increases the blocking potency and essentially compensates for low expression levels.

Along the same lines, we observed that one and the same RBD probe exhibits variable potencies for blocking different events downstream of oncogenic Ras. For instance, the monovalent wild-type unit R1 does not even partially affect matrix metalloproteinase induction by RasG12V, even though it does impinge on proximal Ras effectors like Erk in essentially the same system. The simplest explanation for this and related observations is that distinct cell biological readouts of oncogenic Ras require the action of different Ras effector pathways, or combinations thereof, that are distinctively sensitive to MSOR action. Indeed, the three most well characterized Ras effectors, Raf, PI3K and Ral GDS exhibit a large variance in their thermodynamic affinities for Ras-GTP of up to two orders of magnitude [50]. Taking into account that many other parameters such as steric considerations or subcellular compartmentalization aspects can additionally regulate Ras/effector coupling in vivo, it is well conceivable that the engagement of different effectors by oncogenic Ras may be distinctively sensitive to MSOR action. Indeed, in the mentioned case of *MMP-1* regulation by oncogenic Ras, available evidence suggest that *MMP*-1 expression requires other Ras-sparked signals in addition to Erk, including activation of p38α and likely others [51]. Alternatively, the partial only blockade of a Ras effector pathway like the Raf/MEK/Erk cascade may not suffice to compromise all-or-nothing, switch-like type of threshold-controlled processes [52,53]. Furthermore the final outcome to Ras/Erk pathway activation is subject to regulation by intricate, as yet not fully understood positive and negative feedback loops [54-56] that may add further levels of complexity in settings of incomplete Ras-GTP blockade by MSOR. Taken together, these considerations indicate that the degree of MSOR-mediated inhibition of a proximal downstream effector of Ras such as Erk, will not necessarily translate into the same degree of inhibition of a given Ras-dependent tumorigenic hallmark. At the same time, from a methodological point of view, these considerations indicate that beyond their use as blockers of Ras signaling, MSOR can be instrumental tools for delineating the regulatory and mechanistic properties of the signaling network downstream of Ras.

As mentioned before, the affinity of the individual RBD modules for Ras-GTP is one major parameter that allows adjusting the strength of binding and inhibition. Many RBD point mutants have been described and extensively characterized biochemically and structurally with regard to their interaction with Ras-GTP. For example, replacing arginine 59 for alanine in RBD yields a polypeptide with 29-fold diminished affinity for Ras-GTP, and incorporation of a second mutation (N64D) further reduces affinity by a factor of four [14,16]. In agreement with those properties, the single R59A and double R59A/N64D mutants did not block any of the investigated Ras effects if applied in their monomeric forms (R1(A) and R1(A/D)) but they did inhibit Ras-GTP signaling at all investigated levels once converted to their trivalent counterparts R3(A) and R3(A/D) (Figures 1 and 5). This was a striking observation since it evidenced that even RBD mutants deemed to be biologically inert due to negligible Ras-GTP binding could turn into potent Ras blockers if rendered more avid towards Ras-GTP by oligomerization. These considerations gain further relevance in the light of recent insights into the Ras-dependent activation mechanism of Raf. A wealth of experimental data has recently established that Raf kinases function as homo- and heterodimers [57-60]. Although many details of Raf regulation remain obscure it is evident that only the dimeric form is responsive and sensitive to activation by Ras-GTP [57]. Thus, the oligomeric RBD-based units, as used in the present study may, in essence, reflect and recapitulate aspects of the physiological interaction of Ras-GTP with a Raf dimer.

Aberrant Ras activity due to oncogenic mutations is found with high frequency in different human malignancies and remains one of the most attractive molecular targets for rational cancer treatment [15]. Although different approaches such as DNA vaccination, microRNA targeting Ras and farnesyl-transferase inhibition have been exploited as putative thera-peutic strategies to block oncogenic Ras, they have all not stood the test of time and clinical trials [61]. More recently, various novel structure-guided approaches for targeting oncogenic Ras have been described [29,33,35]. Of note, others have previously exploited the single RBD from c-Raf-1 or other Ras-GTP interacting protein modules in order to suppress oncogenic Ras-induced cell transformation in various experimental settings [62,63]. The MSOR approach described here adds to this panel of Ras inhibitory strategies. As a unique feature, MSOR are amenable to fine-tuning for adjustment of their inhibitory strength. Their potent effect on different pa-rameters of Ras-stimulated cellular transformation in vitro (Figures 2, 4 and 5) provides a solid basis for further studies investigating the performance of MSOR in the context of in vivo tumor growth and progression. However, being genetically encoded, the use of MSOR for treatment of Ras-dependent tumours must await improved gene delivery protocols. Alternatively, however, MSOR could potentially be delivered via alternative routes, taking advantage of specific features of Ras-driven tumours. For example, Ras-positive tumours exhibit strongly enhanced macropinocytosis [64], a property that could be exploited to selectively deliver polypeptides, nanoparticles or other types of drugs into the tumour cells.

2.4 CONCLUSIONS

The data presented herein introduce the multivalent scavengers of oncogenic Ras (MSOR) that can be applied as versatile, adjustable Ras-GTP selective probes. MSOR represent novel tools to potently inhibit the action of oncogenic Ras and can be employed in basic research studies of oncogenic Ras function and studies aiming to block tumor growth and progression.

2.5 MATERIAL AND METHODS

2.5.1 CELL LINES, TRANSFECTION

COS-7 cells and NIH3T3 cells were obtained from the DSMZ (German Collection of Microorganisms and Cell Cultures, Braunschweig, Germany) and cultured in DMEM medium supplemented with 10% FCS and 100 μg/ml Gentamycin. Transfection of COS-7 and NIH3T3 cells with plasmid DNA was performed with Nucleofection^R employing a Nucleofector^R device, "Solution V" and "Program A24" according to directions of the manufacturer (Lonza, Cologne, Germany) or using the Polyfect™ transfection reagent following the directions of the manufacturer (Qiagen, Hilden, Germany).

2.5.2 DNA CONSTRUCTS

Expression constructs for EGFP-fused RBD-mono- and oligomers based on the EGFP-C2 vector (Clontech, Mountain View, CA, USA) as well as plasmids encoding constitutively active RasG12V mutants and HA-tagged Erk2 have been described previously [38,39]. Inducible expression constructs for EGFP and EGFP-MSOR were generated on the basis of the bicistronic *Tet-off* vector pNRTIS-21 [38]. cDNAs encoding EGFP and EGFP-RBD fusions were subcloned as *Eco*RI/*Not*I fragment into pNRTIS-21 by standard molecular biology procedures. The luciferase reporter gene plasmid containing the human *MMP-1* promoter has been described previously [65].

2.5.3 INDUCIBLE MSOR EXPRESSION

COS-7 cells were transiently transfected with constructs encoding inducible, EGFP, mono- or oligovalent EGFP-RBD probes. Expression of these constructs was induced or repressed by culturing the cells in absence or presence of 100 ng/ml Doxycyclin, respectively. Fluorescence microscopy demonstrated that the expression of EGFP-constructs was efficiently suppressed in cultures exposed to Doxycyclin for 72 h.

2.5.4 FLUORESCENCE MICROSCOPY

Visualization of EGFP fluorescence was performed with an Axiovert 135 M fluorescence microscope (Carl Zeiss GmbH, Jena, Germany).

2.5.5 WESTERN BLOT ANALYSIS

Western blot analysis of cell lysates for protein expression and/or protein phosphorylation has been previously described in detail [38].

2.5.6 LUCIFERASE REPORTER GENE ASSAY

5×10^5 NIH3T3 cells were grown in six-well plates (Greiner, Fricken-
hausen, Germany) in 2 ml DMEM/10% FCS to 80–90% confluency.
Cells were transferred to 1 ml of fresh medium and transfected with
plasmids encoding oncogenic Ras and EGFP-coupled RBD-probes.
The next day, cells were transfected simultaneously with 1 µg *firefly* luciferase-coupled *MMP-1*-promoter construct, *MMP-1*-2G/pGL3
[65] and 0.1 µg pRL-TK plasmid encoding *renilla* luciferase (Promega,
Madison, WI). 14 h post transfection, cells were harvested using "reporter lysis buffer" (Promega). *Firefly* and *renilla* luciferase activities
were determined using the Dual-Luciferase Reporter Assay System kit
(Promega, Madison, WI, USA) following the manufacturer's instructions. Luminescence was measured using the Promega GLOMAX[R]
96 Luminometer and reported as relative light units. Relative *MMP-
1*-promoter activation was derived by normalizing the *firefly* luciferase
activity to *renilla* luciferase activity.

2.5.7 SOFT AGAR COLONY FORMATION ASSAY

The soft agar assay to analyze the anchorage independent growth of NI-
H3T3-TM cells was performed as described before [44]. Briefly, NIH3T3-
TM cells, were transfected with constructs encoding EGFP or EGFP-RBD
probes. Subsequently, 2×10^4 transfected cells were suspended in 0.5 ml
DMEM/10% FCS supplemented with 0.4% Seaplaque agarose and seeded per well of a 24-well tissue culture plate (Greiner) on a layer of 0.5
ml DMEM/0.8% Seaplaque agarose. Cultures were fed with 0.2 ml of
DMEM/10% FCS in the presence or absence of 25 ng/ml NGF every 3
days for 2 weeks. Colonies were then stained with p-iodonitrotetrazolium
violet (Sigma, Munich, Germany) and microscopically inspected. Data are
derived from counting the number of colonies in at least ten arbitrarily
selected vision fields.

2.5.8 PROTEASE EXPRESSION ANALYSIS BY CDNA ARRAYS

cDNA microarrays of protease and protease inhibitor sequences on nylon membranes and the synthesis of digoxigenin labeled cDNA have been described previously [66]. Detailed information on the generation of the protease/protease inhibitor probes, their arrangement on the membranes as well as experimental details have been published [67]. In brief, cDNA prepared from COS-7 cells was digoxigenin-labeled and hybridized on a custom oligonucleotide microarray comprising housekeeping genes, positive and negative controls, and genes representing a collection of human intra- and extracellular proteases, and protease inhibitors. Hybridization patterns were subsequently detected by chemiluminescence and analyzed using the AIDA imaging software (Raytest, Straubenhardt, Germany). Average densitometry signals of duplicate spots from K-RasG12V/E1- and K-RasG12V/E1-R3-xpressing cells were corrected for the background and normalized against the respective signal from E1-expressing cells.

2.5.9 CYTOMETRIC CELL ANALYSIS AND SORTING

Cytometric measurements and cell sorting was performed using a FACS Calibur^R instrument (BD Biosciences, Heidelberg, Germany) equipped with a 488 nm laser and the CellQuestPro^R software. For flow cytometric analysis of EGFP expression, cells transfected with constructs encoding EGFP or EGFP-RBD probes were trypsinized and adjusted to a density of $1 \times 10^6/100$ μl, forward scatter (cell size) and sideward scatter (cell granularity) were determined and vital cells were gated. EGFP signals were recorded using a 515–545 nm filter and plotted against the number of events. Sorting of EGFP-positive cells was performed following transfection of 2×10^6 COS-7 cells with pRNTIS 21-derived expression constructs encoding EGFP or EGFP-RBD-probes and subsequent cultivation of cells for 48 h. This procedure routinely yielded an enrichment of EGFP-expressing cells to approximately 90%.

2.5.10 ANNEXIN V STAINING

COS-7 cells were grown in six-well plates (Greiner) to 80% confluency, transfected the next day with plasmids encoding EGFP or EGFP-coupled RBD-probes and then cultured for additional 24 h in fresh culture medium. Cells were detached by trypsin/versene (Gibco®/Life Technologies, Darmstadt, Germany) and collected by centrifugation. The cell pellet was washed twice in $1 \times$ PBS and re-suspended in 220 µl $1 \times$ bindings buffer (BD Biosciences). The sample was divided in two: 100 µl sample were left untreated, the other 100 µl were supplemented with 2.5 µl Annexin V-APC (BD Biosciences). The different preparations were incubated for 5 min at 37°C and then for 25 min at room temperature in the dark. To determine the proportion of dead cells among the EGFP or EGFP-RBD-expressing COS-7 cells Annexin V-APC was measured using the FACS CaliburR instrument (BD Biosciences) and plotted against EGFP. Subsequent propidium iodide (Merck Biosciences, Schwalbach, Germany) staining revealed that approximately 85% of the transfected, dead cells underwent apoptosis.

2.5.11 IN VITRO CELL INVASION ASSAY

COS-7 invasion was studied using polycarbonate Trans-wells (Corning Costar Corp., Cambridge, MA, USA) as previously described [44]. Briefly, cells were cultured in DMEM supplemented with 10% FCS and, optionally, 100 ng/ml Doxycyclin and/or 50 ng/ml EGF for 72 h. 2×10^5 cells were then seeded onto membrane filters coated with Matrigel® (BD Biosciences) and transmigration through the Matrigel® layer was determined after incubation for 24 h. Cell invasion was expressed as the average number of migrated cells per vision field (100× magnification) of at least seven, arbitrarily selected vision fields.

2.5.12 STATISTICS

All data are expressed as the mean S.E.M. SPSS for Windows was used for all statistical analyses. The non-parametric Mann–Whitney (U) test and one-way ANOVA with Newman-Keuls Multiple Comparisons were used to analyze if differences among different experimental groups are statistically significant ($p < 0.05$).

REFERENCES

1. Vigil D, Cherfils J, Rossman KL, Der CJ: Ras superfamily GEFs and GAPs: validated and tractable targets for cancer therapy? Nat Rev Cancer 2010, 10(12):842-857.
2. Stone JC: Regulation and function of the RasGRP family of Ras activators in blood cells. Genes Cancer 2011, 2(3):320-334.
3. Young A, Lyons J, Miller AL, Phan VT, Alarcon IR, McCormick F: Ras signaling and therapies. Adv Cancer Res 2009, 102:1-17.
4. Ahearn IM, Haigis K, Bar-Sagi D, Philips MR: Regulating the regulator: post-translational modification of RAS. Nat Rev Mol Cell Biol 2012, 13(1):39-51.
5. Prior IA, Hancock JF: Ras trafficking, localization and compartmentalized signalling. Semin Cell Dev Biol 2012, 23(2):145-153.
6. Ward AF, Braun BS, Shannon KM: Targeting oncogenic Ras signaling in hematologic malignancies. Blood 2012, 120(17):3397-3406.
7. Pylayeva-Gupta Y, Grabocka E, Bar-Sagi D: RAS oncogenes: weaving a tumorigenic web. Nat Rev Cancer 2011, 11(11):761-774.
8. Rusconi P, Caiola E, Broggini M: RAS/RAF/MEK inhibitors in oncology. Curr Med Chem 2012, 19(8):1164-1176.
9. Wimmer R, Baccarini M: Partner exchange: protein-protein interactions in the Raf pathway. Trends Biochem Sci 2010, 35(12):660-668.
10. Wellbrock C, Karasarides M, Marais R: The RAF proteins take centre stage. Nat Rev Mol Cell Biol 2004, 5(11):875-885.
11. Block C, Janknecht R, Herrmann C, Nassar N, Wittinghofer A: Quantitative structure-activity analysis correlating Ras/Raf interaction in vitro to Raf activation in vivo. Nat Struct Biol 1996, 3(3):244-251.
12. Warne PH, Viciana PR, Downward J: Direct interaction of Ras and the amino-terminal region of Raf-1 in vitro. Nature 1993, 364(6435):352-355.
13. Herrmann C, Martin GA, Wittinghofer A: Quantitative analysis of the complex between p21ras and the Ras-binding domain of the human Raf-1 protein kinase. J Biol Chem 1995, 270(7):2901-2905.
14. Nassar N, Horn G, Herrmann C, Block C, Janknecht R, Wittinghofer A: Ras/Rap effector specificity determined by charge reversal. Nat Struct Biol 1996, 3(8):723-729.
15. Prior IA, Lewis PD, Mattos C: A comprehensive survey of Ras mutations in cancer. Cancer Res 2012, 72(10):2457-2467.
16. Tidyman WE, Rauen KA: The RASopathies: developmental syndromes of Ras/MAPK pathway dysregulation. Curr Opin Genet Dev 2009, 19(3):230-236.
17. Swanson KD, Winter JM, Reis M, Bentires-Alj M, Greulich H, Grewal R, Hruban RH, Yeo CJ, Yassin Y, Iartchouk O, et al.: SOS1 mutations are rare in human malignancies: implications for Noonan Syndrome patients. Genes Chromosomes Cancer 2008, 47(3):253-259.

18. Le LQ, Parada LF: Tumor microenvironment and neurofibromatosis type I: connecting the GAPs. Oncogene 2007, 26(32):4609-4616.

19. Steelman LS, Franklin RA, Abrams SL, Chappell W, Kempf CR, Basecke J, Stivala F, Donia M, Fagone P, Nicoletti F, et al.: Roles of the Ras/Raf/MEK/ERK pathway in leukemia therapy. Leukemia 2011, 25(7):1080-1094.

20. Maurer G, Tarkowski B, Baccarini M: Raf kinases in cancer-roles and therapeutic opportunities. Oncogene 2011, 30(32):3477-3488.

21. Zhu K, Hamilton AD, Sebti SM: Farnesyltransferase inhibitors as anticancer agents: current status. Curr Opin Investig Drugs 2003, 4(12):1428-1435.

22. Lancet JE, Karp JE: Farnesyltransferase inhibitors in hematologic malignancies: new horizons in therapy. Blood 2003, 102(12):3880-3889.

23. Prendergast GC, Oliff A: Farnesyltransferase inhibitors: antineoplastic properties, mechanisms of action, and clinical prospects. Semin Cancer Biol 2000, 10(6):443-452.

24. Bar-Sagi D, McCormick F, Milley RJ, Feramisco JR: Inhibition of cell surface ruffling and fluid-phase pinocytosis by microinjection of anti-ras antibodies into living cells. J Cell Physiol 1987, Suppl 5:69-73.

25. Marais R, Light Y, Mason C, Paterson H, Olson MF, Marshall CJ: Requirement of Ras-GTP-Raf complexes for activation of Raf-1 by protein kinase C. Science 1998, 280(5360):109-112.

26. Mittnacht S, Paterson H, Olson MF, Marshall CJ: Ras signalling is required for inactivation of the tumour suppressor pRb cell-cycle control protein. Curr Biol 1997, 7(3):219-221.

27. Abrams SI, Hand PH, Tsang KY, Schlom J: Mutant ras epitopes as targets for cancer vaccines. Semin Oncol 1996, 23(1):118-134.

28. Orr RM, Dorr FA: Clinical studies of antisense oligonucleotides for cancer therapy. Methods Mol Med 2005, 106:85-111.

29. Hocker HJ, Cho KJ, Chen CY, Rambahal N, Sagineedu SR, Shaari K, Stanslas J, Hancock JF, Gorfe AA: Andrographolide derivatives inhibit guanine nucleotide exchange and abrogate oncogenic Ras function. Proc Natl Acad Sci USA 2013, 110(25):10201-10206.

30. Stites EC, Trampont PC, Ma Z, Ravichandran KS: Network analysis of oncogenic Ras activation in cancer. Science 2007, 318(5849):463-467.

31. Palmioli A, Sacco E, Airoldi C, Di Nicolantonio F, D'Urzo A, Shirasawa S, Sasazuki T, Di Domizio A, De Gioia L, Martegani E, et al.: Selective cytotoxicity of a bicyclic Ras inhibitor in cancer cells expressing K-Ras(G13D). Biochem Biophys Res Commun 2009, 386(4):593-597.

32. Sun Q, Burke JP, Phan J, Burns MC, Olejniczak ET, Waterson AG, Lee T, Rossanese OW, Fesik SW: Discovery of small molecules that bind to K-Ras and inhibit Sos-mediated activation. Angew Chem Int Ed Engl 2012, 51(25):6140-6143.

33. Shima F, Yoshikawa Y, Ye M, Araki M, Matsumoto S, Liao J, Hu L, Sugimoto T, Ijiri Y, Takeda A, et al.: In silico discovery of small-molecule Ras inhibitors that display antitumor activity by blocking the Ras-effector interaction. Proc Natl Acad Sci USA 2013, 110(20):8182-8187.

34. Tanaka T, Rabbitts TH: Interfering with RAS-effector protein interactions prevent RAS-dependent tumour initiation and causes stop-start control of cancer growth. Oncogene 2010, 29(45):6064-6070.
35. Zimmermann G, Papke B, Ismail S, Vartak N, Chandra A, Hoffmann M, Hahn SA, Triola G, Wittinghofer A, Bastiaens PI, et al.: Small molecule inhibition of the *KRAS*-PDEdelta interaction impairs oncogenic *KRAS* signalling. Nature 2013, 497(7451):638-642.
36. Posch C, Moslehi H, Feeney L, Green GA, Ebaee A, Feichtenschlager V, Chong K, Peng L, Dimon MT, Phillips T, et al.: Combined targeting of MEK and PI3K/mTOR effector pathways is necessary to effectively inhibit NRAS mutant melanoma in vitro and in vivo. Proc Natl Acad Sci USA 2013, 110(10):4015-4020.
37. Roberts PJ, Der CJ: Targeting the Raf-MEK-ERK mitogen-activated protein kinase cascade for the treatment of cancer. Oncogene 2007, 26(22):3291-3310.
38. Augsten M, Pusch R, Biskup C, Rennert K, Wittig U, Beyer K, Blume A, Wetzker R, Friedrich K, Rubio I: Live-cell imaging of endogenous Ras-GTP illustrates predominant Ras activation at the plasma membrane. EMBO Rep 2006, 7(1):46-51.
39. Rubio I, Grund S, Song SP, Biskup C, Bandemer S, Fricke M, Forster M, Graziani A, Wittig U, Kliche S: TCR-induced activation of Ras proceeds at the plasma membrane and requires palmitoylation of N-Ras. J Immunol 2010, 185(6):3536-3543.
40. Broggi S, Martegani E, Colombo S: Live-cell imaging of endogenous Ras-GTP shows predominant Ras activation at the plasma membrane and in the nucleus in Saccharomyces cerevisiae. Int J Biochem Cell Biol 2013, 45(2):384-394.
41. Wu RF, Ma Z, Liu Z, Terada LS: Nox4-derived H2O2 mediates endoplasmic reticulum signaling through local Ras activation. Mol Cell Biol 2010, 30(14):3553-3568.
42. Jaitner BK, Becker J, Linnemann T, Herrmann C, Wittinghofer A, Block C: Discrimination of amino acids mediating Ras binding from noninteracting residues affecting raf activation by double mutant analysis. J Biol Chem 1997, 272(47):29927-29933.
43. Heidaran MA, Fleming TP, Bottaro DP, Bell GI, Di Fiore PP, Aaronson SA: Transformation of NIH3T3 fibroblasts by an expression vector for the human epidermal growth factor precursor. Oncogene 1990, 5(8):1265-1270.
44. Cramer A, Kleiner S, Westermann M, Meissner A, Lange A, Friedrich K: Activation of the c-Met receptor complex in fibroblasts drives invasive cell behavior by signaling through transcription factor STAT3. J Cell Biochem 2005, 95(4):805-816.
45. Graziani A, Gramaglia D, Dalla Zonca P, Comoglio PM: Hepatocyte growth factor/scatter factor stimulates the Ras-guanine nucleotide exchanger. J Biol Chem 1993, 268(13):9165-9168.
46. Gherardi E, Birchmeier W, Birchmeier C, Vande Woude G: Targeting MET in cancer: rationale and progress. Nat Rev Cancer 2012, 12(2):89-103.
47. Imler JL, Schatz C, Wasylyk C, Chatton B, Wasylyk B: A Harvey-ras responsive transcription element is also responsive to a tumour-promoter and to serum. Nature 1988, 332(6161):275-278.
48. White LA, Maute C, Brinckerhoff CE: ETS sites in the promoters of the matrix metalloproteinases collagenase (MMP-1) and stromelysin (MMP-3) are auxiliary elements that regulate basal and phorbol-induced transcription. Connect Tissue Res 1997, 36(4):321-335.

49. Wolfman JC, Wolfman A: Endogenous c-N-Ras provides a steady-state anti-apoptotic signal. J Biol Chem 2000, 275(25):19315-19323.

50. Wohlgemuth S, Kiel C, Kramer A, Serrano L, Wittinghofer F, Herrmann C: Recognizing and defining true Ras binding domains I: biochemical analysis. J Mol Biol 2005, 348(3):741-758.

51. Futamura M, Kamiya S, Tsukamoto M, Hirano A, Monden Y, Arakawa H, Nishimura S: Malolactomycin D, a potent inhibitor of transcription controlled by the Ras responsive element, inhibits Ras-mediated transformation activity with suppression of MMP-1 and MMP-9 in NIH3T3 cells. Oncogene 2001, 20(46):6724-6730.

52. Xiong W, Ferrell JE Jr: A positive-feedback-based bistable 'memory module' that governs a cell fate decision. Nature 2003, 426(6965):460-465.

53. Daniels MA, Teixeiro E, Gill J, Hausmann B, Roubaty D, Holmberg K, Werlen G, Hollander GA, Gascoigne NR, Palmer E: Thymic selection threshold defined by compartmentalization of Ras/MAPK signalling. Nature 2006, 444(7120):724-729.

54. Santos SD, Verveer PJ, Bastiaens PI: Growth factor-induced MAPK network topology shapes Erk response determining PC-12 cell fate. Nat Cell Biol 2007, 9(3):324-330.

55. Waters SB, Holt KH, Ross SE, Syu LJ, Guan KL, Saltiel AR, Koretzky GA, Pessin JE: Desensitization of Ras activation by a feedback disassociation of the SOS-Grb2 complex. J Biol Chem 1995, 270(36):20883-20886.

56. Kothe S, Muller JP, Bohmer SA, Tschongov T, Fricke M, Koch S, Thiede C, Requardt RP, Rubio I, Bohmer FD: Features of Ras activation by a mislocalized oncogenic tyrosine kinase: FLT3 ITD signals via K-Ras at the plasma membrane of Acute Myeloid Leukemia cells. J Cell Sci 2013, 126(20):4746-4755.

57. Freeman AK, Ritt DA, Morrison DK: Effects of Raf dimerization and its inhibition on normal and disease-associated Raf signaling. Mol Cell 2013, 49(4):751-758.

58. Lavoie H, Thevakumaran N, Gavory G, Li JJ, Padeganeh A, Guiral S, Duchaine J, Mao DY, Bouvier M, Sicheri F, et al.: Inhibitors that stabilize a closed RAF kinase domain conformation induce dimerization. Nat Chem Biol 2013, 9(7):428-436.

59. Poulikakos PI, Persaud Y, Janakiraman M, Kong X, Ng C, Moriceau G, Shi H, Atefi M, Titz B, Gabay MT, et al.: RAF inhibitor resistance is mediated by dimerization of aberrantly spliced BRAF(V600E). Nature 2011, 480(7377):387-390.

60. Poulikakos PI, Zhang C, Bollag G, Shokat KM, Rosen N: RAF inhibitors transactivate RAF dimers and ERK signalling in cells with wild-type BRAF. Nature 2010, 464(7287):427-430.

61. Downward J: Targeting RAS signalling pathways in cancer therapy. Nat Rev Cancer 2003, 3(1):11-22.

62. Clark GJ, Drugan JK, Terrell RS, Bradham C, Der CJ, Bell RM, Campbell S: Peptides containing a consensus Ras binding sequence from Raf-1 and theGTPase activating protein NF1 inhibit Ras function. Proc Natl Acad Sci USA 1996, 93(4):1577-1581.

63. Fridman M, Tikoo A, Varga M, Murphy A, Nur EKMS, Maruta H: The minimal fragments of c-Raf-1 and NF1 that can suppress v-Ha-Ras-induced malignant phenotype. J Biol Chem 1994, 269(48):30105-30108.

64. Commisso C, Davidson SM, Soydaner-Azeloglu RG, Parker SJ, Kamphorst JJ, Hackett S, Grabocka E, Nofal M, Drebin JA, Thompson CB, et al.: Macropinocyto-

sis of protein is an amino acid supply route in Ras-transformed cells. Nature 2013, 497(7451):633-637.

65. Tsareva SA, Moriggl R, Corvinus FM, Wiederanders B, Schutz A, Kovacic B, Friedrich K: Signal transducer and activator of transcription 3 activation promotes invasive growth of colon carcinomas through matrix metalloproteinase induction. Neoplasia 2007, 9(4):279-291.

66. Fitzgerald JS, Tsareva SA, Poehlmann TG, Berod L, Meissner A, Corvinus FM, Wiederanders B, Pfitzner E, Markert UR, Friedrich K: Leukemia inhibitory factor triggers activation of signal transducer and activator of transcription 3, proliferation, invasiveness, and altered protease expression in choriocarcinoma cells. Int J Biochem Cell Biol 2005, 37(11):2284-2296.

67. Schuler S, Wenz I, Wiederanders B, Slickers P, Ehricht R: An alternative method to amplify RNA without loss of signal conservation for expression analysis with a proteinase DNA microarray in the ArrayTube format. BMC Genomics 2006, 7:144.

There are several supplemental files that are not available in this version of the article. To view this additional information, please use the citation information cited on the first page of this chapter.

CHAPTER 3

DIRECT INHIBITION OF PI3K IN COMBINATION WITH DUAL HER2 INHIBITORS IS REQUIRED FOR OPTIMAL ANTITUMOR ACTIVITY IN HER2+ BREAST CANCER CELLS

BRENT N. REXER, SIPRACHANH CHANTHAPHAYCHITH, KIMBERLY BROWN DAHLMAN, AND CARLOS L. ARTEAGA

3.1 INTRODUCTION

Amplification of the *HER2* oncogene occurs in approximately 25% of human breast cancers and predicts response to therapies targeting human epidermal growth factor receptor 2 (HER2), including trastuzumab, a monoclonal antibody directed against HER2, and lapatinib, a tyrosine kinase inhibitor (TKI) of HER2 and epidermal growth factor receptor (EGFR) [1,2]. HER2 is a member of the ErbB family of receptor tyrosine kinases (RTKs), which form both homo- and heterodimers, resulting in the activation of downstream signaling pathways [3]. In *HER2*-amplified cancers, the heterodimer of HER2 with kinase-deficient HER3 is a major activator of phosphoinositide 3-kinase (PI3K)-Akt signaling, and HER3, when phosphorylated, can directly couple to the p85 subunit of PI3K [4]. *HER2*-amplified tumors show significant reliance on PI3K-Akt signaling [5,6].

This chapter was originally published under the Creative Commons Attribution License. Rexer BN, Chanthaphaychith S, Dahlman KB and Arteaga CL. Direct Inhibition of PI3K in Combination with Dual HER2 Inhibitors is Required for Optimal Antitumor Activity in HER2+ Breast Cancer Cells. Breast Cancer Research *16,R9 (2014). doi:10.1186/bcr3601.*

Importantly, inhibition of PI3K-Akt signaling is believed to be an essential component of the antitumor effect of HER2-directed therapies [7-9].

Alteration of the PI3K-Akt pathway is frequent in human cancers, and among the most frequent alterations are mutations in phosphoinositide 3-kinase catalytic subunit α (*PIK3CA*), the gene encoding the p110α catalytic subunit of PI3K. These mutations cluster in hotspot regions in the helical and kinase domains of p110α [10,11] and confer a gain of function [12]. *PIK3CA* hotspot mutations are found in approximately 25% of breast cancers and can overlap with *HER2* amplification [10,13,14]. The presence of these mutations in *HER2*-amplified cancer cells confers resistance to trastuzumab or lapatinib [7,15-17]. Moreover, aberrant activation of PI3K-Akt signaling by a *PIK3CA* mutation and/or phosphatase and tensin homologue (PTEN) loss is associated with resistance to trastuzumab in patients in some studies [15,18,19].

Recent clinical studies have suggested that targeting HER2-PI3K signaling with combinations of agents that inhibit HER2 by different mechanisms is more effective than a single HER2 inhibitor; combining trastuzumab and lapatinib was more effective than trastuzumab alone in both the metastatic and neoadjuvant settings [20,21]; and combining two HER2 antibodies, trastuzumab and pertuzumab, prolonged survival longer than trastuzumab alone [22]. Preclinical studies have suggested that the HER2/HER3 signaling complex has sufficient buffering capacity to withstand incomplete inhibition of HER2 catalytic activity, even in combination with a PI3K inhibitor, though this capacity can be overcome by fully inactivating HER2 catalytic activity with elevated doses of a TKI that may not be tolerated in clinical practice [23]. Moreover, even so-called dual-targeting of HER2 may not be sufficient to overcome resistance to HER2 inhibition, particularly in the case of *HER2*-amplified cancer with a *PIK3CA* mutation [16,24]. We have previously shown that, once resistance to HER2 inhibitors is established, inhibition of PI3K added to continued HER2 inhibition can overcome resistance [25].

In this work, we show that *PIK3CA*-activating mutations can be acquired during the development of resistance to HER2 inhibitors and that the presence of these mutations uncouples PI3K signaling from HER2. We further demonstrate that adding a PI3K inhibitor to dual-targeting of HER2 is more effective than HER2 targeting alone in a PI3K wild-type tumor and that the combination of HER2 and PI3K targeting is required for tumor regression in a model with *HER2* amplification and *PIK3CA* mutation.

3.2 METHODS

3.2.1 CELL CULTURES, INHIBITOR TREATMENTS AND PROLIFERATION AND APOPTOSIS ASSAYS

BT474, SKBR3, MDA-MB-361, HCC1954 and UACC893 cells were obtained from the American Type Culture Collection (Manassas, VA, USA). SUM190 cells were purchased from Asterand (Detroit, MI, USA). Lapatinib-resistant (LR) cell lines were generated as described previously [25] and cultured in the presence of 1 to 2 µM lapatinib. Lapatinib ditosylate and BIBW2992 were obtained from LC Laboratories (Woburn, MA, USA). BKM120 was obtained from Selleck Chemicals (Houston, TX, USA). Trastuzumab and pertuzumab were obtained from the Vanderbilt University Medical Center outpatient pharmacy. Unless otherwise noted, cells were treated with inhibitors at the following concentrations: lapatinib, 1 µM; trastuzumab, 10 µg/ml; BKM120, 1 µM; and BIBW2992, 1 µM. Cell proliferation was measured using the sulforhodamine B (SRB) reagent. Cells plated in 96-well plates were treated with inhibitors and fixed in 1% trichloroacetic acid after 72-hour treatment. Plates were rinsed with water and air-dried, then stained with 0.4% SRB in 1% acetic acid. Excess stain was removed by washing with 1% acetic acid, and plates were air-dried. Stained cells were solubilized in 10 mM Tris–HCl, pH 7.4, and absorbance at 590 nm was measured in a plate reader. Apoptosis was measured at 24 hours using the Caspase-Glo reagent (Promega, Madison, WI, USA) according to the manufacturer's instructions. For longer-term growth assays, cells were seeded into six-well plates and treated with inhibitors as indicated. Media and inhibitors were replenished twice weekly, and cells were grown for 2 to 3 weeks until confluence in the untreated wells. Cells were fixed and stained in 20% methanol with 0.5% crystal violet and washed with water. Dried plates were imaged on a flatbed scanner.

3.2.2 GENERATION OF PIK3CA MUTANT CELLS

BT474 and SKBR3 parental cells were transduced with an amphotrophic retrovirus (LZRS) with neomycin resistance produced by Phoenix-AM-PHO packaging cells containing C-terminal hemagglutinin (HA)-tagged bovine wild-type *PIK3CA* or E545K or H1047R mutations cloned from the JP1520 retroviral vector, as described previously [26]. To generate LR cell lines, transduced cells were treated with 1 μM lapatinib for approximately 4 to 6 weeks. Wild-type *PIK3CA*-expressing cells did not survive selection. Mutant-expressing, lapatinib-selected cells were maintained in the presence of 1 μM lapatinib.

3.2.3 SNAPSHOT GENOTYPING AND SANGER SEQUENCING

Genomic DNA was isolated from parental and LR BT474 cells and analyzed by the SNaPshot mutational profiling method (SNaPshot; Vanderbilt, NY, USA). This assay involves multiplexed polymerase chain reaction (PCR) and multiplexed single-base primer extension, followed by capillary electrophoresis [27,29]; Vandana G Abramson et al., unpublished data. The current assay was designed to detect 18 somatic point mutations in three genes (Additional file 1: Table S1). Briefly, PCR primers were pooled to amplify the target DNA, and PCR was performed using the following conditions: 95°C (8 minutes), followed by 40 cycles at (95°C (20 seconds), 58°C (30 seconds) and 72°C (1 minute)) and then a final extension at 72°C (3 minutes) (Additional file 2: Table S2). Next, PAGE-purified primers were pooled together, and multiplex single-base extension reactions were performed on ExoSAP-IT-treated (USB/Affymetrix, Santa Clara, CA, USA) PCR products using the following conditions: 96°C (30 seconds), followed by 35 cycles at (96°C (10 seconds), 50°C (5 seconds) and 60°C (30 seconds)) (Additional file 3: Table S3). Extension products were applied to capillary electrophoresis in an ABI 3730 DNA Analyzer (Applied Biosystems, Foster City, CA, USA), and the data were interpreted using ABI GeneMapper software (version 4.0; Applied Biosystems). Human male genomic DNA (Promega) was used as a wild-type control.

Spiking primers were mixed to create a pan-positive control mix for the assay (Additional file 4: Table S4).

To validate the SNaPshot result, exon 9 of *PIK3CA* was amplified by high-fidelity PCR from genomic DNA and sequenced by traditional Sanger sequencing methods. The primer sequences are as follows: 5'-TTCAG-CAGTGTGGTAAAGTTC-3' forward; 5'GAGGCCAATCTTTTAC-CAAGC-3' reverse.

3.2.4 IMMUNOBLOT ANALYSIS AND P85 IMMUNOPRECIPITATION

Cells were treated with inhibitors for 3 hours, and lysates were prepared as described previously [25]. Lysates were resolved on 7.5% acrylamide gels and transferred to Immobilon-FL PVDF (EMD Millipore, Billerica, MA, USA) and incubated in primary antibodies overnight at 4°C. Blots were washed and incubated with infrared fluorescent dye secondary antibody conjugates (LI-COR Biosciences, Lincoln, NE, USA), and blots were imaged using a LI-COR Odyssey scanner. Antibodies from the following sources were used for analysis: pHER2 Y1248 (R&D Systems, Minneapolis, MN, USA); Y877 pHER2 (Epitomics, Burlingame, CA, USA); Y1221/2 pHER2, Y11197 and Y1289 pHER3, S473 pAkt, Akt, S240/44 pS6, pErk1/2, Erk, and p110α PI3K (Cell Signaling Technology, Danvers, MA, USA); p85 N-terminal Src homology 2 (SH2) domain (EMD Millipore); HA (Covance, Princeton, NJ, USA); actin (Sigma-Aldrich, St Louis, MO, USA); HER2 (Thermo Fisher, Pittsburgh, PA, USA); and glyceraldehyde 3-phosphate dehydrogenase (Santa Cruz Biotechnology, Santa Cruz, CA, USA).

To analyze p110α isoforms bound to p85, lysates were prepared from wild-type or mutant-expressing cells and incubated with p85 rabbit antibody (directed against the N-terminal SH2 domain) overnight at 4°C as described previously [30]. Antibody complexes were isolated with Dynabeads Protein A (Life Technologies) and washed. Beads were boiled in SDS sample buffer and resolved on polyacrylamide gels, then immunoblotted with HA, p110α and p85 antibodies. Bands were quantitated using the LI-COR Odyssey scanner and Image Studio software. Mean values from triplicate experiments were compared by analysis of variance (ANOVA).

3.2.5 PHOSPHOPROTEIN ELISA ANALYSIS

PathScan Sandwich ELISA kits for pHER2, pAkt and pS6 were purchased from Cell Signaling Technology and used according to the manufacturer's instructions. Lysates from cells treated with a range of lapatinib doses for 4 hours were prepared, and protein concentration determined by bicinchoninic acid assay (Pierce Biotechnology, Rockford, IL, USA). Lysates were diluted and incubated in enzyme-linked immunosorbent assay (ELISA) plates overnight at 4°C, then washed and developed according to the protocol. Absorbance values were read on a BioTek Epoch microplate optical reader (BioTek, Winooski, VT, USA) and normalized for each experiment to values for untreated cells. ELISA experiments were repeated in triplicate. Mean values ± SEM from the three experiments were used to plot log(inhibitor) vs. response curves using a variable slope model and to determine half-maximal inhibitory concentrations (IC_{50}) using GraphPad Prism software (GraphPad Software, La Jolla, CA, USA). For statistical analysis, IC_{50} values for each phosphoprotein were compared between wild-type and mutant cells by ANOVA.

3.2.6 XENOGRAFT EXPERIMENTS

Animal studies were approved by the Vanderbilt University Medical Center Institutional Animal Care and Use Committee. BT-474 or HCC1954 cells (approximately 5×10^6 in 50% Matrigel) were injected into female athymic nude mice. For BT474 cells, mice were implanted with 60-day, 0.72-mg, slow-release estrogen pellets (Innovative Research of America, Sarasota, FL, USA) on the day prior to injection. After tumors reached ≥ 250 mm^3, mice were randomly assigned to treatment with trastuzumab (30 mg/kg by intraperitoneal injection twice weekly), lapatinib (100 mg/kg by oral gavage daily) and/or BKM120 (30 mg/kg by oral gavage daily). Tumors were measured with calipersn and tumor volume in cubic millimeters was calculated by the formula length/$(2 \times \text{width}^2)$. Mean tumor volumes for each treatment group are displayed on $\log_2$ scale. Kaplan-Meier curves were constructed using the time point at which tumor volumes exceeded 100 mm^3.

3.3 RESULTS

3.3.1 ACQUIRED PIK3CA MUTATION IN LAPATINIB-RESISTANT CELLS

We previously showed that LR cells exhibit continued activation of PI3K signaling despite inhibition of the HER2 tyrosine kinase [25]. To identify mechanisms associated with maintenance of PI3K signaling in these drug-resistant cells, we profiled them using the SNaPshot assay [27–29; Vandana G Abramson et al., unpublished] for a panel of mutations in *PIK3CA, AKT* and *PTEN* (Additional file 1: Table S1). Four of the six LR cell lines were derived from cells with preexisting hotspot *PIK3CA* mutations. None of the resistant cell lines acquired *AKT* or *PTEN* mutations. However, BT-474 LR cells acquired an E542K *PIK3CA* mutation (Figure 1A). We confirmed the presence of this mutation in cDNA by Sanger sequencing. Parental cells did not have any detectable E542K mutant sequence. The mutation persisted in LR cells even after culture in the absence of lapatinib for more than 2 weeks (Figure 1B). The E542 mutation is unique to BT474 cells. Other *HER2*-amplified cell lines known to contain *PIK3CA* mutations have either an E545K or H1047R mutation [10,14].

3.3.2 PIK3CA MUTATION PARTIALLY UNCOUPLES HER2 INHIBITION FOR PHOSPHOINOSITIDE 3-KINASE SIGNALING

We observed that the four *HER2*-amplified cell lines that contain *de novo* *PIK3CA* mutations develop resistance to lapatinib more rapidly than *PIK3CA* wild-type cells in chronic cell culture. Thus, because the BT474 cells acquired a *PIK3CA* mutation upon the development of lapatinib resistance, we reasoned that the gain of function conferred by this mutation in p110α uncouples downstream PI3K signaling from HER2 and thereby blunts the inhibitory effect of lapatinib. We first evaluated response to lapatinib in parental and LR BT474 cells. Resistant cells were cultured in the absence of lapatinib for at least 2 weeks until recovery of HER2 phosphorylation was observed. After treatment of cells with increasing doses of lapatinib, we observed a similar inhibition of HER2 phosphorylation in parental and resistant cells, but the resistant cells with an acquired E542K mutation showed only very limited inhibition of PI3K signaling as measured by Akt and S6 phosphorylation (Figure 1C).

We next transduced parental BT474 and SKBR3 cells with wild-type, E545K or H1047R *PIK3CA* retroviral constructs. Expression of the ectopic p110α construct was verified by Western blot analysis for a C-terminal HA-tag (Additional file 5: Figure S1). We then evaluated the inhibition of pHER2 and PI3K signaling with a range of lapatinib doses. Similarly to BT474 LR cells with acquired E542K mutation, cells with the other hotspot mutations showed a blunted inhibitor response of PI3K signaling to lapatinib with persistent PI3K-Akt signaling (Figure 2A).

To better quantify the signaling output of mutant *PIK3CA* in cells in which HER2 was inhibited, we measured pHER2 by ELISA in cells treated with lapatinib and compared the IC_{50} for HER2 inhibition to the IC_{50} for PI3K inhibition as measured by Akt S473 and S6 S240/244 phosphorylation. We first compared BT474 and SKBR3 cells (without endogenous hotspot mutations) infected with wild-type, E545K or H1047R p110α retroviral constructs. As expected, IC_{50} data for inhibition of HER2 by lapatinib were similar between wild-type and *PIK3CA* mutant expressing cells (Figure 2B). However, both cell lines expressing either mutant *PIK3CA* isoform showed a significant increase in the IC_{50} for both Akt and S6 phosphorylation (Figure 2B and Additional file 6: Figure S2). For cells expressing ectopic mutant PI3K, the increase in IC_{50} was typically two- to threefold higher for the mutant cells than for the wild-type cells (Table 1). Interestingly, we observed a similar increase in the IC_{50} for pAkt and pS6 in the LR BT474 cells with an acquired E542K mutation (Additional file 7: Figure S3). These data suggest that mutant p110α uncouples PI3K signaling from HER2.

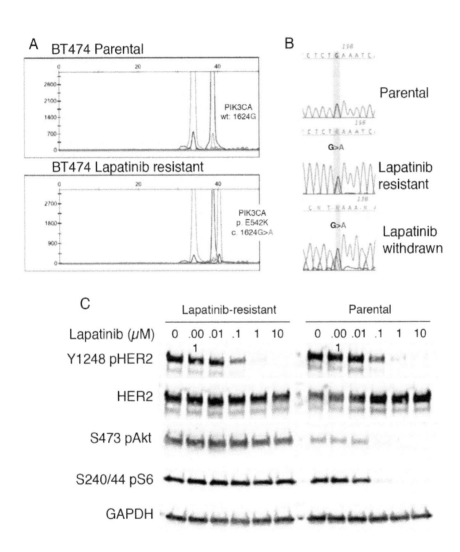

FIGURE 1: BT474 lapatinib-resistant cells acquire E542K *PIK3CA* mutation. (A) SNaPshot analysis of genomic DNA isolated from parental and lapatinib-resistant (LR) cells reveals a distinct peak corresponding to a G1624A nucleotide change in resistant cells. (B) Sanger sequencing of exon 9 of *PIK3CA* confirms the presence of the G1624A nucleotide change in LR cells. (C) Lysates from parental and LR cells treated with increasing doses of lapatinib, ranging from 0.001 to 10 μM, were analyzed by immunoblotting with the indicated antibodies.

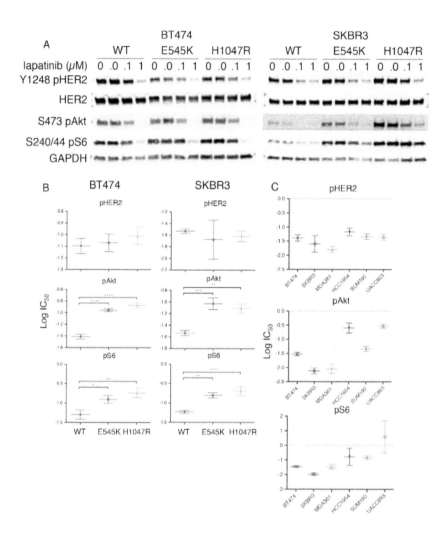

FIGURE 2: *PIK3CA* mutation uncouples phosphoinositide 3-kinase signaling from HER2 inhibition by lapatinib. (A) BT474 and SKBR3 cells infected with wild-type, E545K or H1047R constructs were treated with lapatinib at the indicated doses, and lysates were analyzed by immunoblotting with the indicated antibodies. (B) Lysates from *PIK3CA* wild-type or mutant expressing cells treated with a range of lapatinib doses (0.0016 to 5 μM) were analyzed by ELISA for pHER2, pAkt and pS6. Half-maximal concentration (IC_{50}) values were calculated, and the mean log IC_{50} ± SEM values for three replicate dose–inhibitor curves are shown. *P<0.05, **P<0.01, ***P<0.001 and ****P<0.0001. (C) HER2+ cell lines with wild-type *PIK3CA* (BT474 or SKBR3) or with a *PIK3CA* mutation (MDA361, HCC1954, SUM190 or UACC893) were treated with varying lapatinib doses and analyzed as described in (B). Mean log IC_{50} values from three replicates ± SEM are shown. Mean IC_{50} data are shown in Table 1.

TABLE 1: Mean half-maximal inhibitory concentrations for lapatinib inhibition of HER2 and signaling proteins in *PIK3CA* mutant cells[a]

	HER2	Akt	Increase over WT	S6	Increase over WT
BT474	0.0411	0.0300		0.0347	
BT474 LR	0.0204	0.3925	13.1	0.7863	22.7
BT474 wt	0.0798	0.0387		0.0506	
BT474 E545K	0.0852	0.1098	2.8	0.1251	2.5
BT474 H1047R	0.0978	0.1336	3.4	0.1829	3.6
SKBR3	0.0254	0.0077		0.0106	
SKBR3 wt	0.0297	0.0293		0.0593	
SKBR3 E545K	0.0211	0.0936	3.2	0.1563	2.6
SKBR3 H1047R	0.0241	0.0770	2.6	0.2028	3.4
MDA361	0.0160	0.0092	0.5	0.0354	1.6
HCC1954	0.0681	0.2527	13.4	0.1678	7.4
SUM190	0.0465	0.0464	2.5	0.1461	6.5
UACC893	0.0444	0.2920	15.5	4.0040	177.0

[a]*WT, Wild type. Half-maximal inhibitory concentration (IC_{50}) values for inhibition of HER2, Akt and S6 phosphorylation by lapatinib treatment were determined by enzyme-linked immunosorbent assay (dose–response curves are shown in Additional file 6: Figure S2, Additional file 7: Figure S3 and Additional file 8: Figure S4). Mean IC_{50} values derived from at least three separate experiments are shown. For phosphoinositide 3-kinase signaling proteins Akt and S6, IC_{50} values from cells with PIK3CA mutations were compared to average values for cells without hotspot mutations (BT474 and SKBR3).*

We next tested whether a similar partial uncoupling of PI3K signaling from HER2 activation was occurring in cells with endogenous PI3K hotspot mutations. We measured the IC_{50} values for HER2 and PI3K inhibition in four cell lines with endogenous *PIK3CA* mutations and compared them to the data for BT474 and SKBR3 cells. Again, we observed similar magnitudes of inhibition of HER2, but cell lines with endogenous PI3K mutations, particularly H1047R, showed an increased IC_{50} for pAkt and, in one cell line, for pS6 (Figure 2C and Additional file 8: Figure S4). There was a trend toward increased IC_{50} for inhibition of S6 phosphorylation by lapatinib, but the difference was not statistically significant. We observed

in the dose–response curves that the level of the remaining S6 phosphorylation, even at maximal (5 μM) doses of lapatinib, was higher than that of
wild-type cells (Additional file 8: Figure S4). This finding confirms that
cell lines with endogenous PI3K mutations have a blunted inhibitory response and explains in part how resistance to lapatinib may emerge more
quickly in *PIK3CA* mutant cells.

3.3.3 LAPATINIB-RESISTANT CELLS PREFERENTIALLY UTILIZE H1047R MUTANT PIK3CA

We reasoned that cells that developed resistance to lapatinib by exploiting the uncoupling of PI3K from HER2 inhibition to allow escape from
that inhibition would rely on the gain of function conferred by the mutant
p110α isoform to allow for continued signaling in the presence of lapatinib. Thus, after selection for resistance, we hypothesized that the pool of
p110α in complex with p85 and available for PI3K signaling would contain more mutant p110α than that available before selection for resistance.
To test this hypothesis, BT474 and SKBR3 cells expressing ectopic HA-
tagged E545K or H1047R p110α were selected for lapatinib resistance
for 4 to 6 weeks. After an initial period of slower growth, the lapatinib-
selected cells proliferated. We observed continued inhibition of HER2 by
lapatinib, but recovery of Akt and S6 phosphorylation (Figure 3A). In LR
cells, PI3K-Akt signaling was still dependent on p110α catalytic activity,
as treatment with the PI3K inhibitor BKM120 abolished S473 Akt phosphorylation in both parental and LR ectopic *PIK3CA* mutant–expressing
cells (Figure 3B).

   We hypothesized that the gain of function conferred by the mutant
p110α isoform in the LR cells would result in the preferential engagement
of the mutant isoform for PI3K signaling under the selective pressure of
HER2 inhibition. As an indirect measure of this utilization of mutant PI3K,
we tested whether the PI3K signaling complex in the LR cells would be
enriched for the mutant p110α isoform. We immunoprecipitated the pool
of p110α in a complex with p85 using a p85 antibody and evaluated the
level of mutant isoform present before and after lapatinib selection by immunoblotting for the HA-tag. In BT474 cells, we observed that more of

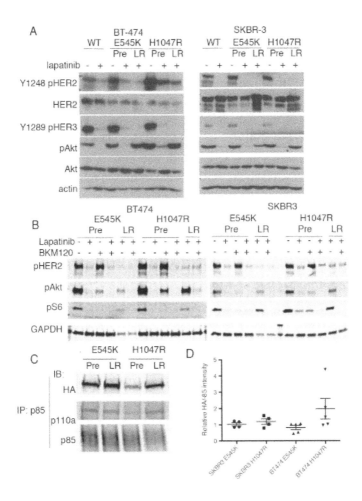

FIGURE 3: Lapatinib-resistant *PIK3CA* mutant cell lines show reactivation of phosphoinositide 3-kinase and utilize mutant p110α for phosphoinositide 3-kinase signaling. (A) Cells expressing E545K or H1047R were selected for lapatinib resistance. Unselected cells (Pre) were treated with lapatinib for 3 hours and analyzed along with lapatinib-resistant (LR) cells in the presence of lapatinib by immunoblotting with the indicated antibodies. (B) Lapatinib-sensitive and LR cells were treated with lapatinib and/or BKM120 for 3 hours, and lysates were analyzed by immunoblotting. (C) Phosphoinositide 3-kinase (PI3K) was immunoprecipitated from *PIK3CA* mutant–expressing cells before and after selection for lapatinib resistance with a p85 antibody. Complexes were separated by SDS-PAGE, and expression of mutant p110α was determined by immunoblot analysis for the hemagglutinin (HA)-tag. (D) HA band intensity detected by infrared fluorescence was quantified using a LI-COR Odyssey imaging system. Mean HA intensity normalized to p85 intensity from at least four replicate experiments is shown (bars = SEM).

the H1047R mutant p110α was detected in a complex with p85 in resistant cells than in parental cells (Figures 3C and 3D). We found a similar increase in HA expression in p85 immunoprecipitates from SKBR3 cells expressing H1047R p110α, but this increase did not persist after normalizing for levels of immunoprecipitated p85 (Figure 3D and Additional file 9: Figure S5). We did not observe any increase in levels of the E545K mutant in a complex with p85 in either cell line. The E545K mutation has been proposed to alter interaction with p85 and so may have a different mechanism of action than the H1047R catalytic site mutation [12,31]. This may explain why levels of E545K mutant p110α did not increase in PI3K from LR cells.

3.3.4 HER2 INHIBITOR RESISTANCE CONFERRED BY PHOSPHOINOSITIDE 3-KINASE MUTATION IS SUSCEPTIBLE TO P110α INHIBITION

The data we compiled in this study suggest that the gain of function conferred by *PIK3CA* mutations partially uncouples PI3K signaling from HER2 and allows for eventual escape from HER2 inhibition, but that signaling remains susceptible to inhibitors of p110α. This idea is in agreement with previous reports that lapatinib resistance from PI3K mutations or PTEN loss could be overcome by combining lapatinib with a dual PI3K/mammalian target of rapamycin (PI3K/mTOR) inhibitor [16]. Another possibility is that the gain of function of mutant PI3K can amplify low levels of HER2 signaling that remain after single-inhibitor treatment of HER2+ breast cancer [23], but the combination of dual HER2 inhibitor blockade may reduce this low level of signaling and thus inhibit even *PIK-3CA* mutant cancers. This idea is consistent with recent clinical data indicating that the combination of trastuzumab and lapatinib or trastuzumab and pertuzumab is more effective than single HER2 inhibitor therapy [20-22]. To test whether additional HER2 blockade could overcome resistance to single-agent lapatinib or trastuzumab conferred by PI3K mutation, we treated LR *PIK3CA* mutant cells with lapatinib and trastuzumab and with or without the PI3K inhibitor BKM120. We found that the addition of trastuzumab did not inhibit the already low level of HER2 phosphorylation or further diminish the minimally detectable HER3 phosphorylation in these cells, nor did it further decrease PI3K-Akt signaling through Akt or S6 phosphorylation (Figure 4A). This low level of signaling appears to be sufficient for cell proliferation, as ectopic PI3K expression allowed for continued proliferation of cells in the presence of both trastuzumab and lapatinib (Additional file 10: Figure S6). Only the addition of the PI3K inhibitor resulted in PI3K pathway inhibition (Figure 4A). We also tested the addition of pertuzumab, a monoclonal antibody against HER2 that recognizes a different epitope than trastuzumab, and BIBW2992, a covalent inhibitor of EGFR/HER2 RTKs. None of these additional HER2 inhibitors resulted in any decrease in cell proliferation or increase in apoptosis in LR cells; only the addition of PI3K inhibition resulted in decreased proliferation and the induction of apoptosis (Figures 4B and 4C).

3.3.5 HER2 INHIBITORS IN COMBINATION WITH PHOSPHOINOSITIDE 3-KINASE INHIBITORS CAN PREVENT OUTGROWTH OF RESISTANT TUMORS

Because we did not find any benefit associated with additional HER2 inhibitors once resistance to an HER2 inhibitor was established in cell culture, we sought to test whether addition of a PI3K inhibitor to the combination of lapatinib and trastuzumab would prevent the outgrowth of resistant tumors, both with and without PI3K mutations. We first used BT474 cells (without PI3K hotspot mutations). As previously reported [32], trastuzumab and lapatinib together induced BT474 tumor regression. Treatment with BKM120 alone was able to block tumor growth, but did not induce any tumor regression. Combination of a single inhibitor of HER2 with a PI3K inhibitor appeared to be somewhat effective at inducing tumor regression, with a magnitude of benefit similar to the trastuzumab/lapatinib combination. Only the combination of all three drugs was sufficient to induce a robust regression in tumor growth, however, such that all of the mice in the

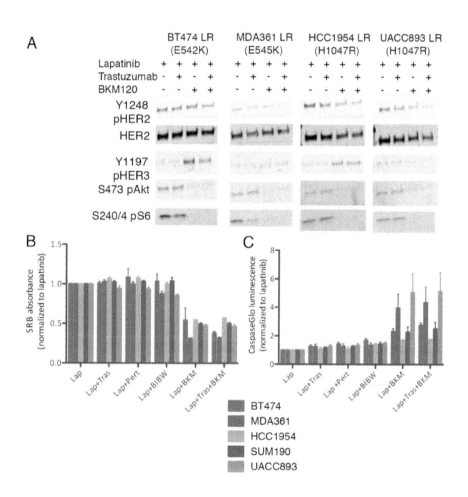

FIGURE 4: Blockade of phosphoinositide 3-kinase, but not additional HER2 blockade, inhibits phosphoinositide 3-kinase signaling in lapatinib-resistant cells with *PIK3CA* mutation. (A) Lapatinib-resistant cells in the presence of lapatinib were treated with trastuzumab overnight, with BKM120 or with the combinations as indicated, and lysates were analyzed by immunoblotting. (B) LR cell lines were plated in 96-well plates. Twenty-four hours after plating, cells were treated with inhibitors as indicated and fixed after seventy-two hours of treatment (lap=lapatinib, tras=trastuzumab, pert=pertuzumab and BKM=BKM120). Cells were stained with sulforhodamine B (SRB). Mean SRB absorbance normalized to cells treated with lapatinib only is shown (duplicates from two separate experiments; bars=SEM). (C) Cells were treated as described in (B), and apoptosis was measured by luminescence after 24-hour treatment with the Caspase-Glo reagent. Mean luminescence (duplicates from two experiments) normalized to lapatinib only is displayed (bars=SEM).

treatment group showed complete regression of tumor growth after about 3 weeks of treatment (Figure 5B). At the end of 4 weeks of treatment, only residual tumors in the BKM120-only or dual-therapy groups were available for analysis, and we observed blockade of PI3K-Akt signaling by the PI3K inhibitor in these residual tumors (Additional file 11: Figure S7). Although the combination of all three drugs was better able to induce tumor regression than two-drug combinations, there were some mice in each group with complete tumor regression with dual- or triple-therapy. To determine whether the response to all three drugs was more durable than two-drug combinations, we followed mice after 28 days of treatment for tumor regrowth. Tumors recurred more rapidly in the trastuzumab/lapatinib combination–treated group, although all groups had at least one mouse with recurrent tumor growth. In all cases where significant recurrence was observed, retreatment with the combination of all three drugs appeared to be effective (black arrows in Figure 5C).

We next tested the combination of HER2 and PI3K inhibition in a xenograft model of HCC1954 cells with endogenous an H1047R *PIK3CA* mutation. These tumors showed continued growth in the presence of trastuzumab or lapatinib as single agents, further suggesting that *PIK3CA* mutations confer resistance to HER2 inhibitors (Figure 5D). In these tumors, the combination of trastuzumab and lapatinib was unable to induce tumor regression as it did in BT474 xenografts. BKM120 alone was as effective at inhibiting growth as dual HER2 inhibition, but the combination of HER2 and PI3K inhibitors was required to induce tumor regression. Unlike treatment of HER2+/*PIK3CA* wild-type tumors, no complete regressions were observed in any treatment group, though most tumors in the triple-therapy group remained only barely palpable at the end of treatment. After 28 days, treatment was stopped and mice were again followed for tumor recurrence, which eventually developed in about half of the mice (Figure 5E). Retreatment with all three inhibitors was able to reinduce tumor regression, suggesting that tumors remain sensitive to the combination of all three drugs. These results imply that a longer duration of treatment would be necessary to induce complete regression in tumors with the *PIK3CA* mutation.

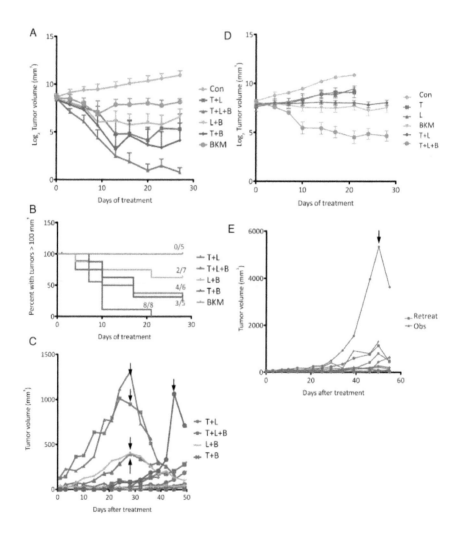

FIGURE 5: Phosphoinositide 3-kinase inhibition combined with HER2 inhibition is more effective at inducing tumor regression than HER2 inhibition alone. (A) Mice were injected with BT474 cells, and, after tumor formation, mice were treated with trastuzumab (T), lapatinib (L) or BKM120 (B) in the combinations indicated. Tumor growth was measured twice weekly. The $\log_2$ tumor volume data is shown over the 28-day treatment course (bars=SEM). (B) Kaplan-Meier plot showing the number of mice in each treatment group whose tumors regressed to below 100 mm^3 in volume during treatment. The number of responses and the total number in each group are displayed above the curve. (C) After 28 days of treatment, mice whose tumors had regressed were followed for recurrence. Tumor volume for individual mice is plotted with the color according to the initial treatment group. Arrows indicate retreatment with T+L+B. (D) Mice were injected with HCC1954 cells (H1047R *PIK3CA* mutant), and, after tumor formation, mice were treated with trastuzumab (T), lapatinib (L) or BKM120 (B) alone or in the combinations indicated. Tumor growth was measured twice weekly. The $\log_2$ tumor volume data are shown over the 28-day treatment course (bars=SEM). Control, T and L alone mice were killed at day 21 of treatment because of excessive tumor volume. (E) After 28 days of treatment, mice from the T+L+B group whose tumors had regressed were followed for recurrence. Tumor volumes for individual mice are plotted. Red indicates mice whose tumors regrew, and those retreated with T+L+B are indicated by the arrow.

3.4 DISCUSSION

The importance of the PI3K–Akt axis in oncogenic signaling is becoming increasingly apparent, especially in the case of HER2+ breast cancer, where inhibition of PI3K signaling is critical for the antitumor action of HER2 inhibitors and activating mutations in the PI3K pathway can confer resistance to HER2 inhibitors. We show in our present study that acquisition of a hotspot *PIK3CA* mutation is a mechanism of acquired resistance to lapatinib and that *PIK3CA* mutations partially uncouple PI3K from HER2 to allow for the development and maintenance of resistance. Further, targeting of PI3K itself, in combination with maximal HER2 blockade with both an antibody and a TKI, is more effective than HER2 targeting alone for HER2 tumors without *PIK3CA* mutations and is required for HER2 tumors with *PIK3CA* mutations.

We and others have found that both helical and catalytic domain mutations of *PIK3CA* can confer resistance to HER2 inhibitors [7,16,24,33]. In our biochemical assays, ectopic expression of either mutation appeared to uncouple HER2 inhibition from PI3K signaling to a similar degree (Figure 2), and cells expressing either mutation showed reactivation of PI3K upon the development of resistance [25]. When we assayed resistant cells for the proportion of mutant vs. wild-type p110α in the PI3K signaling complex compared to sensitive cells, however, we observed an increase in utilization of the H1047R mutant isoform but not the E545K isoform. We also did not observe the same degree of uncoupling of downstream signaling in a cell line with endogenous E545K compared with H1047R expressing cell lines (Figure 2). This is consistent with different proposed mechanisms for these mutations, whereby the helical domain mutant may function primarily to abolish the normal regulatory inhibition of PI3K, whereas the increased catalytic activity conferred by the kinase domain mutation may be required by the mutant cells still under the selective pressure of HER2 inhibition [12,31]. In either case, catalytic inhibition of p110α was effective at blocking downstream signaling for both mutations.

This catalytic inhibition of p110α is emerging as an attractive possible therapeutic option, with a number of inhibitors currently in preclinical and clinical development [34]. Several studies have investigated the potential importance of PI3K inhibition as a therapeutic strategy in *HER2*-amplified breast cancer, but these studies have uncovered feedback loops and other factors that may limit the use of PI3K inhibitors as single agents. A p110-specific inhibitor, GDC-0941, inhibited the growth of *HER2*-amplified cells in culture, but the combination of GDC-0941 with trastuzumab was required for tumor growth inhibition in mice [7]. Interestingly, the combination of GDC-0941 with antibody inhibitors of HER2 (trastuzumab and pertuzumab) appeared to be more effective than GDC-0941 in combination with a TKI [35]. In HER2+ cells with *PIK3CA* mutations, low doses of lapatinib are ineffective, but the cells are susceptible to dual PI3K/mTOR inhibitors such as BEZ235 [16] or INK-128 [36]. In the latter study, however, combination of the HER2 TKI with the PI3K inhibitor was not able to induce tumor regression, whereas combination of a PI3K p110 inhibitor (BKM120) with trastuzumab did result in tumor regression, albeit in a wild-type *PIK3CA* model. In a transgenic HER2+/*PIK3CA* H1047R mutant mouse model, tumors that were resistant to the combinations of trastuzumab and lapatinib or trastuzumab and pertuzumab could be inhibited by BKM120 alone or in combination with HER2 inhibitors [24]. Those observations are in agreement with our present findings that BKM120 alone did not induce tumor regression, but did result in tumor regression when combined with HER2 inhibitors. This increased efficacy of HER2 and PI3K inhibitor combinations may be partly explained by increased Erk signaling and feedback upregulation of HER3 after PI3K inhibitor treatment alone, whereas targeting HER2 in combination with PI3K inhibition could overcome these compensatory mechanisms [37,38].

Our results support the clinical testing of combinations of PI3K inhibition with maximal HER2 inhibition for HER2+ breast cancer. In biomarker studies from the recent CLEOPATRA clinical trial, although the combination of trastuzumab and pertuzumab was superior to trastuzumab alone, regardless of *PIK3CA* mutation status, the magnitude of benefit was less for those tumors with *PIK3CA* mutations than for wild-type tumors [39]. There is a potential for overlapping toxicities when combining multiple inhibitors, so important questions regarding selection of patients and the sequence and timing of combination therapies remain to be addressed. Our data suggest that *PIK3CA* mutation can be acquired during HER2 inhibitor treatment and that the presence of a *PIK3CA* mutation requires PI3K blockade in addition to HER2 blockade. Despite several lines of preclinical evidence that either *PIK3CA* mutation can confer HER2 inhibitor resistance, a robust correlation of the occurrence of *PIK3CA* mutations with outcomes after trastuzumab or lapatinib therapy in patients is still lacking. A recent study of PTEN expression and *PIK3CA* mutation in HER2+ patients with either recurrent disease after trastuzumab treatment or progression of metastatic disease while on trastuzumab therapy showed a significantly increased frequency of either PTEN loss or *PIK3CA* mutation compared to untreated HER2+ tumors, but *PIK3CA* mutation status alone did not appear to be significantly enriched in the clinically trastuzumab-refractory cohort [19]. In addition to determining whether *PIK3CA* mutation predicts for the eventual development of resistance, there is a critical need to understand how frequently *PIK3CA* mutations occur after patients develop resistance to HER2 inhibitor treatment. Another important question is whether these combinations need to be given together as part of the initial treatment or whether direct PI3K inhibition can be added after therapeutic resistance develops. In all of our models of resistance, we found reactivation of PI3K signaling and susceptibility to PI3K inhibition. We also found that additional HER2 blockade after the development of lapatinib resistance was ineffective and that PI3K inhibition was essential to overcoming resistance. This suggests that a potentially better-tolerated sequential approach might be effective, and, indeed, sequential additive treatment with HER2 inhibitors appears to have some clinical efficacy, though these patients eventually develop resistance to combined HER2 blockade [21]. In our xenograft models, however, we found that up-front combination of both HER2 and PI3K blockade was most effective at inducing regression and preventing tumor regrowth, regardless of PI3K mutation status. In the case of *PIK3CA* mutation, this combination was required to induce tumor regression. This suggests that the combination of HER2 and PI3K therapies up front, if tolerable, might provide optimal treatment for patients with *PIK3CA* mutations and also have potential benefits for patients with *PIK3CA* wild-type tumors.

3.5 CONCLUSIONS

Our results show that the gain of function conferred by *PIK3CA* mutations contributes to resistance to HER2 inhibitors by partially uncoupling PI3K signaling from HER2 inhibition. They also show that, though maximal HER2 blockade is insufficient to overcome the effects of the mutation, addition of a PI3K inhibitor to HER2 inhibitors can reverse or prevent resistance. This suggests that this combination may be an effective strategy to overcome resistance in patients that warrants clinical testing.

REFERENCES

1. Geyer CE, Forster J, Lindquist D, Chan S, Romieu CG, Pienkowski T, Jagiello-Gruszfeld A, Crown J, Chan A, Kaufman B, Skarlos D, Campone M, Davidson N, Berger M, Oliva C, Rubin SD, Stein S, Cameron D: Lapatinib plus capecitabine for HER2-positive advanced breast cancer. N Engl J Med 2006, 355:2733-2743.
2. Slamon DJ, Leyland-Jones B, Shak S, Fuchs H, Paton V, Bajamonde A, Fleming T, Eiermann W, Wolter J, Pegram M, Baselga J, Norton L: Use of chemotherapy plus a monoclonal antibody against HER2 for metastatic breast cancer that overexpresses HER2. N Engl J Med 2001, 344:783-792.
3. Yarden Y, Sliwkowski MX: Untangling the ErbB signalling network. Nat Rev Mol Cell Biol 2001, 2:127-137.

4. Holbro T, Beerli RR, Maurer F, Koziczak M, Barbas CF 3rd, Hynes NE: The ErbB2/ ErbB3 heterodimer functions as an oncogenic unit: ErbB2 requires ErbB3 to drive breast tumor cell proliferation. Proc Natl Acad Sci USA 2003, 100:8933-8938.

5. Brachmann SM, Hofmann I, Schnell C, Fritsch C, Wee S, Lane H, Wang S, Garcia-Echeverria C, Maira SM: Specific apoptosis induction by the dual PI3K/mTor inhibitor NVP-BEZ235 in HER2 amplified and PIK3CA mutant breast cancer cells. Proc Natl Acad Sci USA 2009, 106:22299-22304.

6. O'Brien C, Wallin JJ, Sampath D, GuhaThakurta D, Savage H, Punnoose EA, Guan J, Berry L, Prior WW, Amler LC, Belvin M, Friedman LS, Lackner MR: Predictive biomarkers of sensitivity to the phosphatidylinositol 3' kinase inhibitor GDC-0941 in breast cancer preclinical models. Clin Cancer Res 2010, 16:3670-3683.

7. Junttila TT, Akita RW, Parsons K, Fields C, Lewis Phillips GD, Friedman LS, Sampath D, Sliwkowski MX: Ligand-independent HER2/HER3/PI3K complex is disrupted by trastuzumab and is effectively inhibited by the PI3K inhibitor GDC-0941. Cancer Cell 2009, 15:429-440.

8. Ritter CA, Perez-Torres M, Rinehart C, Guix M, Dugger T, Engelman JA, Arteaga CL: Human breast cancer cells selected for resistance to trastuzumab in vivo over-express epidermal growth factor receptor and ErbB ligands and remain dependent on the ErbB receptor network. Clin Cancer Res 2007, 13:4909-4919.

9. Yakes FM, Chinratanalab W, Ritter CA, King W, Seelig S, Arteaga CL: Herceptin-induced inhibition of phosphatidylinositol-3 kinase and Akt Is required for antibody-mediated effects on p27, cyclin D1, and antitumor action. Cancer Res 2002, 62:4132-4141.

10. Saal LH, Holm K, Maurer M, Memeo L, Su T, Wang X, Yu JS, Malmström PO, Mansukhani M, Enoksson J, Hibshoosh H, Borg A, Parsons R: PIK3CA mutations correlate with hormone receptors, node metastasis, and ERBB2, and are mutually exclusive with PTEN loss in human breast carcinoma. Cancer Res 2005, 65:2554-2559.

11. Samuels Y, Wang Z, Bardelli A, Silliman N, Ptak J, Szabo S, Yan H, Gazdar A, Powell SM, Riggins GJ, Willson JK, Markowitz S, Kinzler KW, Vogelstein B, Velculescu VE: High frequency of mutations of the PIK3CA gene in human cancers. Science 2004, 304:554.

12. Zhao L, Vogt PK: Helical domain and kinase domain mutations in p110α of phosphatidylinositol 3-kinase induce gain of function by different mechanisms. Proc Natl Acad Sci USA 2008, 105:2652-2657.

13. Koboldt DC, Fulton RS, McLellan MD, Schmidt H, Kalicki-Veizer J, McMichael JF, Fulton LL, Dooling DJ, Ding L, Mardis ER, Wilson RK, Ally A, Balasundaram M, Butterfield YS, Carlsen R, Carter C, Chu A, Chuah E, Chun HJ, Coope RJ, Dhalla N, Guin R, Hirst C, Hirst M, Holt RA, Lee D, Li HI, Mayo M, Moore RA, Mungall AJ, Cancer Genome Atlas Network, et al.: Comprehensive molecular portraits of human breast tumours. Nature 2012, 490:61-70.

14. Stemke-Hale K, Gonzalez-Angulo AM, Lluch A, Neve RM, Kuo WL, Davies M, Carey M, Hu Z, Guan Y, Sahin A, Symmans WF, Pusztai L, Nolden LK, Horlings H, Berns K, Hung MC, van de Vijver MJ, Valero V, Gray JW, Bernards R, Mills GB,

Hennessy BT: An integrative genomic and proteomic analysis of PIK3CA, PTEN, and AKT mutations in breast cancer. Cancer Res 2008, 68:6084-6091.

15. Berns K, Horlings HM, Hennessy BT, Madiredjo M, Hijmans EM, Beelen K, Linn SC, Gonzalez-Angulo AM, Stemke-Hale K, Hauptmann M, Beijersbergen RL, Mills GB, van de Vijver MJ, Bernards R: A functional genetic approach identifies the PI3K pathway as a major determinant of trastuzumab resistance in breast cancer. Cancer Cell 2007, 12:395-402.

16. Eichhorn PJA, Gili M, Scaltriti M, Serra V, Guzman M, Nijkamp W, Beijersbergen RL, Valero V, Seoane J, Bernards R, Baselga J: Phosphatidylinositol 3-kinase hyperactivation results in lapatinib resistance that is reversed by the mTOR/phosphatidylinositol 3-kinase inhibitor NVP-BEZ235. Cancer Res 2008, 68:9221-9230.

17. Serra V, Markman B, Scaltriti M, Eichhorn PJA, Valero V, Guzman M, Botero ML, Llonch E, Atzori F, Di Cosimo S, Maira M, Garcia-Echeverria C, Parra JL, Arribas J, Baselga J: NVP-BEZ235, a dual PI3K/mTOR inhibitor, prevents PI3K signaling and inhibits the growth of cancer cells with activating PI3K mutations. Cancer Res 2008, 68:8022-8030.

18. Nagata Y, Lan KH, Zhou X, Tan M, Esteva FJ, Sahin AA, Klos KS, Li P, Monia BP, Nguyen NT, Hortobagyi GN, Hung MC, Yu D: PTEN activation contributes to tumor inhibition by trastuzumab, and loss of PTEN predicts trastuzumab resistance in patients. Cancer Cell 2004, 6:117-127.

19. Chandarlapaty S, Sakr RA, Giri D, Patil S, Heguy A, Morrow M, Modi S, Norton L, Rosen N, Hudis C, King TA: Frequent mutational activation of the PI3K-AKT pathway in trastuzumab-resistant breast cancer. Clin Cancer Res 2012, 18:6784-6791.

20. Baselga J, Bradbury I, Eidtmann H, Di Cosimo S, de Azambuja E, Aura C, Gómez H, Dinh P, Fauria K, Van Dooren V, Aktan G, Goldhirsch A, Chang TW, Horváth Z, Coccia-Portugal M, Domont J, Tseng LM, Kunz G, Sohn JH, Semiglazov V, Lerzo G, Palacova M, Probachai V, Pusztai L, Untch M, Gelber RD, Piccart-Gebhart M, NeoALTTO Study Team: Lapatinib with trastuzumab for HER2-positive early breast cancer (NeoALTTO): a randomised, open-label, multicentre, phase 3 trial. Lancet 2012, 379:633-640.

21. Blackwell KL, Burstein HJ, Storniolo AM, Rugo HS, Sledge G, Aktan G, Ellis C, Florance A, Vukelja S, Bischoff J, Baselga J, O'Shaughnessy J: Overall survival benefit with lapatinib in combination with trastuzumab for patients with human epidermal growth factor receptor 2–positive metastatic breast cancer: final results from the EGF104900 Study. J Clin Oncol 2012, 30:2585-2592.

22. Baselga J, Cortés J, Kim SB, Im SA, Hegg R, Im YH, Roman L, Pedrini JL, Pienkowski T, Knott A, Clark E, Benyunes MC, Ross G, Swain SM, CLEOPATRA Study Group: Pertuzumab plus trastuzumab plus docetaxel for metastatic breast cancer. N Engl J Med 2012, 366:109-119. Amin DN, Sergina N, Ahuja D, McMahon M, Blair JA, Wang D, Hann B, Koch KM, Shokat KM, Moasser MM: Resiliency and vulnerability in the HER2-HER3 tumorigenic driver. Sci Transl Med 2010, 2:16ra7.

23. Hanker AB, Pfefferle AD, Balko JM, Kuba MG, Young CD, Sánchez V, Sutton CR, Cheng H, Perou CM, Zhao JJ, Cook RS, Arteaga CL: Mutant PIK3CA accelerates HER2-driven transgenic mammary tumors and induces resistance to combinations of anti-HER2 therapies. Proc Natl Acad Sci USA 2013, 110:14372-14377.

24. Rexer BN, Ham AJL, Rinehart C, Hill S, de Matos Granja-Ingram N, González-Angulo AM, Mills GB, Dave B, Chang JC, Liebler DC, Arteaga CL: Phosphoproteomic mass spectrometry profiling links Src family kinases to escape from HER2 tyrosine kinase inhibition. Oncogene 2011, 30:4163-4174.

25. Isakoff SJ, Engelman JA, Irie HY, Luo J, Brachmann SM, Pearline RV, Cantley LC, Brugge JS: Breast cancer–associated PIK3CA mutations are oncogenic in mammary epithelial cells. Cancer Res 2005, 65:10992-11000.

26. Dias-Santagata D, Akhavanfard S, David SS, Vernovsky K, Kuhlmann G, Boisvert SL, Stubbs H, McDermott U, Settleman J, Kwak EL, et al.: Rapid targeted mutational analysis of human tumours: a clinical platform to guide personalized cancer medicine. EMBO Mol Med 2010, 2:146-158.

27. Su Z, Dias-Santagata D, Duke M, Hutchinson K, Lin YL, Borger DR, Chung CH, Massion PP, Vnencak-Jones CL, Iafrate AJ, Pao W: A platform for rapid detection of multiple oncogenic mutations with relevance to targeted therapy in non-small-cell lung cancer. J Mol Diagn 2011, 13:74-84.

28. Lovly CM, Dahlman KB, Fohn LE, Su Z, Dias-Santagata D, Hicks DJ, Hucks D, Berry E, Terry C, Duke M, Su Y, Sobolik-Delmaire T, Richmond A, Kelley MC, Vnencak-Jones CL, Iafrate AJ, Sosman J, Pao W: Routine multiplex mutational profiling of melanomas enables enrollment in genotype-driven therapeutic trials. PLoS One 2012, 7:e35309.

29. Engelman JA, Jänne PA, Mermel C, Pearlberg J, Mukohara T, Fleet C, Cichowski K, Johnson BE, Cantley LC: ErbB-3 mediates phosphoinositide 3-kinase activity in gefitinib-sensitive non-small cell lung cancer cell lines. Proc Natl Acad Sci USA 2005, 102:3788-3793.

30. Huang CH, Mandelker D, Schmidt-Kittler O, Samuels Y, Velculescu VE, Kinzler KW, Vogelstein B, Gabelli SB, Amzel LM: The structure of a human p110α/p85α complex elucidates the effects of oncogenic PI3Kα mutations. Science 2007, 318:1744-1748.

31. Garrett JT, Olivares MG, Rinehart C, Granja-Ingram ND, Sánchez V, Chakrabarty A, Dave B, Cook RS, Pao W, McKinely E, Manning HC, Chang J, Arteaga CL: Transcriptional and posttranslational up-regulation of HER3 (ErbB3) compensates for inhibition of the HER2 tyrosine kinase. Proc Natl Acad Sci USA 2011, 108:5021-5026.

32. Chakrabarty A, Rexer BN, Wang SE, Cook RS, Engelman JA, Arteaga CL: H1047R phosphatidylinositol 3-kinase mutant enhances HER2-mediated transformation by heregulin production and activation of HER3. Oncogene 2010, 29:5193-5203.

33. Miller TW, Rexer BN, Garrett JT, Arteaga CL: Mutations in the phosphatidylinositol 3-kinase pathway: role in tumor progression and therapeutic implications in breast cancer. Breast Cancer Res 2011, 13:224.

34. Yao E, Zhou W, Lee-Hoeflich ST, Truong T, Haverty PM, Eastham-Anderson J, Lewin-Koh N, Gunter B, Belvin M, Murray LJ, Friedman LS, Sliwkowski MX, Hoeflich KP: Suppression of HER2/HER3-mediated growth of breast cancer cells with combinations of GDC-0941 PI3K inhibitor, trastuzumab, and pertuzumab. Clin Cancer Res 2009, 15:4147-4156.

35. García-García C, Ibrahim YH, Serra V, Calvo MT, Guzmán M, Grueso J, Aura C, Pérez J, Jessen K, Liu Y, Rommel C, Tabernero J, Baselga J, Scaltriti M: Dual

mTORC1/2 and HER2 blockade results in antitumor activity in preclinical models of breast cancer resistant to anti-HER2 therapy. Clin Cancer Res 2012, 18:2603-2612.

36. Chakrabarty A, Sánchez V, Kuba MG, Rinehart C, Arteaga CL: Feedback upregulation of HER3 (ErbB3) expression and activity attenuates antitumor effect of PI3K inhibitors. Proc Natl Acad Sci USA 2012, 109:2718-2723.

37. Serra V, Scaltriti M, Prudkin L, Eichhorn PJ, Ibrahim YH, Chandarlapaty S, Markman B, Rodriguez O, Guzman M, Rodriguez S, Gili M, Russillo M, Parra JL, Singh S, Arribas J, Rosen N, Baselga J: PI3K inhibition results in enhanced HER signaling and acquired ERK dependency in HER2-overexpressing breast cancer. Oncogene 2011, 30:2547-2557.

38. Baselga J, Cortés J, Im SA, Kiermaier A, Ross G, Swain SM: Biomarker analyses in CLEOPATRA: a phase III, placebo-controlled study of pertuzumab in HER2-positive, first-line metastatic breast cancer (MBC). Cancer Res 2012, 72:Abstract nr S5-1.

There are several supplemental files that are not available in this version of the article. To view this additional information, please use the citation information cited on the first page of this chapter.

CHAPTER 4

WNT3A EXPRESSION IS ASSOCIATED WITH MMP-9 EXPRESSION IN PRIMARY TUMOR AND METASTATIC SITE IN RECURRENT OR STAGE IV COLORECTAL CANCER

MYUNG AH LEE, JIN-HEE PARK, SI YOUNG RHYU, SEONG-TAEK OH, WON-KYOUNG KANG, AND HEE-NA KIM

4.1 BACKGROUND

Colorectal cancer (CRC) is the 3rd most common cancer in Korea and is becoming more common in Asia and Western countries [1]. As new anti-cancer agents and new technology for local treatment have been developed, the survival rates of patients with CRC have been increasing, even for patients with stage IV cancer. However, distant organ metastasis eventually develops in stage IV cancer, and leads to fatal organ failure. To improve survival outcomes, it is important to control cancer invasion and metastasis, but the mechanisms by which CRC becomes metastasis are not yet known.

This chapter was originally published under the Creative Commons Attribution License. Lee MA, Park J-H, Rhyu SY, Oh S-T, Kang W-K, and Kim H-N. Wnt3a Expression is Associated with MMP-9 Expression in Primary Tumor and Metastatic Site in Recurrent or Stage IV Colorectal Cancer. BMC Cancer 14,125 (2014). doi:10.1186/1471-2407-14-125.

The wnt/β-catenin pathway is known to play an important role in maintaining cell homeostasis in normal cells and in embryologic cell development. In CRC, mutation in APC or β-catenin, component of the wnt signaling pathway, are well-known oncogenic factors in familial and some sporadic CRC cases. Recently, many researchers have suggested that the wnt signaling pathway is also involved in controlling cancer cell invasion and metastasis by interacting with the tumor microenvironment or other signal pathways [2-4]. However, most studies investigating the effects of the wnt pathway on CRC have focused on β-catenin, and the role of wnt3a (the initiator of wnt/β-catenin pathway) is not well understood.

The non-canonical pathway via wnt5a is another of the identified wnt signaling pathways. The non-canonical pathway plays a role in embryonic cell motility, but the role of the non-canonical pathway in cancer is unknown. Katoh has suggested that the non-canonical pathway is involved in tumor cell invasion and metastasis [5], but studies investigating the role of wnt5a expression in cancer have been limited and controversial. Several authors have reported that the wnt5a expression is associated with higher grade, poor differentiation or poor clinical outcomes, but others have suggested that wnt5a expression antagonizes the wnt/β-catenin pathway and inhibits oncogenesis [6]. Dejimek et al. suggested that wnt5a expression in stage II colon cancer is associated with good prognosis, and another study reported that wnt5a methylation is associated with microsatellite instability and BRAF mutation [7,8]. Considering these data, the exact role of wnt5a in cancer is unclear.

MMP-9 and VEGF expression are commonly studied in cancer research, and their expression is associated with invasion and metastasis. One study reported that in vitro inhibition of DKK-1 signaling inhibits the MMP-9 expression [9], but MMP-9 expression decreased after wnt3a treatment in another study [10]. Many studies investigating the role of MMP-9 in CRC have been conducted using cell lines or animal models. Few trials have reported that MMP-9 expression can be used as a prognostic factor for disease recurrence in human tissue [11]. Thus, the association between the wnt signaling pathway and MMP-9 and VEGFR-2 expression in CRC remains unclear.

Previous experiments investigating the role of the wnt proteins in metastasis have been conducted at the primary site of stages I-III CRC because diseased tissue can be easily obtained through surgery. Studies focusing on the relationships between the primary tissue and tissue from the metastatic site have rarely been reported. Recently, resection of metastatic lesions has become a treatment option for patients with stage IV CRC. In the present study, we evaluated the expression of wnt3a, wnt5a, MMP-9, and VEGFR-2 in human tissue from primary and metastatic sites of stage IV advanced CRC patients to identify associations between these proteins. We also analyzed the concordance between the primary and metastatic lesions, and aimed to identify the potential prognostic markers of survival outcome.

4.2 METHODS

4.2.1 PATIENTS

This study included eighty-three patients with colon or rectal cancer who had resection for both a primary mass and metastatic lesions resected in a single procedure at Seoul St. Mary's Hospital between January 2000 and December 2006. All patients had colorectal cancer with organ metastasis at the initial diagnosis. Clinical records and pathological reports were reviewed retrospectively. This study was approved by the institutional review board of Seoul St. Mary's hospital (KC10SIMI0621).

4.2.2 TISSUE MICROARRAY (TMA) AND IMMUNOHISTOCHEMICAL STAINING

Core biopsies with 3.0 mm diameter were taken from representative areas of tumor tissue. Each patient had biopsy samples taken from the primary tumor mass and its adjacent mesenchyme as well as the metastatic site and its adjacent mesenchyme. Tissue cores from each specimen were assembled on a recipient paraffin block with a precision instrument (Micro Digital Co. Korea), following previously established methods [12].

We performed immunohistochemical staining on 5 μm sections of the TMA blocks. Each paraffin section was deparaffinized for 1 hour at 60°C in

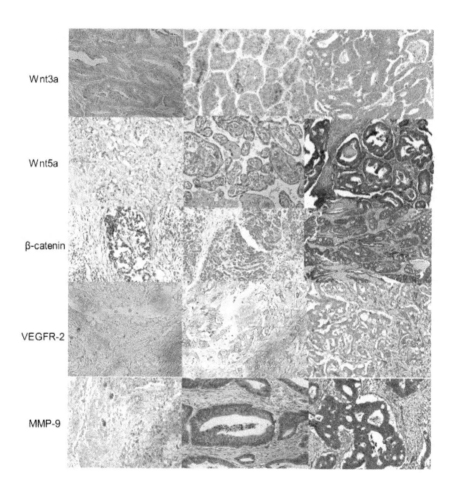

FIGURE 1: Immunohistochemical staining for the wnt3a, wnt5a, β-catenin, MMP-9, and VEGFR-2 (magnification x 200).

xylene and rehydrated in serial graded ethanol before being stored overnight in citrate buffer (0.01 M, pH 6.0) at 75°C for antigen retrieval. Endogenous peroxidase activity was blocked with 0.3% hydrogen peroxide in methanol. Sections were incubated for 1 hour at room temperature with the following primary antibodies at the specified dilutions: Wnt3a (Abcam, Cambridge, UK) diluted 1:100, Wnt5a (Abcam, Cambridge, UK) diluted 1:50, β-catenin (Abcam, Cambridge, UK) diluted 1:100, MMP-9 (Cell Signaling, Danvers, MA, USA) diluted 1:100, and VEGFR-2 (Cell Signaling, Danvers, MA, USA) diluted 1:200. Immunohistochemical staining was performed using the rabbit or mouse DAKO ChemMate TM EnVision TM system and a Peroxidase/DAB kit (DAKO). Sections were then counterstained with Mayer hematoxylin and dehydrated, cleared and mounted.

The results were interpreted by two independent pathologists who were blinded to the specific diagnosis and prognosis for each case. The staining intensity was scored on a three-tiered scale: score 0 = less than 10% of cells positive; 1 = 10–49% positive; and 2 = more than 50% of cells positive. The criterion for positive staining was more than 1+ of tumor cells that showed distinct nuclear or cytoplasmic staining (Figure 1).

4.2.3 STATISTICAL ANALYSIS

Continuous and categorical variables were compared using the Student's t test and chi-square test. All statistical analyses were performed using SPSS (version 13.0) and p value under 0.05 was considered statistically significant.

4.3 RESULTS

4.3.1 PATIENT CHARACTERISTICS AND PROTEIN EXPRESSION

Of the 83 patients, 46 were male and the median age was 60 years. Liver metastasis was the most common (57.8%), followed by peritoneum metastasis (26.5%). Thirteen patients had no lymph node involvement even with stage IV disease, and two patients had T2 disease. The patient characteristics are summarized in Table 1.

TABLE 1: Patients' characteristics

		Number
Age	Years	60 (27–78)
Sex	M : F	46 : 37
Location	Colon	51 (61.4%)
	Rectum	32 (38.6%)
T	2	2 (2.4%)
	3	24 (28.9%)
	4	57 (68.7%)
N	0	13 (15.7%)
	1	26 (31.3%)
	2	44 (53.0%)
Metastasis	Liver	48 (57.8%)
	Peritoneum	22 (26.5%)
	Lung	4 (4.8%)
	Ovary	3 (3.6%)
	Others	6 (7.2%)

Wnt3a, wnt5a, MMP-9 and β-catenin were expressed in more than 50% of the primary tumors, but VEGFR-2 was not. These protein expression levels were slightly decreased in the tissue taken from the metastatic sites (Table 2). We analyzed the concordance rate of the protein expression between the primary tumor and metastatic site. The concordance rates of wnt3a, wnt5a and β-catenin expression were high; all of the rates were in the range of 76.2% to 79.4%, and MMP-9 expression had a 68.3% concordance rate. However, VEGFR-2 was expressed in 67.4% of the metastatic sites when there was no expression in the primary tumors, with only a 40.0% concordance rate between primary tumor and metastatic sites.

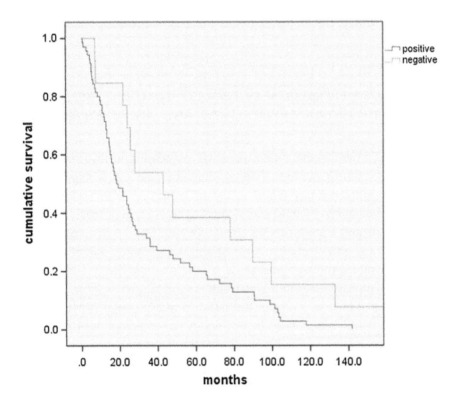

FIGURE 2: Overall survival according to β-catenin expression. β-catenin-expressing group in primary tumor showed poorer survival outcome than non-expressing group (18.4 vs. 42.9 months, p=0.05).

TABLE 2: Immunohistochemical staining for primary tumors and metastatic site

	Wnt3a	Wnt5a	β-catenin	MMP-9	VEGFR-2
Primary tumor	51 (61.4%)	58 (69.9%)	70 (84.3%)	45 (54.2%)	30 (36.1%)
Metastatic site	38 (45.8%)	39 (47.0%)	44 (53.0%)	28 (33.7%)	22 (26.5%)
Concordance rate	79.4%	76.2%	79.4%	68.3%	40.0%

4.3.2 WNT3A EXPRESSION IN PRIMARY TUMOR CORRELATED WITH LYMPH NODE INVOLVEMENT AND MMP-P EXPRESSION

We analyzed the association between wnt expression in the primary tumor and the clinicopathologic findings, including the T & N stage. Wnt3a expression in the primary tumor was significantly correlated with lymph node involvement (p=0.038) and MMP-9 expression in primary, adjacent mesenchyme and metastatic sites (p=0.038, 0.022 and 0.004, respectively). There was no association between wnt5a expression and other findings, but wnt5a expression did show a correlation tendency with lymph node and lymphatic invasion. This result is summarized in Table 3.

Analysis of liver or peritoneal metastasis, wnt expression, MMP expression, and VEGFR-2 expression in the primary tumor did not show any associations but venous invasion was associated with liver metastasis (p=0.047). We also performed immunohistochemical staining of the adjacent mesenchymal tissue, but there was no association between wnt3a, wnt5a, MMP, or VEGFR-2 expression and liver or peritoneal metastasis (Table 4).

In survival analysis, patients with positive β-catenin staining in primary tumors showed significantly poorer survival outcomes than those with no staining (18.4 months vs. 42.9 months, respectively, p=0.05, Figure 2). There were no other prognostic factors for survival.

4.4 DISCUSSION

The wnt/β-catenin signaling pathway is an emerging target for cancer research; studies indication the pathway may play a role in cancer invasion and progression by interacting with the tumor microenvironment and oncogenesis [13]. Wnt/β-catenin has been widely studied as a prognostic factor for CRC, and it may also be involved in the mechanism of cancer invasion [14,15]. However, research has focused on aberrant nuclear β-catenin rather than wnt3a, despite the role of wnt3a as a major initiating factor in the wnt/β-catenin pathway. In the present study, we analyzed wnt3a and wnt5a to determine their roles in cancer progression. We selected the patients who had had both their primary tumor and the metastatic sites resected to identify differences in the protein expression between the primary and the metastatic sites. Expression of wnt3a was very high in tissues from both primary tumors and metastatic sites but was higher at the primary site with a concordance rate higher than 70%. Wnt expression at the metastatic site was rare if the primary tumor tested negative for both wnt3a and wnt5a. This result suggests that wnt could be expressed initially when CRC develops, not newly emerged as the cancer progresses. This result is consistent with the previous studies on the oncogenic role of the wnt signaling pathway in various cancers [16,17]. Recently, Boutros reported that sustained wnt activity through wnt3a and Evi/Wls/GPR177 can be important for the proliferation in colon cancer cell, independently from APC or β-catenin mutation [18]. It suggests that the upper stream factor of wnt signaling pathway may play an important role in the cancer progression.

It is well understood that MMP-9 overexpression is a key factor in degradation of the extracellular matrix, an essential step in tumor invasion and metastasis; this role has been observed in human tissue and cell line studies of CRC [19,20]. In the present study, wnt3a expression was significantly correlated with MMP-9 expression in the primary tumor, mesenchyme and metastatic site. In previous study, inhibition of the wnt/β-catenin pathway decreased the level of MMP-9 mRNA in embryonic neural stem cells [9]. These data suggest that the wnt/β-catenin signaling pathway may have an effect on MMP-9 expression, with a role in cancer invasion and metastasis. However, in one mouse study, wnt3a stimulation was shown to inhibit MMP-2 and MMP-9 expression in mesenchymal stem cells; the investigator suggested that regulation of wnt3a could be different in mice and human [13]. Recently, some studies have reported that inhibition of β-catenin by some agents can also inhibit MMP-2 or MMP-9 expression

[21,22]. The previous data suggest that the wnt/β-catenin pathway may play a role in cancer invasion and metastasis through MMP-9 expression.

TABLE 3: The association of wnt expression in primary tumor and other pathologic findings

		Wnt3a		Wnt5a	
		No.	p	No.	p
T	2	1		0	
	3	13	0.626	17	0.092
	4	37		41	
N	0	8		9	
	1	21	0.038	19	0.910
	2	22		30	
Grade	Well	6		6	
	Moderate	37	0.268	45	0.945
	Poorly	8		7	
Lymphatic invasion	(+)	46	0.482	54	0.088
Venous invasion	(+)	16	0.594	20	0.249
Perineural invasion	(+)	34	0.060	37	0.136
Primary tumor	β-catenin	46	0.063	52	0.054
	MMP-9	34	0.004	33	0.306
	VEGFR-2	21	0.166	24	0.102
Adjacent	β-catenin	39	0.299	37	0.289
Mesemchyme	MMP-9	32	0.022	32	0.359
	VEGFR-2	15	0.205	17	0.158
Metastatic site	β-catenin	29	0.382	32	0.598
	MMP-9	23	0.031	20	0.511
	VEGFR-2	16	0.396	15	0.364

VEGFR is also a topic of interest in cancer proliferation and metastasis research. Many studies have associated the wnt/β-catenin signaling pathway with VEGFR activity [23,24]. In this study, however, wnt expression was not correlated with VEGFR-2 expression. In addition, VEGFR-2 expression was relatively low in our study, particularly at the metastatic site. In other experiments, VEGFR-1 expression was regulated independent of the wnt/β-catenin pathway, and VEGFR-2 did not show any significant association with lymph node or lymphovascular invasion [25,26]. Based on these data, the mechanism by which VEGFR is involved in metastasis should be explored independently of the wnt signaling pathway.

TABLE 4: The association between the protein expression and liver or peritoneal seeding

		Liver		Peritoneum	
		No.	p	No.	p
T	2	1		0	
	3	18	0.130	4	0.264
	4	29		18	
N	0	6		2	
	1	17	0.508	5	0.243
	2	25		15	
Grade	Well	4		3	
	Moderate	39	0.749	17	0.419
	Poorly	5		2	
Lymphatic invasion	(+)	44	0.304	20	0.556
Venous invasion	(+)	19	0.047	5	0.230
Perineural invasion	(+)	27	0.353	15	0.223
Primary tumor	Wnt3a	30	0.498	13	0.493
	Wnt5a	34	0.506	12	0.062
	β-catenin	39	0.277	18	0.470
	MMP-9	26	0.584	10	0.238
	VEGFR-2	15	0.196	9	0.385
Adjacent mesenchyme	Wnt3a	18	0.498	11	0.151
	Wnt5a	29	0.298	15	0.412
	β-catenin	36	0.453	15	0.347
	MMP-9	25	0.510	12	0.533
	VEGFR-2	13	0.431	5	0.494

Wnt5a is a key initiating factor in the non-canonical pathway, but its role in cancer is not known. Kato has suggested that the non-canonical pathway may be involved in cancer cell invasion [5]. In previous study, wnt5a expression showed aggressive behavior in breast or gastric cancer [27,28]. However, wnt5a has also been associated with a good prognosis or tumor suppression by inhibiting the wnt/β-catenin pathway in CRC [7,29]. In our study, wnt5a showed no correlation with pathologic findings or invasion related protein expression, but showed higher expression in the primary and metastatic tumor sites. The data do not show an antagonistic relationship between wn3a and wnt5a in the present study. To determine the role of wnt5a in CRC, further analysis of other signaling pathways is warranted.

Theoretically, wnt3a expression is directly associated with β-catenin expression. However, previous studies have reported that β-catenin can be independently, aberrantly expressed without altering wnt3a in CRC [14,15] and could not be differentiated from the β-catenin that is activated by wnt3a. This is the reason β-catenin was higher than wnt3a expression in our study. In survival analysis of our study, β-catenin expression was significantly correlated with poor survival outcome, independently of wnt3a expression. It has previously been shown that the β-catenin expression can be independent prognostic marker for CRC patients.

As a prognostic factor for overall survival, β-catenin expression was significantly correlated only with the survival outcome. Known prognostic factors, such as lymph node involvement or lymphovascular invasion, did not show any significance in our survival analysis. We analyzed the stage IV patients with metastasis in the present study; these factors could have less of effect on the survival status in stage IV patients than in stage II or III CRC patients.

There is a limitation in our study. We could not determine whether the wnt and MMP-9 expression levels are prognostic or predictive factors because we performed the present study in stage IV CRC patients. According to the objective of this study, we enrolled the patients who underwent surgery for primary and metastatic sites; thus, patients with early stages of CRC were not included. Therefore, a comparative study would be required to determine whether wnt and MMP-9 expression levels are prognostic factors for the recurrence of distant metastasis.

In summary, wnt3a and wnt5a expression is high in primary and metastatic tumors in CRC with a high concordance rate. The wnt3a expression is highly correlated with MMP-9 expression, but not with VEGFR-2, and we did not determine the role of wnt5a. To investigate the mechanism of invasion and metastasis, further studies of the wnt/β-catenin pathway and MMP-9 should be performed, and another approach for evaluating VEGFR or wnt5a should be explored.

4.5 CONCLUSIONS

Wnt3a and wnt5a are highly expressed in colorectal cancer both in primary and metastatic sites with a higher than 50% concordance rate. The wnt3a expression is significantly associated with MMP-9 expression, the metastasis related protein, but is not related with VEGFR-2 expression, and other metastatic related protein.

REFERENCES

1. Jung KW, Park SH, Won YJ, Kong YJ, Lee JY, Park EC, Lee JS: Prediction of cancer incidence and mortality in Korea. Cancer Res Treat 2011, 43(1):12-18.
2. Polarkis P: Wnt signaling and cancer. Genes Dev 2000, 14:1837-1851.
3. Saif MW, Chu E: Biology of colorectal cancer. Cancer J 2010, 16(3):196-201.
4. Huang D, Du X: Crosstalk between tumor cells and microenvironment via wnt pathway in colorectal cancer dissemination. J Gastroenterol 2008, 14(12):1823-1827.
5. Katoh M: WNT/PCP signaling pathway and human cancer (review). Oncol Repub 2005, 14(6):1583-1588.
6. McDonald SL, Silver A: The opposing roles of wnt-5a in cancer. Br J Can 2009, 101:209-214.
7. Dejimek J, Dejimek A, Säfholm A, Sjölander A, Andersson T: Wnt-5a protein expression in primary dukes B colon cancers identifies a subgroup of patients with good prognosis. Cancer Res 2005, 65(20):9142-9146.
8. Rawson JB, Mrkonjic M, Daftary D, Dicks E, Buchanan DD, Younghusband HB, Parfrey PS, Young JP, Pollett A, Green RC, Gallinger S, McLaughlin JR, Knight JA,

Bapat B: Promoter methylation of Wnt5a is associated with microsatellite instability and BRAF V600E mutation in two large populations of colorectal cancer patients. Br J Cancer 2011, 104(12):1906-1912.

9. Ingraham CA, Park GC, Makarenkova HP, Crossin K: Matrix Metalloproteinase (MMP)-9 induced by wnt signaling increases the proliferation and migration of embryonic neural stem cells at low O2 levels. J Bio Chem 2011, 286(20):17649-17657.

10. Karow M, Popp T, Egea V, Ries C, Jochum M, Neth P: Wnt signaling in muse mesenchymal stem cells: impact on proliferation, invasion and MMP expression. J Cell Mol Med 2009, 13(88):2506-2520.

11. Bendardaf R, Buhmeida A, Hilska M, Laato M, Syrjänen S, Syrjänen K, Collan Y, Pyrhönen S: MMP-9 (gelatinase B) expression is associated with disease-free survival and disease-specific survival in colorectal cancer patients. Cancer Invest 2010, 28(1):38-43.

12. Lee MA, Park KS, Lee HJ, Jung JH, Kang JH, Hong YS, Lee KS, Kim DG, Kim SN: Survivin expression and its clinical significance in pancreatic cancer. BMC Cancer 2005, 5:127-132.

13. Neth P, Ciccarella M, Egea V, Hoelters J, Jochum M, Ries C: Wnt signaling regulates the invasion capacity of human mesenchymal stem cells. Stem Cells 2006, 24(8):1892-1903.

14. Elzagheid A, Buhmeida A, Korkeila E, Collan Y, Syrjanen K, Pyrhonen S: Nuclear beta-catenin expression as a prognostic factor in advanced colorectal carcinoma. World J Gastroenterol 2008, 14(24):3866-3871.

15. Suzuki H, Masuda N, Shimura T, Araki K, Kobayashi T, Tsutsumi S, Asao T, Kuwano H: Nuclear beta-catenin expression at the invasive front and in the vessels predicts liver metastasis in colorectal carcinoma. Anticancer Res 2008, 28(3B):1821-1830.

16. Kahlil S, Tan GA, Giri DD, Zou XK, Howe LR: Activation status of Wnt/β-catenin signaling in normal and neoplastic breast tissues: relationship to HER2/neu expression in human and mouse. PLos One 2012, 7(3):e33421.

17. Moyes LH, McEwan H, Radulescu S, Pawlikowski J, Lamm CG, Nixon C, Sansom OJ, Going JJ, Fullarton GM, Adams PD: Activation of Wnt signalling promotes development of dysplasia in Barrett's oesophagus. J Pathol 2012, 228(1):99-112.

18. Voloshanenko O, Erdmann G, Dubashi TD, Augustin I, Metzig M, Moffa G, Hundsrucker C, Ken G, Sandmann T, Anchang B, Demir K, Boehm C, Leible S, Ball CR, Glimm H, Spang R, Boutros M: Wnt secretion is required to maintain high levels of wnt activity in colon cancer cells. Nat Commun 2013, 4:2610.

19. Cheung LW, Leung PC, Wong AS: Gonadotropin-releasing hormone promotes ovarian cancer cell invasiveness through c-Jun NH2-terminal kinase-mediated activation of matrix metalloproteinase (MMP)-2 and MMP-9. Cancer Res 2006, 66(22):10902-10910.

20. Roh SA, Choi EY, Cho DH, Jang SJ, Kim SY, Kim YS, Kim JC: Growth and invasion of sporadic colorectal adenocarcinomas in terms of genetic change. J Korean Med Sci 2010, 25:353-360.

21. Song KS, Li G, Kim JS, Jing K, Kim TD, Kim JP, Seo SB, Yoo JK, Park HD, Hwang BD, Lim K, Yoon WH: Protein-bound polysaccharide from Phellinus linteus inhibits tumor growth, invasion, and angiogenesis and alters Wnt/β-catenin in SW480 human colon cancer cells. BMC Cancer 2011, 11:307.

22. Singh T, Katiyar SK: Honokiol Inhibits Non-Small Cell Lung Cancer Cell Migration by Targeting PGE2-Mediated Activation of β-Catenin Signaling. PLos One 2013, 8(3):e60749.

23. Naik S, Dotharger RS, Marasa J, Lewis CL, Piwnica-Worms D: Vascular endothelial growth factor receptor-1 is synthetic lethal to aberrant beta catenin activation in colon cancer. Clin Cancer Res 2009, 15(24):7529-7537.

24. Zeitlin BD, Ellis LM, Nor JE: Inhibition of vascular endothelial growth factor receptor-1/wnt beta catenin cross talk leads to tumor cell death. Clin Cancer Res 2009, 15(24):7453-7455.

25. Yoshihara T, Takahashi-Yanaga F, Shiraishi F, Morimoto S, Watanabe Y, Hirata M, Hoka S, Sasaguri T: Anti-angiogenic effects of differentiation-inducing factor-1 involving VEGFR-2 expression inhibition independent of the Wnt/β-catenin signaling pathway. Mol Cancer 2010, 16(9):245.

26. Kin JU, Bae BN, Kim HJ, Park KM: Prognostic significance of epidermal growth factor receptor and vascular endothelial growth factor receptor in colorectal adenocarcinoma. APMIS 2011, 119(7):449-459.

27. Kurayoshi M, Oue N, Yamamoto H, Kishida M, Inoue A, Asahara T, Yasui W, Kikuchi A: Expression of Wnt-5a is correlated with aggressiveness of gastric cancer by stimulating cell migration and invasion. Cancer Res 2006, 66(21):10439-10448.

28. Pukrop T, Klemm F, Hagemann T, Gradl D, Schulz M, Siemes S, Trümper L, Binder C: Wnt5a signaling is critical for macrophage-induced invasion of breast cancer cell lines. Proc Natl Acad Sci U S A 2006, 103(14):5454-5459.

29. Ying J, Li H, Yu J, Ng KM, Poon FF, Wong SC, Chan AT, Sung JJ, Tao Q: WNT5A exhibits tumor-suppressive activity through antagonizing the Wnt/beta-catenin signaling, and is frequently methylated in colorectal cancer. Clin Cancer Res 2008, 14(1):55-61.

WNT5A PROMOTES MIGRATION OF HUMAN OSTEOSARCOMA CELLS BY TRIGGERING A PHOSPHATIDYLINOSITOL-3 KINASE/ AKT SIGNALS

AILIANG ZHANG, SHUANGHUA HE, XIAOLIANG SUN, LIANGHUA DING, XINNAN BAO, AND NENG WANG

5.1 INTRODUCTION

Osteosarcoma, characterized by a high malignant and metastatic potential, principally affects children and adolescents [1]. Progression of disease is inexorable and response to therapy can be unrewarding: fewer than 50% of patients live beyond 10 years, and there are no reliable predictors to guide the choice or intensity of therapy [1,2]. Several improvements in understanding the molecular pathology of metastatic osteosarcoma have been achieved in the last several years [1]. However, the molecular mecha-

*This chapter was originally published under the Creative Commons Attribution License. Zhang A, He S, Sun X, Ding L, Bao X, and Wang N. Wnt5a Promotes Migration of Human Osteosarcoma Cells by Triggering a Phosphatidylinositol-3 Kinase/Akt Signals. Cancer Cell International **14**,15 (2014). doi:10.1186/1475-2867-14-15*

nisms underlying this malignancy are still largely unknown. For this reason, elucidating the signaling pathways involved in the metastatic cascade has become a key goal for developing novel effective therapeutics aimed at reducing osteosarcoma mortality rates.

Wnt signaling regulates several developmental and oncogenic processes in both insects and vertebrates [3]. Signals triggered by Wnt are classified into two groups, a canonical β-catenin pathway and non-canonical pathways. In the canonical pathway, Wnt signals are mediated by disheveled (Dvl), which inhibits glycogen synthase kinase-3β (GSK-3β) activity [4]. The accumulated β-catenin enters the nucleus, forms complexes as a co-factor with Tcf/Lef transcription factors, and then triggers transcription of a set of target genes, which ultimately leads to regulation of cell proliferation and cell fate as well as cell transformation [5]. Compared to the canonical pathway, the non-canonical Wnt pathways are not fully understood, especially in mammals, while its contributions to cell polarity in *Drosophila* (a planar cell polarity pathway, a PCP pathway) and convergent extension movement in *Xenopus* and zebrafish are demonstrated [6-9]. Wnt5a has been originally classified into the non-canonical Wnt group and reported to trigger Ca^{2+} pathways that subsequently activate protein kinase C (PKC) and Ca^{2+}/Calmodulin kinase II [10,11]. Wnt5a can activate other intracellular protein kinases such as extracellular signal-regulated kinase (ERK), Akt and c-Jun N-terminal kinase (JNK) [12,13].

It has been reported that trimeric G protein mediated activation of phospholipase C and phosphodiesterase is involved in Wnt-induced gene expression [14]. Phosphatidylinositol-3 kinases (PI3Ks) are a family of enzymes involved in cellular functions including cell growth, proliferation, differentiation, motility, survival and intracellular trafficking, all of which can potentially influence the development of cancer. The serine/threonine kinase Akt is a major effector of the PI3K pathway and is activated by many polypeptide growth factors [15]. In this study, we hypothesized that the PI3K/Akt signaling pathway might mediate Wnt5a-induced osteosarcoma cell migration.

5.2 MATERIALS AND METHODS

5.2.1 CELL CULTURE

Human osteosarcoma cell line MG-63 purchased from Cells Resource Center of Shanghai Institutes for Biological Sciences, Chinese Academy of Sciences (Shanghai, China) and were cultured in Dulbecco-modified Eagle's medium (DMEM) supplemented with 10% fetal bovine serum (FBS).

5.2.2 SMALL INTERFERING RNA (SIRNA)

For gene knockdown, siRNA duplexes specific for Akt (Cell Signaling) were transfected into MG-63 cells by using Lipofectamine 2000 reagent (Invitrogen, Carlsbad, CA) in serum-free OPTI-MEM according to the manufacturer's instructions. Knockdown efficiency was evaluated 48 h after transfection by measuring protein levels in cell lysates through using immunoblotting.

5.2.3 IMMUNOBLOTTING ANALYSIS

Subconfluent cells were washed twice with PBS, and then lysed with ice-cold RIPA lysis buffer (50 mmol/L Tris, 150 mmol/L NaCl, 1% Triton X-100, 1% sodium deoxycholate, 0.1% SDS, 1 mmol/L sodium orthovan-adate, 1 mmol/L sodium fluoride, 1 mmol/L EDTA, 1 mmol/L PMSF, and 1% cocktail of protease inhibitors) (pH7.4). The lysates were then clarified by centrifugating at 12,000 g for 20 min at 4°C. The protein extracts were separated by SDS-PAGE. The immunoblotting procedure was performed as described [16] and the following antibodies were used: mouse anti-β-actin antibody (KangChen Bio-tech, Shanghai, China), rabbit anti-Akt antibody, rabbit anti-phospho-Akt (p-Ser473) antibody, rabbit anti-PI3K p85 antibody, rabbit anti-phospho-PI3K p85 (Tyr458) (Cell Signaling Technology, Danvers, MA), mouse anti-β-catenin antibody, rabbit anti-phospho-β-catenin (p-Ser33) antibody (Santa Cruz Biotechnology, Santa Cruz, CA). Protein bands were detected by incubating with horseradish peroxidase-conjugated antibodies (Santa Cruz Biotechnology, Santa Cruz, CA) and visualized with ECL reagent (Thermo Scientific, Rockford, IL).

5.2.4 WOUND HEALING ASSAY

MG-63 cells were plated onto 96-well cell culture clusters (Costar) and grown to confluence, and then serum-starved for 24 h. The monolayer cells were scratched manually with a plastic pipette tip, and after two washes with PBS, the wounded cellular monolayer was allowed to heal for 10 h in DMEM containing 100 ng/ml recombinant Wnt5a (rWnt5a) (R&D Systems). Photographs of central wound edges per condition were taken at time 0 and at the indicated time points using digital camera (Nikon, Tokyo, Japan).

5.2.5 CELL MIGRATION ASSAYS

Cell migration was assessed in a modified Boyden chamber (Costar), in which two chambers were separated by a polycarbonate membrane (pore diameter, 8.0 μm). MG-63 cells were grown to subconfluence in tissue culture plates and then detached; thereafter, they were centrifuged and rendered into single cell suspensions in serum-free culture medium supplemented with 5 μg/mL BSA. The suspensions containing 5×10^4 cells were added to wells with a membrane placed in the bottom. Medium containing indicated Wnt5a was added to the upper and lower compartment of the Boyden chamber. The cells were allowed to migrate for the indicated periods of time at 37°C in this assay. Thereafter, the medium was discarded, stationary cells were removed with a cotton-tipped applicator and the membranes were cut out of the chamber and stained with 0.5% crystal violet. The response was evaluated in a light microscope by counting the number of cells that had migrated into the membrane.

5.2.6 STATISTICAL ANALYSIS

All experiments here were repeated at least three times, with independent treatments, each of which showed essentially the same results. The data were analyzed using Student's t-test by SPSS statistical software package. All the results were expressed as mean ± SD. For all analyses a two-sided p value of less than 0.05 was deemed statistically significant. This study had been approved by institutional ethics committee of Changzhou No. 1 People's Hospital.

5.3 RESULTS

5.3.1 WNT5A STIMULATES OSTEOSARCOMA CELL MIGRATION IN VITRO

To assess the effect of Wnt5a on osteosarcoma cell migration, we treated MG-63 cells with different doses of recombinant Wnt5a (rWnt5a), and measured the migration rate by wound healing assays and Boyden chamber assays. We found that Wnt5a had a potent stimulatory effect on MG-63 cell migration (Figure 1A, 1B). An approximately 2-fold increase in cell migration was observed in cells treated with 100 ng/ml rWnt5a (Figure 1A, 1B). Nevertheless, low concentration of Wnt5a (50 ng/ml) has no stimulative effect on MG-63 cell migration (Figure 1A). Accordingly, 100 ng/ml rWnt5a was used for the remaining studies hereafter to identify the mechanism that accounts for the changes in the migration of MG-63 cells.

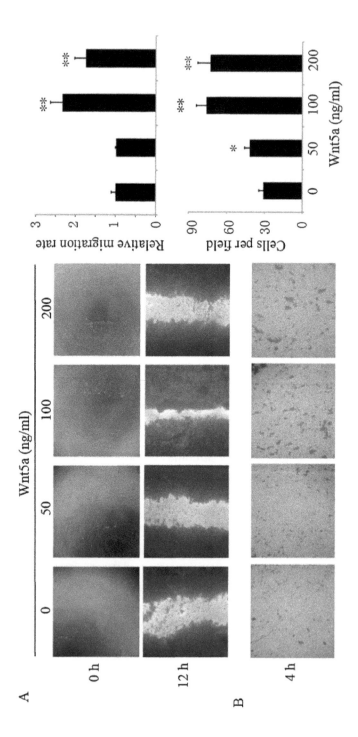

FIGURE 1: Effect of Wnt5a on the migration of osteosarcoma cells. Relative cell migration rate was determined by using wound healing assay (A) and Boyden chamber assays (B) in MG-63 cells incubated in the absence (0 ng/ml) or presence of 50, 100, and 200 ng/ml Wnt5a for 10 h. *, **: p<0.05, p<0.01 in the cultures with Wnt5a relative to the cultures without Wnt5a, respectively. Data were presented as mean ± SD of 5 determinations.

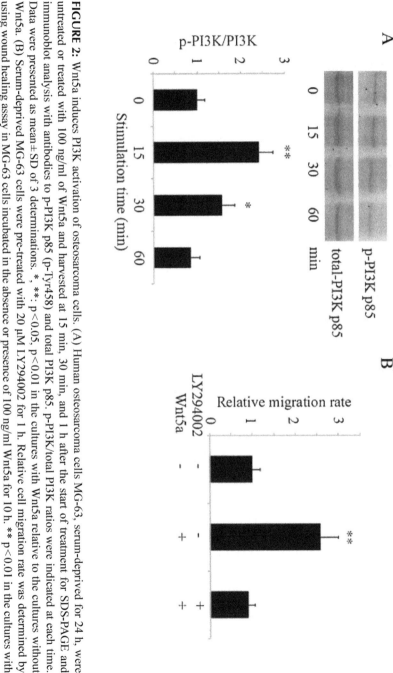

FIGURE 2: Wnt5a induces PI3K activation of osteosarcoma cells. (A) Human osteosarcoma cells MG-63, serum-deprived for 24 h, were untreated or treated with 100 ng/ml of Wnt5a and harvested at 15 min, 30 min, and 1 h after the start of treatment for SDS-PAGE and immunoblot analysis with antibodies to p-PI3K p85 (p-Tyr458) and total PI3K p85. p-PI3K/total PI3K ratios were indicated at each time. Data were presented as mean±SD of 3 determinations. *, **: $p<0.05$, $p<0.01$ in the cultures with Wnt5a relative to the cultures without Wnt5a. (B) Serum-deprived MG-63 cells were pre-treated with 20 μM LY294002 for 1 h. Relative cell migration rate was determined by using wound healing assay in MG-63 cells incubated in the absence or presence of 100 ng/ml Wnt5a for 10 h. ** $p<0.01$ in the cultures with Wnt5a relative to the cultures without Wnt5a. Data were presented as mean±SD of 5 determinations.

5.3.2 WNT5A INDUCES PI3K AND AKT PHOSPHORYLATIONS

Wnt5a-triggered signals in human osteosarcoma cells have remained completely unknown. To address the question, we first tried to identify the downstream signals triggered by Wnt5a in MG-63 cells. We first detected the phosphorylated-PI3K p85 (p-Tyr458), which represents the PI3K activation state. Human osteosarcoma cells, serum-starved for 24 h, were treated with 100 ng/ml of rWnt5a. The cells were harvested at 15 min, 30 min and 1 h after the start of Wnt5a treatment, followed by SDS-PAGE and immunoblot analyses. PI3K showed visible signs of basal phosphorylation and elevated phosphorylation at 15 min after stimulation with rWnt5a and continued to be elevated at least until 30 min after the start of treatment with Wnt5a (Figure 2A).

The most established activator of Akt is PI3K, therefore we sought to determine whether Akt activation was triggered by Wnt5a. The same assays were performed to detect the phosphorylated-Akt (p-Ser473), which represents the Akt activation state. Akt also showed visible signs of basal phosphorylation and elevated phosphorylation at 15 min after stimulation with rWnt5a and continued to be elevated at least until 1 h after the start of treatment with Wnt5a (Figure 3A).

5.3.3 WNT5A PROMOTES CELL MIGRATION VIA PI3K PATHWAY

The finding that Wnt5a could induce PI3K/Akt phosphorylation in MG-63 cells prompted us to determine whether PI3K/Akt activation was required for Wnt5a-mediated cell migration. Wnt5a-induced cell migration was largely abolished by pre-treatment with 20 μM LY294002, the PI3K-specific inhibitor (Figure 2B), suggesting that PI3K activation is required for Wnt5a-induced MG-63 cell migration.

To demonstrate the involvement of PI3K in Wnt5a-induced activation of Akt, we tested the effect of LY294002 for Akt activation. Human osteosarcoma cells, serum-starved for 24 h and pretreated with 20 μM LY294002 for 1 h, were incubated with 100 ng/ml of Wnt5a. The cells were harvested 15 min after the start of Wnt5a treatment and the cell lysates were subjected to SDS-PAGE and immunoblot analysis.

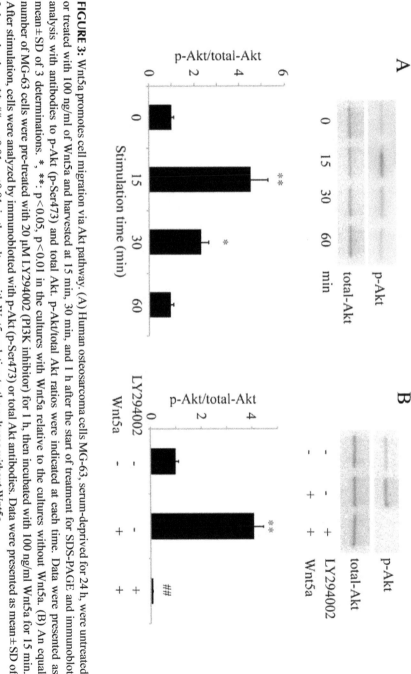

FIGURE 3: Wnt5a promotes cell migration via Akt pathway. (A) Human osteosarcoma cells MG-63, serum-deprived for 24 h, were untreated or treated with 100 ng/ml of Wnt5a and harvested at 15 min, 30 min, and 1 h after the start of treatment for SDS-PAGE and immunoblot analysis with antibodies to p-Akt (p-Ser473) and total Akt. p-Akt/total Akt ratios were indicated at each time. Data were presented as mean±SD of 3 determinations. *, **: p<0.05, p<0.01 in the cultures with Wnt5a relative to the cultures without Wnt5a. (B) An equal number of MG-63 cells were pre-treated with 20 μM LY294002 (PI3K inhibitor) for 1 h, then incubated with 100 ng/ml Wnt5a for 15 min. After stimulation, cells were analyzed by immunoblotted with p-Akt (p-Ser473) or total Akt antibodies. Data were presented as mean±SD of 3 determinations. **, ##: p<0.01, p<0.01, in the cultures with Wnt5a relative to the cultures without Wnt5a.

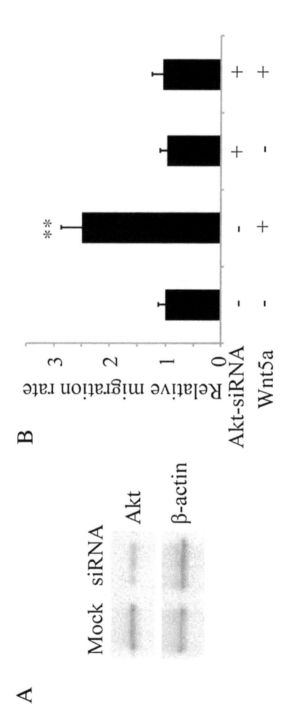

FIGURE 4: Effect of Akt in Wnt5a-induced cell migration. (A) Effect of Akt siRNA on the expression of Akt. MG-63 cells were transiently transfected with Akt siRNA pool, or mock for 48 h. Cells were analyzed by immunoblotted with total Akt antibody. (B) Effect of Akt siRNA on Wnt5a-stimulated cell migration. Cells were transiently transfected with Akt siRNA pool, and stimulated with 100 ng/ml Wnt5a or not for 10 h. Relative cell migration rate was determined by using wound healing assay. ** p<0.01 in the cultures with Wnt5a relative to the cultures without Wnt5a. Data were presented as mean±SD of 5 determinations.

The Wnt5a-induced activation of Akt was mostly blocked by pretreatment of LY294002 (Figure 3B). These data indicate that PI3K mediates Wnt5a-induced activation of Akt.

5.3.4 AKT ACTIVATION REGULATES OSTEOSARCOMA CELL MIGRATION

To analyze the role of endogenous Akt activation on Wnt5a-induced cell migration, we knocked down Akt expression by using siRNA, which reduced the protein level of Akt by approximately 60%, as assessed by immunoblotting (Figure 4A) and significantly reduced Wnt5a-induced migration of MG-63 cells (Figure 4B). Taken together, these experiments demonstrated that Akt activation was required for Wnt5a-induced MG-63 cell migration.

5.3.5 WNT5A DOES NOT ALTER THE TOTAL EXPRESSION AND PHOSPHORYLATION OF β-CATENIN

Mikels and Nusse reported that purified Wnt5a inhibits Wnt3a protein-induced canonical Wnt signaling in a dose-dependent manner, not by influencing β-catenin levels but by downregulating β-catenin-induced reporter gene expression in HEK293 cells or mouse L cells [17]. To assess the effect of Wnt5a on the total expression and phosphorylation of β-catenin, we treated MG-63 cells with different doses of rWnt5a, and measured the total expression and phosphorylation by immunoblotting assays. We found that the total expression and phosphorylation of β-catenin had not been altered under Wnt5a stimulation for less than 1 hour (Figure 5).

5.4 DISCUSSION

Wnt5a is a prototypic ligand that activates a β-catenin independent pathway in Wnt signaling [3]. Mice with the disrupted Wnt5a gene also exhibit various developmental abnormalities such as dwarfism, facial abnormalities, shortened limbs and tails, dysmorphic ribs and vertebrae, absence of the genital tubercle, and abnormal distal lung morphogenesis, indicating that it plays a crucial role in the development of various organs [18]. Wnt5a has been demonstrated to exert differential effects on cancer development [5]. Wnt5a can promote cancer progression and metastasis in malignant melanoma, breast cancer, and gastric cancer [19-23]. However, some evidence supports the hypothesis that Wnt5a acts as a tumor suppressor in certain experimental systems [24-28]. The first observation in the present study was that Wnt5a induced the migration of MG-63 osteosarcoma cells. Our finding suggest that Wnt5a acts as a migratory stimulator in osteosarcoma cells.

Kinases show the comprehensive functions triggered by Wnt5a in a variety of cells. It has been reported that Wnt5a may promote tumor progression through inducing actin reorganization and increasing cell motility via activating the protein kinase C (PKC) [19,29]. Wnt5a activates extracellular regulated protein kinase (ERK) in endothelial cells [30]. Wnt5a prevents apoptosis caused by serum-deprivation in osteoblasts by activating both Akt and ERK [12], and promotes cell adhesion in mammary epithelial cells in a PI3K/Akt-dependent manner [31]. Furthermore, Wnt5a triggers Akt phosphorylation via PI3K, but not ERK or PKC phosphorylation in human dermal fibroblasts [14]. These tanglesome viewpoints of kinases in Wnt5a signaling indicate that phosphorylated signaling molecules triggered by Wnt5a in these cases determine the diverse roles of these kinases. In this study, we demonstrated that Wnt5a promoted MG-63 osteosarcoma cell migration by activating PI3K/Akt. We found that Wnt5a induced PI3K (p-Tyr458) and Akt phosphorylation (p-Ser473) rapidly and transiently in MG-63 cells. Blocking Akt signaling by chemical inhibition of PI3K or Akt siRNA completely abolished Wnt5a-induced cell migration, indicating that PI3K/Akt activation participated in the regulation of the MG-63 cell migration.

The activation of PI3K/Akt signaling is thought to play a central role in cell viability-promoting phenotypes such as inhibition of cell death and cell cycle progression [14]. It has been shown that the PI3K/Akt signaling could be activated not only by Wnt5a but also by Wnt3a. Wnt3a prevents the serum deprivation-triggered apoptosis of osteoblasts by activating the canonical β-catenin pathway, ERK-, and/or PI3K-mediated uncharacterized pathways [12]. In this study, we have showed that Wnt5a promotes cell migration of human osteosarcoma cells in the PI3K/Akt-dependent manner.

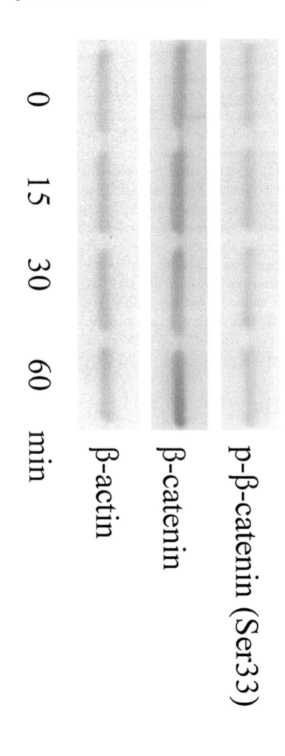

0   15   30   60   min

p-β-catenin (Ser33)

β-catenin

β-actin

FIGURE 5: Wnt5a does not alter the total expression and phosphorylation of β-catenin. MG-63 were untreated or treated with 100 ng/ml of Wnt5a and harvested at 15 min, 30 min, and 1 h after the start of treatment for SDS-PAGE and immunoblot analysis with antibodies to p-β-catenin (p-Ser33) and total β-catenin. The total expression and phosphorylation of β-catenin were not been altered by Wnt5a stimulation.

These results indicated that PI3K/Akt played crucial and complicated roles in canonical and non-canonical Wnt signalings in various cell and tissue types. Besides mentions above, PI3K/Akt signaling presents wide crosstalks referred to instances in which one or more components of one signal transduction pathway affect another. High-mobility group box 1 (HMGB1) activates the FAK/PI3K/mTOR signaling cascade and regulates tumor-associated cell migration through the interaction with BTB domain [32]. CXCR4 drives the metastatic phenotype in breast cancer through induction of CXCR2, and activation of MEK and PI3K pathways [33]. Akt signaling is involved in fucoidan-induced inhibition of growth and migration of human bladder cancer cells [34]. These studies provide us a complicated and crossed network around the PI3K/Akt signaling.

In summary, we present the first direct evidence here that Wnt5a promotes osteosarcoma cell migration via PI3K/Akt signaling. These findings elucidate a molecular pathway linking Wnt5a signaling to PI3K/Akt in cell motility. This result will contribute to further understanding of biological roles of Wnt5a/PI3K/Akt in cell migration of osteosarcoma and other cancers.

REFERENCES

1. Marulanda GA, Henderson ER, Johnson DA, Letson GD, Cheong D: Orthopedic surgery options for the treatment of primary osteosarcoma. Cancer Control 2008, 15(1):13-20.
2. Mirabello L, Troisi RJ, Savage SA: Osteosarcoma incidence and survival rates from 1973 to 2004: data from the surveillance, epidemiology, and end results program. Cancer 2009, 115(7):1531-1543.
3. Logan CY, Nusse R: The Wnt signaling pathway in development and disease. Annu Rev Cell Dev Biol 2004, 20:781-810.
4. Widelitz R: Wnt signaling through canonical and non-canonical pathways: recent progress. Growth Factors 2005, 23(2):111-116.
5. Klaus A, Birchmeier W: Wnt signalling and its impact on development and cancer. Nat Rev Cancer 2008, 8(5):387-398.
6. Moon RT, Campbell RM, Christian JL, McGrew LL, Shih J, Fraser S: Xwnt-5A: a maternal Wnt that affects morphogenetic movements after overexpression in embryos of Xenopus laevis. Development 1993, 119(1):97-111.
7. Wallingford JB, Rowning BA, Vogeli KM, Rothbacher U, Fraser SE, Harland RM: Dishevelled controls cell polarity during Xenopus gastrulation. Nature 2000, 405(6782):81-85.

8. Heisenberg CP, Tada M, Rauch GJ, Saude L, Concha ML, Geisler R, Stemple DL, Smith JC, Wilson SW: Silberblick/Wnt11 mediates convergent extension movements during zebrafish gastrulation. Nature 2000, 405(6782):76-81.

9. Tada M, Smith JC: Xwnt11 is a target of Xenopus Brachyury: regulation of gastrulation movements via Dishevelled, but not through the canonical Wnt pathway. Development 2000, 127(10):2227-2238.

10. Veeman MT, Axelrod JD, Moon RT: A second canon. Functions and mechanisms of beta-catenin-independent Wnt signaling. Dev Cell 2003, 5(3):367-377.

11. Kuhl M, Sheldahl LC, Park M, Miller JR, Moon RT: The Wnt/Ca2+ pathway: a new vertebrate Wnt signaling pathway takes shape. Trends Genet 2000, 16(7):279-283.

12. Almeida M, Han L, Bellido T, Manolagas SC, Kousteni S: Wnt proteins prevent apoptosis of both uncommitted osteoblast progenitors and differentiated osteoblasts by beta-catenin-dependent and -independent signaling cascades involving Src/ERK and phosphatidylinositol 3-kinase/AKT. J Biol Chem 2005, 280(50):41342-41351.

13. Yamanaka H, Moriguchi T, Masuyama N, Kusakabe M, Hanafusa H, Takada R, Takada S, Nishida E: JNK functions in the non-canonical Wnt pathway to regulate convergent extension movements in vertebrates. EMBO Rep 2002, 3(1):69-75.

14. Kawasaki A, Torii K, Yamashita Y, Nishizawa K, Kanekura K, Katada M, Ito M, Nishimoto I, Terashita K, Aiso S, et al.: Wnt5a promotes adhesion of human dermal fibroblasts by triggering a phosphatidylinositol-3 kinase/Akt signal. Cell Signal 2007, 19(12):2498-2506.

15. Datta SR, Brunet A, Greenberg ME: Cellular survival: a play in three Akts. Genes Dev 1999, 13(22):2905-2927.

16. Zhang A, Zhang J, Sun P, Yao C, Su C, Sui T, Huang H, Cao X, Ge Y: EIF2alpha and caspase-12 activation are involved in oxygen-glucose-serum deprivation/restoration-induced apoptosis of spinal cord astrocytes. Neurosci Lett 2010, 478(1):32-36.

17. Mikels AJ, Nusse R: Purified Wnt5a protein activates or inhibits beta-catenin-TCF signaling depending on receptor context. PLoS Biol 2006, 4(4):e115.

18. Yamaguchi TP, Bradley A, McMahon AP, Jones S: A Wnt5a pathway underlies outgrowth of multiple structures in the vertebrate embryo. Development 1999, 126(6):1211-1223.

19. Weeraratna AT, Jiang Y, Hostetter G, Rosenblatt K, Duray P, Bittner M, Trent JM: Wnt5a signaling directly affects cell motility and invasion of metastatic melanoma. Cancer Cell 2002, 1(3):279-288.

20. Zhu Y, Tian Y, Du J, Hu Z, Yang L, Liu J, Gu L: Dvl2-dependent activation of Daam1 and RhoA regulates Wnt5a-induced breast cancer cell migration. PloS One 2012, 7(5):e37823.

21. Zhu Y, Shen T, Liu J, Zheng J, Zhang Y, Xu R, Sun C, Du J, Chen Y, Gu L: Rab35 is required for Wnt5a/Dvl2-induced Rac1 activation and cell migration in MCF-7 breast cancer cells. Cell Signal 2013, 25(5):1075-1085.

22. Kurayoshi M, Oue N, Yamamoto H, Kishida M, Inoue A, Asahara T, Yasui W, Kikuchi A: Expression of Wnt-5a is correlated with aggressiveness of gastric cancer by stimulating cell migration and invasion. Cancer Res 2006, 66(21):10439-10448.

23. Bittner M, Meltzer P, Chen Y, Jiang Y, Seftor E, Hendrix M, Radmacher M, Simon R, Yakhini Z, Ben-Dor A, et al.: Molecular classification of cutaneous malignant melanoma by gene expression profiling. Nature 2000, 406(6795):536-540.

24. Kremenevskaja N, von Wasielewski R, Rao AS, Schofl C, Andersson T, Brabant G: Wnt-5a has tumor suppressor activity in thyroid carcinoma. Oncogene 2005, 24(13):2144-2154.

25. Dejmek J, Dejmek A, Safholm A, Sjolander A, Andersson T: Wnt-5a protein expression in primary dukes B colon cancers identifies a subgroup of patients with good prognosis. Cancer Res 2005, 65(20):9142-9146.

26. Liu XH, Pan MH, Lu ZF, Wu B, Rao Q, Zhou ZY, Zhou XJ: Expression of Wnt-5a and its clinicopathological significance in hepatocellular carcinoma. Dig Liver Dis 2008, 40(7):560-567.

27. Safholm A, Tuomela J, Rosenkvist J, Dejmek J, Harkonen P, Andersson T: The Wnt-5a-derived hexapeptide Foxy-5 inhibits breast cancer metastasis in vivo by targeting cell motility. Clin Cancer Res 2008, 14(20):6556-6563.

28. Jonsson M, Dejmek J, Bendahl PO, Andersson T: Loss of Wnt-5a protein is associated with early relapse in invasive ductal breast carcinomas. Cancer Res 2002, 62(2):409-416.

29. Witze ES, Litman ES, Argast GM, Moon RT, Ahn NG: Wnt5a control of cell polarity and directional movement by polarized redistribution of adhesion receptors. Science 2008, 320(5874):365-369.

30. Masckauchan TN, Agalliu D, Vorontchikhina M, Ahn A, Parmalee NL, Li CM, Khoo A, Tycko B, Brown AM, Kitajewski J: Wnt5a signaling induces proliferation and survival of endothelial cells in vitro and expression of MMP-1 and Tie-2. Mol Biol Cell 2006, 17(12):5163-5172.

31. Dejmek J, Dib K, Jonsson M, Andersson T: Wnt-5a and G-protein signaling are required for collagen-induced DDR1 receptor activation and normal mammary cell adhesion. Int J Canc Suppl J Int Canc Suppl 2003, 103(3):344-351.

32. Ko YB, Kim BR, Nam SL, Yang JB, Park SY, Rho SB: High-mobility group box 1 (HMGB1) protein regulates tumor-associated cell migration through the interaction with BTB domain. Cell Signal 2014, 26(4):777-783.

33. Sobolik T, Su YJ, Wells S, Ayers GD, Cook RS, Richmond A: CXCR4 drives the metastatic phenotype in breast cancer through induction of CXCR2, and activation of MEK and PI3K pathways. Mol Biol Cell 2014, mbc.E13-07-0360.

34. Cho TM, Kim WJ, Moon SK: AKT signaling is involved in fucoidan-induced inhibition of growth and migration of human bladder cancer cells. Food Chem Toxicol 2014, 64:344-352.

CHAPTER 6

FOXA1 PROMOTES TUMOR CELL PROLIFERATION THROUGH AR INVOLVING THE NOTCH PATHWAY IN ENDOMETRIAL CANCER

MEITING QIU, WEI BAO, JINGYUN WANG, TINGTING YANG, XIAOYING HE, YUN LIAO, AND XIAOPING WAN

6.1 BACKGROUND

Endometrial cancer (EC) is one of the most common gynecologic malignancies. The incidence of EC has markedly increased in recent years. EC is broadly classified into two groups [1]; type I ECs are linked to estrogen excess, hormone-receptor positivity, and favorable prognoses, whereas type II, primarily serous tumors, are more common in older women and have poorer outcomes [2]. Primary treatment, including surgery and radiation, cannot provide sufficient tumor control, especially in high-grade, undifferentiated tumors with deep muscle infiltration. Endocrine treatment, including medroxyprogesterone acetate or tamoxifen, is sometimes useful to improve the outcome. However, patients with type II EC and even some patients with type I EC are refractory to traditional endocrine treatment [3]. Thus, a new treatment is needed to achieve a better response.

*This chapter was originally published under the Creative Commons Attribution License. Qiu M, Bao W, Wang J, Yang T, He X, Liao Y, and Wan X. FOXA1 Promotes Tumor Cell Proliferation Through AR Involving the Notch Pathway in Endometrial Cancer. BMC Cancer **14,**78 (2014). doi:10.1186/1471-2407-14-78.*

Several studies have shown that the majority of ECs also express another hormone receptor, androgen receptor (AR) [4,5]. The results of immunohistochemical analysis indicate that, compared with endometrial glandular epithelial cells in normal cycling endometrium, more epithelial cells express AR in ECs [4]. Moreover, in female mice, in contrast to AR−/− uteri, AR+/+ uteri have uterine hypertrophy and endometrial growth [6]. It thus is very important to examine the possible actions and metabolism mediated by AR in human EC.

Forkhead box A1 (FOXA1) is a transcription factor that belongs to the forkhead family consisting of the winged-helix DNA-binding domain and the N-terminal and C-terminal transcriptional domains. FOXA1 is expressed in various organs, including breast, liver, pancreas, and prostate, and can influence the expression of a large number of genes associated with metabolic processes, regulation of signaling, and the cell cycle [7,8]. FOXA1 has been identified as a "pioneer factor" that binds to chromatin-packaged DNA and opens the chromatin for binding of additional transcription factors, including AR [9]. FOXA1 also binds directly to AR and regulates transcription of prostate-specific genes in prostate cancer [10]. Recent global gene expression studies of prostate cancer and triple-negative breast cancer have shown that high FOXA1 expression, which correlates positively with AR level, promotes tumor proliferation [11,12]. Thus, FOXA1 expression is considered a predictor of poor survival in prostate cancer and triple-negative breast cancer. However, the interaction between FOXA1 and AR in EC remains unclear.

An aberrant Notch pathway has been documented in various cancer types and has been associated with tumorigenesis [13-15]. The Notch pathway is initiated by ligand binding, which is followed by intramembranous proteolytic cleavage of the Notch1 receptor to release an active form of the Notch intracellular domain (NICD). The NICD subsequently translocates to the nucleus and acts as a transcriptional activator to enhance the expression of target genes such as Hairy-enhancer of split1 (Hes1) [16]. Abnormal activation of the Notch pathway promotes proliferation in a variety of cancer cell types, including EC [15,17].

In the present study, we investigated the dependency of AR on FOXA1 expression in tissue paraffin sections, in multiple cellular contexts, and on tumor-bearing nude mice. Here we show, for the first time, that FOXA1 activates the Notch pathway through AR and that AR is required for FOXA1-enhanced cell proliferation in EC.

TABLE 1: The relationship between protein expression and clinicopathological features in EC

Parameter	n	FOXA1		p	AR		p
		High	Low		High	Low	
Age							
≤55	31	21	10	0.916	19	12	0.493
>55	45	31	14		31	14	
FIGO stage							
I–II	64	41	23	0.059	41	23	0.464
III–IV	12	11	2		9	3	
Pathological type							
Adenocarcinoma	63	47	16	0.469	41	22	0.774
Papillary serous carcinoma	13	10	3		9	4	
Histological grade							
G1	35	19	16	0.038	18	17	0.040
G2	22	17	5		18	4	
G3	6	6	0		5	1	
Lymph node metastasis							
Positive	7	6	1	0.271	5	2	0.86
Negative	66	43	23		45	21	
Depth of myometrial invasion							
≤1/2	47	28	19	0.035	29	18	0.339
>1/2	29	24	5		21	8	
ERα expression							
Positive	61	43	18	0.434	41	20	0.598
Negative	15	9	6		9	6	
P53 expression							
Positive	20	14	6	0.86	15	5	0.312
Negative	56	38	18		35	21	

6.2 METHODS

6.2.1 PATIENTS AND TISSUES

A total of 57 normal endometrial samples, 11 atypical hyperplasias, and 76 EC specimens obtained from Chinese female patients who underwent surgical treatment from 2011 to 2013 at the Shanghai Jiao Tong University Affiliated International Peace Maternity & Child Health Hospital (Shanghai, China) were available for examination in this study. Tissues were embedded in paraffin. Two independent pathologists verified the histological diagnosis of all collected tissues. No patient had received neoadjuvant therapy or endocrine therapy before the surgery. The clinicopathological characteristics of EC patients are presented in Table 1. The samples of EC, atypical hyperplasias and normal endometrial tissues were collected after written informed consent from the patients. The Human Investigation Ethical Committee of the International Peace Maternity & Child Health Hospital Affiliated Shanghai Jiao Tong University approved this study.

6.2.2 IMMUNOHISTOCHEMICAL STAINING

Staining was performed on paraffin-embedded specimens using primary antibodies as follows: anti-FOXA1 (1:200; Abcam, Cambridge, MA, USA) and anti-AR (1:50; Abcam). The percentage of positively stained cells was rated as follows: 0 point=0%, 1 point=1% to 25%, 2 points=26% to 50%, 3 points=51% to 75%, and 4 points=greater than 75%. The staining intensity was rated in the following manner: 0 points=negative staining, 1 point=weak intensity, 2 points=moderate intensity, and 3 points=strong intensity. Then, immunoreactivity scores for each case were obtained by multiplying the values of the two parameters described above. The average score for all of five random fields at 200× magnification was used as the histological score (HS). Tumors were categorized into two groups based on the HS: low-expression group (HS=0–5) and high-expression group (HS=6–12).

6.2.3 CELL CULTURE AND EXPERIMENTAL SETUP

The human endometrial cell lines AN3CA, RL95-2, and HEC-1B were obtained from the Chinese Academy of Sciences Committee Type Culture Collection cell bank. These three cell lines were grown in Dulbecco's modified Eagle's medium (DMEM)/F12 (HyClone, Waltham, MA, USA) supplemented with 10% fetal bovine serum (Gibco, Carlsbad, CA, USA) in a humidified atmosphere of 5% CO2 at 37°C. The human endometrial cell line MFE-296 was purchased from Sigma (St. Louis, MO, USA). The MFE-296 cell line was grown in high-glucose DMEM (4.5 g/L glucose) (HyClone) supplemented with 10% fetal bovine serum in a humidified atmosphere of 5% CO_2 at 37°C.

To investigate the impact of FOXA1 on the AR-mediated transcription, the AR pathway agonist 5α-dihydrotestosterone (DHT) (Dr. Ehrenstorfer, Augsburg, Germany) and the AR pathway blocker flutamide (Sigma) were purchased and dissolved in 100% ethanol for storage. In this study they were diluted with phenol red–free DMEM/F12 (Gibco) immediately before each experiment, with the final concentration of ethanol at 0.1%. DHT was added into the cell culture media at concentrations of 10^{-9} to 10^{-7} M for different periods (0–48 h). To block the activation of AR-mediated transcription, flutamide (10^{-6} M) was added into the media 30 min before DHT. Vehicle contained 0.1% absolute ethanol/phenol red–free DMEM/F12.

6.2.4 STABLE TRANSFECTION

To stably knock down endogenous FOXA1 expression, MFE-296 cells were grown to 30% confluency in 6-well culture plates and then infected with lentivirus carrying an shRNA targeting FOXA1 (shFOXA1) or a negative control vector (NC; LV3-pGLV-h1-GFP-puro vector, D03004; GenePharma, Shanghai, China) at a multiplicity of infection of 70 in the presence of polybrene (8 μg/mL). After 48 h of infection at 37°C, the medium was replaced with fresh medium and incubated further for 72 h before analysis using quantitative RT-PCR (qRT-PCR) and western blotting for FOXA1 expression. The shRNA sequences used were 5'-GAGA-

GAAAAAAUCAACAGC-3′ (shFOXA1) and 5′-TTCTCCGAACGTGT-CACGT-3′ (NC).

6.2.5 TRANSIENT TRANSFECTION

The plasmid PWP1/GFP/Neo-AR containing transfection-ready AR cDNA (exAR) and its negative control PWP1/GFP/Neo were gifts from Doctor Yuyang Zhao at Shanghai First People's Hospital. MFE-296 cells stably transfected with shFOXA1 or NC were transiently cotransfected with PWP1/GFP/Neo-AR (exAR) or its negative control (NC). The plasmid pCMV/3FLAG/Neo-FOXA1 containing transfection-ready FOXA1 cDNA (exFOXA1) (GenBank: BC033890) and a pure pCMV/3FLAG/Neo (NC) were purchased from Genechem (Product code: GOSE33403; Shanghai, China). AN3CA cells were transiently transfected with exFOXA1 or NC or cotransfected with a siRNA targeting AR (siAR) (Genephama Biotech, Shanghai, China) or its negative control (NC) in Opti-MEM (Invitrogen, Carlsbad, CA, USA) using Lipo2000 (Invitrogen). The siRNA targeting FOXA1 (siFOXA1) and its negative control (NC) were purchased from Genephama Biotech (Shanghai, China). AN3CA cells were transiently transfected with exAR or NC or cotransfected with siFOXA1 or NC in Opti-MEM (Invitrogen) using Lipo2000 (Invitrogen). The transfection solution was removed from the cells and replaced with standard medium after 8 h. The sequences of the siRNA oligos used were: siAR: sense: 5′-AUGUCAACUCCAGGAUGCUTT-3′, antisense: 5′-AGCAUCCUG-GAGUUGACAUTT-3′; siFOXA1: sense: 5′-GAGAGAAAAAAUCAA-CAGC-3′, antisense: 5′-GCUGUUGAUUUUUUCUCUC-3′.

6.2.6 QRT-PCR

Total RNA was extracted from cultured cells by Trizol Reagent (Invitrogen). RNA was converted to cDNA with the one-step Prime Script RT reagent kit (TaKaRa, Dalian, China), and the cDNA was analyzed by real-time PCR using SYBR Premix Ex Taq (TaKaRa) in an Eppendorf Mastercycler® realplex. Each sample was assayed in triplicate in each of three independent experiments. All values are expressed as mean±standard deviation. The following primers were used: FOXA1: sense: 5'-AGGTGTGTATTCCAGACCCG-3', antisense: 5'-TTGACG-GTTTGGTTTGTGTG-3'; AR: sense: 5'-CCTGGCTTCCGCAACTTA-CAC-3', antisense: 3'-GGACTTGTGCATGCGGTACTCA-5'; MYC: sense: 5'-AAAGGCCCCCAAGGTAGTTA-3', antisense: 5'-TTTCCG-CAACAAGTCCTCTT-3'; XBP1: sense: 5'-CCTTGTAGTTGAGAAC-CAGG-3', antisense: 5'-GGGGCTTGGTATATATGTGG-3'; UHRF1: sense: 5'-AAGGTGGAGCCCTACAGTCTC-3', antisense: 5'-CACTT-TACTCAGGAACAACTGGAAC-3'; and ZBTB16: sense: 5'-CCAGCA-GATTCTGGAGTATGCA-3', antisense: 5'-GCATACAGCAGGTCATC-CAAGTC-3'.

6.2.7 WESTERN BLOTTING

Total protein was extracted using a RIPA kit (Beyotime, Shanghai, China) containing a 1% dilution of the protease inhibitor PMSF (Beyotime). Protein concentrations were determined by the enhanced BCA Protein Assay kit (Beyotime). Equal amounts of protein in each lane were separated by 8% SDS-PAGE and transferred to a PVDF membrane (Millipore, Billerica, MA, USA). After blocking the membrane in blocking buffer (5% milk powder in 20 mM Tris–HCl pH 7.5, 500 mM NaCl, 0.1% (v/v) Tween 20), the membrane was incubated with primary antibodies against FOXA1 (1:1000; Abcam), AR (1:2000; Cell Signaling Technology, Danvers, MA, USA), Notch1 (1:2000; Epitomics, Burlingame, CA, USA), Hes1 (1:2000; Epitomics), and β-actin (1:2000, Cell Signaling Technology) at 4°C overnight. Peroxidase-linked secondary anti-rabbit or anti-mouse antibodies were used to detect the bound primary antibodies.

6.2.8 CO-IMMUNOPRECIPITATION

Total protein was extracted from cells treated or not treated with 10^{-7} M DHT for 24 h (described in the Cell culture and experimental setup section). After protein quantification, 500 μg of each cell lysate was added to

10 μl of anti-FOXA1 (Epitomics) and shaken at 4°C overnight, then added to 30 μl of Protein A+G Agarose (Beyotime), shaken at 4°C for 4 h, centrifuged at 2500 × g for 5 min, and washed with a RIPA kit (Beyotime) to collect the immunoprecipitate-bound agarose beads. Each immunoprecipitate was denatured with 20 μl of 1× SDS-PAGE loading buffer at 100°C for 5 min. Each supernatant was subjected to SDS-PAGE (8% acrylamide). It is important to note that FOXA1 (51 kDa) is close in size to IgG (55 kDa). To avoid detecting IgG protein left from the immunoprecipitation process and FOXA1 protein from the same species in the western blot at the same time, we used anti-FOXA1 from mouse in western blotting, whereas anti-FOXA1 from rabbit was used for immunoprecipitation. Primary antibodies against AR (1:2000) and FOXA1 (1:500; Santa Cruz Biotechnology, Dallas, TX, USA) were used for western blotting. Other steps were as described in the Western blotting section.

6.2.9 CHROMATIN IMMUNOPRECIPITATION (CHIP)-PCR

Chromatin immunoprecipitation (ChIP) assays were performed as previously described [18] using anti-FOXA1 antibody (ab23738, Abcam), anti-AR antibody (sc-7305, Santa Cruz Biotechnology). FOXA1-AR overlapping binding sites were identified by Chip-seq as previously depicted [19] and by qRT-PCR using SYBR Premix Ex Taq (Takara). Enrichment was calculated using the comparative Ct method, and was analyzed for specificity, linearity range, and efficiency in order to accurately evaluate the occupancy (percentage of immunoprecipitation/input). IgG was used as negative control. The primers used included: MYC pro: sense: 5′-CCCCC-GAATTGTTTTCTCTT-3′, antisense: 5′-TCTCATCCTTGGTCCCT-CAC-3′; MYC enh-1: sense: 5′-AGACAGAGGCAGGGTGGAG-3′, antisense: 3′-CCCAGGTAAACAGCCAATGT-5′; MYC enh-2a: sense: 5′-CCGTTCCGTGTCTAACCACT-3′, antisense: 5′-ATGAAACTC-GGGGAGTGTTG-3′; MYC enh-2b: sense: 5′-AGCGTTCTCTTTGC-CAGAAA-3′, antisense: 3′-GGCAAAGCTTCACAGAGGAC-5′; MYC enh-2c: sense: 5′-CACACAAGAAGAGCAAACTGAAG-3′, antisense: 5′-TGAGGATTGTTAGGAATCTCTGG-3′.

6.2.10 MTT ASSAY

Cells (3×10^3 cells/well) were plated in 96-well plates. Then, 20 µl of 3-(4,5-dimethylthiazol-2-yl)-2,5-diphenyltetrazolium bromide (MTT, 5 mg/ml; Sigma) was added to each well and subsequently incubated at 37°C for 4 h. The absorbance at 490 nm was then measured using a microplate reader. Cells incubated with culture medium were used as a control group. Each sample was assayed in triplicate.

6.2.11 COLONY-FORMATION ASSAYS

Cell lines were trypsinized to generate a single-cell suspension, and 120 cells/well (MFE-296 cells) or 200 cells/well (AN3CA cells) were seeded into 6-well plates. Dishes were returned to the incubator for 14 days, and the colonies were fixed with methanol for 30 min at room temperature and then stained with 0.5% crystal violet for 1 h.

6.2.12 CELL MIGRATION AND INVASION ASSAYS

Cells were trypsinized, centrifuged, and resuspended in serum-free medium. Cells were then plated at a density of 1×10^5/well (for the migration assay) or 2×10^5/well (for the invasion assay) in invasion chambers (8 µm pore size; BD Biosciences, California, USA) with or without matrigel coating for invasion and migration assays. Complete medium (600 µl) was added to the lower chamber as a chemoattractant. After incubation for 5 h (MFE-296) or 24 h (AN3CA) for the migration assay, or after incubation for 24 h (MFE-296) or 48 h (AN3CA) for the invasion assay, cells were fixed with 4% paraformaldehyde for 1 h. Cells on the apical side of each insert were removed by mechanical scraping. Cells that migrated to the basal side of the membrane were stained with 0.5% crystal violet and counted at 200× magnification. The migration and invasion assays were repeated at least three times.

6.2.13 XENOGRAFT TUMOR–FORMATION ASSAYS

Female athymic mice of 4 weeks of age were obtained from the Shanghai Experimental Animal Center of the Chinese Academy of Science. Our animal research was carried out in strict accordance with the recommendations in the Guideline for the Care and Use of Laboratory Animals of China. The protocol was approved by the Committee on the Ethics of Animal Experiments of the Obstetrical and Gynecological Hospital affiliated Fudan University (Permit Number: SYXK (hu) 2008–0064). All efforts were made to minimize animal suffering.

To establish a nude mouse model bearing EC, uninfected MFE-296 cells (MFE-296), stable MFE-296 cells infected with lentivirus carrying shFOXA1 (MFE-296/shFOXA1) or vector alone (MFE-296/NC) were used. All mice were randomly divided into three groups of four mice. Each mouse was given a unilateral subcutaneous injection of 1×10^7 cells. Tumor measurement began one week after injection and was conducted weekly using digital calipers. The tumors were removed and weighed after 42 days. Tumor volume was calculated as follows: tumor volume $(cm^3) = (the longest diameter) \times (the shortest diameter)2 \times 0.5$.

6.2.14 IMMUNOHISTOCHEMICAL STAINING OF MOUSE TUMOR SAMPLES

Tumor samples from xenografted mice were collected and fixed according to routine procedures. Histological staining was then performed on the tissue sections of the paraffin-embedded tumors using the streptavidin-biotin-peroxidase method. Primary antibodies were as follows: anti-FOXA1 (1:200; Abcam), anti-AR (1:50, Abcam), anti-Notch1 (1:100; Epitomics,), anti-Hes1 (1:250; Epitomics), anti-Ki67 (1:100; Boster, Wuhan, China), and anti-PCNA (1:100; Boster). The sections were then counterstained with hematoxylin and eosin (H&E).

6.2.15 STATISTICS

Measured data were assessed by unpaired Student's t-test or one-way ANOVA for multiple comparisons, and χ^2 test for 2×2 tables was used to compare the categorical data. $p < 0.05$ was considered significant.

6.3 RESULTS

6.3.1 EXPRESSION OF FOXA1 AND AR IN ENDOMETRIAL TISSUES AND THE CLINICOPATHOLOGICAL SIGNIFICANCE IN EC SPECIMENS

We assessed relative FOXA1 and AR levels in EC samples, atypical hyperplasias, and normal endometrial tissue samples using immunohistochemistry. FOXA1 was higher in atypical hyperplasias and even higher in EC compared with normal endometrial tissues ($p = 0.005$) (Figure 1, Additional file 1: Table S1). Notably, the expression of AR was also significantly higher in EC ($p = 0.033$) (Figure 1, Additional file 2: Table S2). The results also showed that FOXA1 expression correlated positively with AR expression ($p = 0.003$) (Table 2). Correlation analysis between FOXA1 and pathological grade of EC showed that FOXA1 expression was higher in G3 tumors (6/6) compared with either G2 (17/22) or G1 (19/35) tumors ($p = 0.038$) (Table 1). Significantly higher FOXA1 expression was also found in tumors that displayed a greater depth of myometrial invasion ($p = 0.035$). Finally, our results also indicated that AR was much higher in G3 and G2 tumors compared to G1 ($p = 0.040$) (Table 1). These results suggested that FOXA1 expression, which correlated with AR expression, had a connection with the development of EC and risk-associated clinical features of the disease.

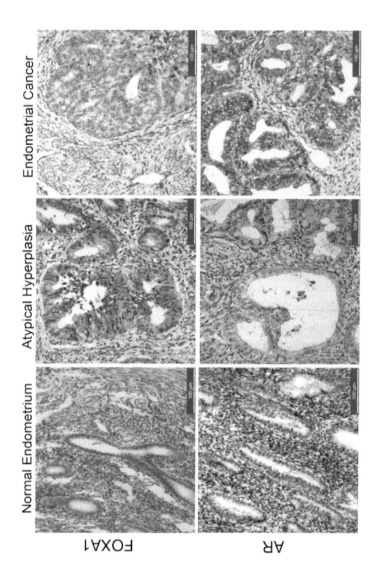

FIGURE 1: Immunohistochemical staining of FOXA1 and AR in normal endometrium, atypical hyperplasias, and endometrial cancer. FOXA1 and AR expression in normal endometrium, atypical hyperplasias and endometrial cancer. (Immunohistochemical staining, ×200).

TABLE 2: Immunohistochemical analysis of protein expression in different endometrial tissues

	Normal endometrium		Atypical hyperplasia		Endometrial cancer		*p
	Negative	Positive	Negative	Positive	Negative	Positive	
FOXA1	32	25	4	7	24	52	
n	57		11		76		
AR	30	27	5	6	26	50	0.003
n	57		11		76		

The relationship between FOXA1 and AR expression was assessed by χ^2 test.

6.3.2 FOXA1 AFFECTS AR EXPRESSION IN HUMAN EC CELLS

We used western blotting to examine FOXA1 and AR expression in EC cells. FOXA1 was upregulated in MFE-296 cells compared with KLE, HEC-1B, and AN3CA cells. Furthermore, the AR level was also markedly higher in MFE-296 cells than in the other three EC cell lines (Figure 2A).

We next manipulated FOXA1 expression and examined its influence on AR expression. AN3CA cells were transiently transfected with a FOXA1 plasmid to overexpress FOXA1 (AN3CA/exFOXA1) or with control vector (AN3CA/NC). Moreover, to knock down FOXA1 expression, MFE-296 cells were stably transfected with FOXA1 shRNA (MFE-296/shFOXA1) or control vector (MFE-296/NC) (Figure 2B). AR expression was then analyzed by qRT-PCR and western blotting, which showed that the AR level was significantly enhanced by FOXA1 overexpression and reduced by FOXA1 depletion (Figure 2C–H). Together, the data suggested that FOXA1 affected the AR level in EC cells.

6.3.3 FOXA1 EXPRESSION AFFECTS AR TARGET GENE EXPRESSION IN HUMAN EC CELLS

We next examined whether the FOXA1 level impacted the expression of AR target genes in EC cells. MFE-296 cells were hormone deprived and

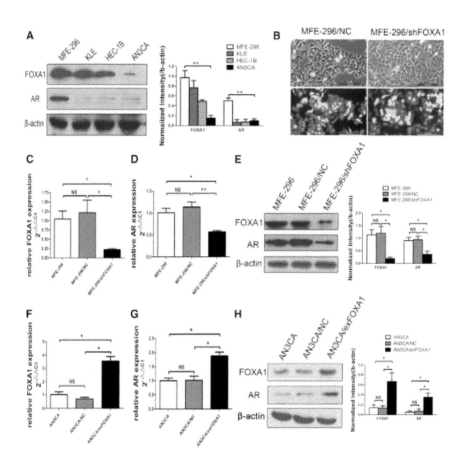

FIGURE 2: FOXA1 affects the expression of AR in human EC cells. A: FOXA1 and AR expression in the indicated EC cell lines as determined were measured by western blotting (Left), and further quantified by densitometry of triplicate experiments (Right). β-actin was used as a loading control. B: Stable transfection of MFE-296 cells with negative control vector (MFE-296-NC) or shFOXA1 (MFE-296-shFOXA1). By comparing the cells in white light (the upper panels) with the cells in green fluorescence (the lower panels), the percentage of transfected/fluorescing cells was estimated at >85%. Magnification, ×400. C: Quantification of FOXA1 mRNA by qRT-PCR in untransfected MFE-296 (MFE-296), MFE-296 transfected with shRNA control plasmid (MFE-296/NC), and MFE-296 transfected with shFOXA1 (MFE-296/shFOXA1). D: Quantification of AR mRNA by qRT-PCR in MFE-296, MFE-296/NC, and MFE-296/shFOXA1 cells. E: FOXA1 and AR expression in MFE-296, MFE-296/NC and MFE-296/shFOXA1 cells were measured by western blotting (Left), and further quantified by densitometry of triplicate experiments (Right). F: Quantification of FOXA1 mRNA by qRT-PCR in untransfected AN3CA (AN3CA), AN3CA transfected with control plasmid (AN3CA/NC), and AN3CA

transfected with FOXA1 expression plasmid (AN3CA/exFOXA1). G: Quantification of AR mRNA by qRT-PCR in AN3CA, AN3CA/NC, and AN3CA/exFOXA1 cells. H: AR and FOXA1 expression in AN3CA, AN3CA/NC and AN3CA/exFOXA1 cells were measured by western blotting (Left), and further quantified by densitometry of triplicate experiments (Right). *p<0.05, **p<0.01, NS p>0.05 compared with NC.

treated with vehicle or the AR pathway agonist DHT [20], and then the expression of AR target genes (XBP1, MYC, ZBTB16, and UHRF1) [12] was evaluated by qRT-PCR. This analysis confirmed that the expression of these four genes increased after treatment with DHT. Furthermore, the dose-response study ($0-10^{-7}$ M DHT) and time-response study (0–48 h) indicated that 10^{-7} M DHT and 24 h of incubation elicited the strongest expression of AR and its target genes (Figure 3A and 3B). These data confirmed that these four genes were downstream of the AR-mediated transcription in EC cells. To partially confirm the promoting effect of DHT on AR-mediated transcription at the protein level, AR expression was examined by western blotting; DHT acted as an agonist, whereas the addition of the AR antagonist flutamide [21] reduced the DHT-enhanced expression of AR in MFE-296 cells (Figure 3C).

To investigate whether FOXA1 influences AR-mediated transcription, we transfected hormone-deprived EC cells with shFOXA1, exFOXA1, or the appropriate negative control vector and then treated them with vehicle or DHT for 24 h. In MFE-296/NC cells, DHT caused a ≥10-fold increase in the expression of the four AR-regulated genes compared with the MFE-296/NC cells treated with vehicle. When FOXA1 expression was knocked down in MFE-296 cells transfected with shFOXA1, however, the expression of these genes was not as markedly increased, and their expression decreased by 8- to 20-fold after treatment with DHT (Figure 3D). Moreover, we found that the increase in the expression of AR and AR-regulated genes was remarkably greater by DHT in the AN3CA/exFOXA1 cells compared with the AN3CA/NC cells (Figure 3E). Our findings indicated that FOXA1 expression globally affected AR-mediated transcription, with all of the four AR-regulated genes requiring FOXA1 for appropriate AR-mediated regulation.

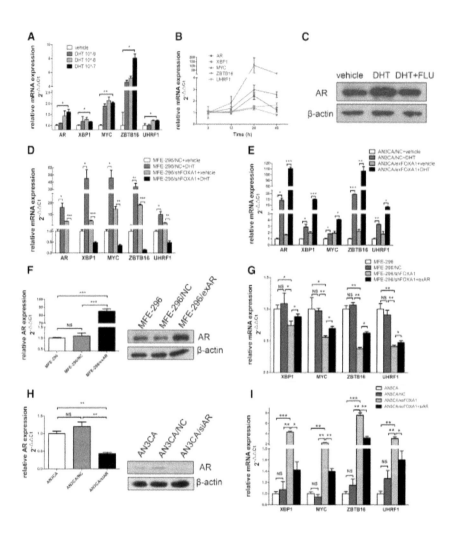

FIGURE 3: FOXA1 affects AR-mediated transcription. A: MFE-296 cells were treated with DHT (10^{-9} to 10^{-7} M) or vehicle (control) for 24 h. qRT-PCR was used to assess the levels of AR, XBP1, MYC, ZBTB16, and UHRF1 mRNA. The levels of each mRNA are shown relative to the level expressed in the vehicle sample. B: Quantification of AR, XBP1, MYC, ZBTB16, and UHRF1 mRNA by qRT-PCR in MFE-296 cells treated with 10^{-7} M DHT for 0–48 h. C: Western blotting analysis of AR in MFE-296 cells treated with vehicle, 10^{-7} M DHT, or 10^{-7} M DHT plus 10^{-6} M flutamide (DHT + FLU) for 24 h. β-actin was used as a loading control. D: MFE-296/NC and MFE-296/shFOXA1 cells were treated with 10^{-7} M DHT or vehicle for 24 h followed by qRT-PCR analysis of AR, XBP1, MYC, ZBTB16, and UHRF1 mRNA. E: AN3CA/NC and AN3CA/exFOXA1 cells were treated with 10^{-7}

M DHT or vehicle for 24 h followed by qRT-PCR analysis of AR, XBP1, MYC, ZBTB16, and UHRF1 mRNA. F: Quantification of AR expression by qRT-PCR and western blotting in untransfected MFE-296 cells (MFE-296) and MFE-296 cells transfected with NC (MFE-296/NC) or exAR (MFE-296/exAR). G: Expression of XBP1, MYC, ZBTB16, and UHRF1 mRNA in untransfected MFE-296 cells and MFE-296 cells transfected with NC, shFOXA1, or shFOXA1 and exAR was measured by qRT-PCR. H: Quantification of AR expression by qRT-PCR and western blotting in untransfected AN3CA cells (AN3CA) and AN3CA cells transfected with NC (AN3CA/NC) or siAR (AN3CA/siAR). I: Expression of XBP1, MYC, ZBTB16, and UHRF1 mRNA in untransfected AN3CA cells and AN3CA cells transfected with NC, exFOXA1, or exFOXA1 and siAR was measured by qRT-PCR. *p < 0.05, **p < 0.01, ***p < 0.001, NS p > 0.05.

6.3.4 FOXA1 PROMOTES AR TARGET GENE EXPRESSION BY INTERACTION WITH AR

FOXA1 can target a series of transcription factors representing anywhere from several to hundreds of genes. To address whether the effects of FOXA1 on AR downstream targets are primarily through upregulating AR, rather than upregulating AR downstream targets directly, we used untransfected MFE-296 cells (MFE-296) and MFE-296 cells transfected with shFOXA1 (MFE-296/shFOXA1), NC (MFE-296/NC), or shFOXA1 together with exAR (MFE-296/shFOXA1 + exAR). qRT-PCR and western blotting analysis confirmed that transfection of exAR resulted in overexpression of AR (Figure 3F). qRT-PCR verified that MFE-296/shFOXA1 cells exhibited substantial decreases in the four AR targets compared with MFE-296/NC cells (Figure 3G). Furthermore, cotransfection with exAR rescued the inhibited expression of the target genes caused by FOXA1 downregulation in MFE-296/shFOXA1 cells (Figure 3G). In addition, we used untransfected AN3CA cells (AN3CA) and AN3CA cells transfected with NC (AN3CA/NC), exFOXA1 (AN3CA/exFOXA1), or exFOXA1 together with siAR (AN3CA/exFOXA1 + siAR). qRT-PCR and western blotting analysis confirmed that transfection with siAR resulted in silencing of AR (Figure 3H). Overexpression of FOXA1 increased the expression of the four AR target genes. Moreover, cotransfection with siAR partially reversed the FOXA1-induced overexpression (Figure 3I). These results verified that AR downregulation attenuated the effect of FOXA1 on AR-mediated transcription and suggested that FOXA1 might promote AR downstream targets at least in part through AR.

6.3.5 FOXA1 AND AR ARE FOUND IN THE SAME PROTEIN COMPLEX

To investigate whether FOXA1 affects AR-mediated transcription through binding to AR, we performed co-immunoprecipitation experiments. We used nuclear lysates from MFE-296 cells to conduct immunoprecipitation with anti-FOXA1. FOXA1 co-immunoprecipitated with AR, whereas immuno-precipitation with the isotype IgG control did not pull down AR or FOXA1 (Figure 4A), indicating that FOXA1 interacted with AR in MFE-296 cells. We also performed the co-immunoprecipitation experiment in AN3CA cells, which has low level of AR. As shown in Figure 4B, AR could be immuno-precipitated by anti-FOXA1 in the presence of DHT but not in its absence. This result indicated that FOXA1 and AR interacted physically. It is likely that FOXA1 affects AR-mediated transcription via binding with AR.

We further examined whether FOXA1 and AR could bind to the five putative FOXA1-AR binding regions, including the promoter and en-hancer regions upstream of the TSS (transcription start sites) of AR tar-get genes such as MYC (Figure 4C). Our ChIP assays showed that both FOXA1 and AR could bind to all the five putative FOXA1-AR-binding regions in MFE-296 cells. Moreover, both FOXA1 and AR bound most greatly to the Enh-1 (enhancer 1) region among the five binding regions (Figure 4D). Our ChIP data together with our co-immunoprecipitation data suggested that FOXA1 forming protein complex with AR might bind to FOXA1-AR overlapping binding regions upstream of MYC, leading to MYC activation in EC cells.

6.3.6 AR IS REQUIRED FOR FOXA1-ENHANCED NOTCH PATHWAY ACTIVATION OF EC CELLS

Pathway analysis in liver cancer shows that FOXA1/AR dual target genes are most involved in the cellular growth/proliferation pathway [22]. Notch pathway activation appears to affect proliferation in many cancers. In EC, the Notch pathway has also been shown to be involved in cell proliferation [17]. Thus, we considered that the interaction between FOXA1 and AR might be related with the Notch pathway. We used western blot analysis to assess the levels of Notch1 and the Notch pathway target protein, Hes1, in MFE-296/shFOXA1 and AN3CA/exFOXA1 cells after exAR or siAR cotransfection, respectively. Cotransfection with exAR rescued the decreased expression of Notch1 and Hes1 caused by FOXA1 downregulation in MFE-296/shFOXA1 cells (Figure 4E). Furthermore, cotransfection with siAR attenuated the increased expression of Notch1 and Hes1 caused by upregulation of FOXA1 in AN3CA/exFOXA1 cells (Figure 4F). These results suggested that the effects of FOXA1 on Notch pathway activation were mediated by AR. In order to determine whether AR was required for FOXA1-enhanced Notch pathway activation, we over-expressed AR expression in AN3CA cells, which has low level of AR. We assessed the levels of Notch1 and Hes1 in untransfected AN3CA cells (AN3CA) as well as AN3CA cells transfected with NC (AN3CA/NC), exAR (AN3CA/exAR), or exAR together with siFOXA1 (AN3CA/exAR + siFOXA1). AN3CA/exAR cells exhibited a substantial increase in AR expression as compared to AN3CA/NC cells, accompanied by over-expression of Notch1 and Hes1 (Additional file 3: Figure S1). Furthermore, cotransfection with siFOXA1 did not rescue the activation of Notch1 and Hes1 caused by AR upregulation in AN3CA/exAR cells (Additional file 3: Figure S1). These results suggested a mechanism, where AR might be a necessary medium in FOXA1-enhanced Notch pathway activation in AN3CA cells.

6.3.7 FOXA1 PROMOTES PROLIFERATION OF HUMAN EC CELLS

To examine the role of FOXA1 in EC cell proliferation, we assessed the effect of FOXA1 in colony-forming and MTT assays. In the colony-forming assay, MFE-296 cell transfected with shFOXA1 showed significantly decreased colony-forming ability when compared with MFE-296 cells transfected with NC (Figure 5A). Moreover, upregulation of FOXA1 in AN3CA cells showed increased colony-forming ability compared with NC cells (Figure 5B). In the MTT assay, downregulation of FOXA1 in MFE-296 cells resulted in poor cell viability (Figure 5C), and upregula-

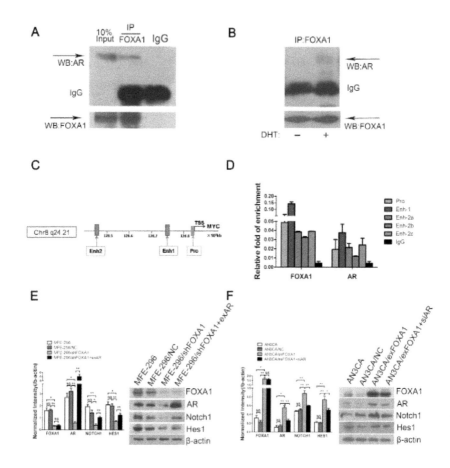

FIGURE 4: FOXA1 affects AR-mediated transcription via binding with AR and activates the Notch pathway. A: Co-immunoprecipitation (IP) of FOXA1 with AR in MFE-296 cells. WB: western blotting. B: Co-immunoprecipitation of FOXA1 with AR in AN3CA cells treated with 10^{-7} M DHT or vehicle. C: Schematic representation of the MYC locus. FOXA1-binding sites and AR-binding sites upstream of the TSS of MYC were predicted by ChIP-seq analysis. ChIP-PCR assays were performed using anti-FOXA1 antibody or anti-AR antibody. Pro: promoter region; Enh-1: enhancer 1 region; End-2: enhancer 2 region; TSS: transcription starting sites. D: Immunoprecipitated DNA fragments in ChIP-PCR assays were examined by qRT-PCR. Each sample was assayed in triplicate in each of three independent experiments. IgG was used as negative control. Primers were designed specifically for the promoter region (Pro), the enhancer 1 region (Enh-1), and the three putative FOXA1-AR binding sites within enhancer 2 region (Enh-2a, Enh-2b, and Enh-2c) according to the study [19]. E: Protein levels of FOXA1, AR, Notch1, and Hes1 in untransfected MFE-296 cells (MFE-296) and MFE-296 cells transfected with NC (MFE-296/NC), shFOXA1 (MFE-296/shFOXA1), or shFOXA1 and exAR (MFE-296/

shFOXA1+exAR) were measured by western blotting (Right), and further quantified by densitometry of triplicate experiments (Left). F: Protein levels of FOXA1, AR, Notch1, and Hes1 in untransfected AN3CA cells (AN3CA) and AN3CA cells transfected with NC (AN3CA/NC) , exFOXA1 (AN3CA/exFOXA1), or exFOXA1 and siAR (AN3CA/exFOXA1+siAR) were measured by western blotting (Right), and further quantified by densitometry of triplicate experiments (Left). β-actin was used as a loading control. *p<0.05, **p<0.01 and NS p>0.05.

tion of FOXA1 in AN3CA cells caused increased cell viability (Figure 5D). These data indicated that FOXA1 promoted cell proliferation.

6.3.8 AR IS REQUIRED FOR FOXA1-ENHANCED PROLIFERATION OF EC CELLS

To directly address whether the effects of FOXA1 in promoting EC cell proliferation can be attributed to its activation of AR, a rescue experiment in MFE-296 cells was performed. In the colony-forming assay, cotransfection with exAR rescued the decreased rate of cell growth caused by FOXA1 downregulation in shFOXA1 cells (Figure 5E). The MTT assay also showed that cotransfection with exAR rescued the inhibition of cell viability caused by FOXA1 downregulation in shFOXA1 cells (Figure 5F). The similarity of results from the colony-forming and MTT assays suggested that the effects of FOXA1 in mediating cell proliferation of EC cells were mediated through AR.

6.3.9 AR IS NOT REQUIRED FOR FOXA1-ENHANCED MIGRATION AND INVASION OF EC CELLS

Our immunohistochemistry results revealed that patients with myometrial invasion displayed higher FOXA1 expression. With this observation in mind, we hypothesized that functional expression of FOXA1 might induce tumor metastasis in EC. To explore the role of FOXA1 in the regulation of metastatic function and to determine whether AR is involved in FOXA1-

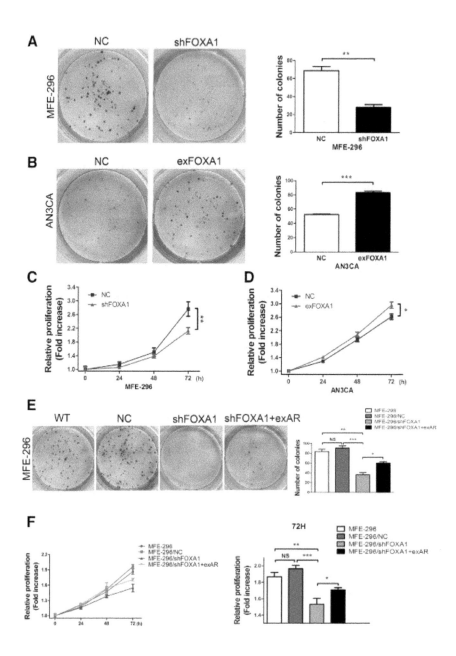

FIGURE 5: FOXA1 promotes proliferation of human EC cells by affecting AR-mediated transcription. A: Proliferation in MFE-296 cells transfected with NC or shFOXA1 was assessed by the colony-forming assay (Left) and further quantified in the number of colonies of triplicate experiments (Right). B: Proliferation in AN3CA cells transfected with NC or exFOXA1 was assessed by the colony-forming assay (Left) and further quantified in the number of colonies of triplicate experiments (Right). C: Assessment of proliferation by the MTT assay in MFE-296 cells transfected with NC or shFOXA1. D: Assessment of proliferation by the MTT assay in AN3CA cells transfected with NC or exFOXA1. E: Left: Colony-formation assay of untransfected MFE-296 cells (WT) and MFE-296 cells transfected with NC, shFOXA1, or shFOXA1 and exAR. Right: Graphical representation of the fold change in the number of colonies in untransfected MFE-296 cells (MFE-296) and MFE-296 cells transfected with NC (MFE-296/NC), shFOXA1 (MFE-296/shFOXA1), or shFOXA1 and exAR (MFE-296/shFOXA1 + exAR). F: Proliferation of MFE-296, MFE-296/NC, MFE-296/shFOXA1, or MFE-296/shFOXA1 + exAR cells was assessed by MTT assay. The right panel reiterates the data in the left panel at 72 h. $*p < 0.05$, $**p < 0.01$, $***p < 0.001$ and NS $p > 0.05$.

mediated regulation of metastatic function, we examined the migration and invasion ability of MFE-296/shFOXA1 and AN3CA/exFOXA1 cells after exAR or siAR cotransfection using transwell migration and invasion assays. MFE-296/shFOXA1 cells displayed a decreased rate of migration compared to MFE-296/NC cells (Figure 6A). However, cotransfection of MFE-296/shFOXA1 cells with exAR (MFE-296/shFOXA1 + exAR) did not rescue the migration to the levels observed in MFE-296/NC or untransfected cells (MFE-296) (Figure 6A). Furthermore, AN3CA/exFOXA1 cells exhibited a high migration rate as compared with AN3CA/NC cells, but cotransfection with siAR (AN3CA/exFOXA1 + siAR) did not significantly attenuate the migration rate (Figure 6B).

Consistent with these findings, the invasion rate was significantly reduced in MFE-296/shFOXA1 cells, but the reduction was not reversed upon transfection with exAR (Figure 6C). Likewise, the invasion rate was enhanced in AN3CA/exFOXA1 cells, but this enhancement was not attenuated upon transfection with siAR (Figure 6D). These results demonstrated a functional role for FOXA1 in mediating migration and invasion in EC cells and suggested a mechanism (distinct from that for EC cell proliferation) by which AR might not contribute to FOXA1-mediated metastasis of EC.

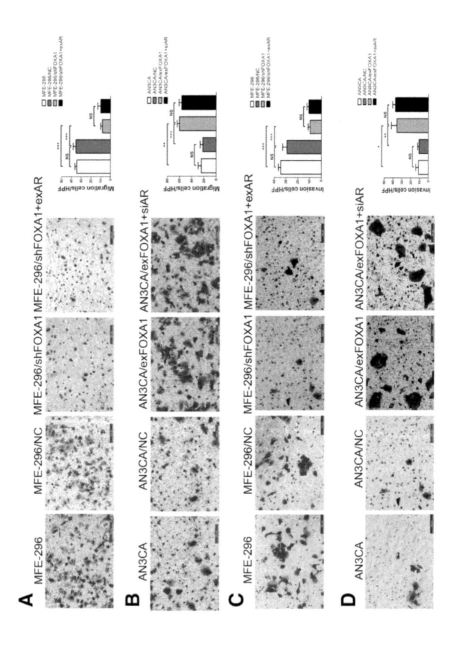

FIGURE 6: FOXA1 induces migration and invasion in EC cells. A: Cell migration of MFE-296, MFE-296/NC, MFE-296/shFOXA1 and MFE-296/shFOXA1 + exAR cells was assessed by the transwell migration analysis (Left). The mean ± SD number of migrated cells of three independent experiments was showed in the right panel. The abbreviation "HPF" on the y axis means one high power field. B: Cell migration of AN3CA, AN3CA/NC, AN3CA/exFOXA1 and AN3CA/exFOXA1 + siAR cells were subjected to transwell migration analysis (Left). The mean ± SD number of migrated cells of three independent experiments was showed in the right panel. C: Cell invasion of MFE-296, MFE-296/NC, MFE-296/shFOXA1 and MFE-296/shFOXA1 + exAR cells was assessed by the transwell invasion analysis (Left). The mean ± SD number of invased cells of three independent experiments was shown in the right panel. D: Cell invasion of AN3CA, AN3CA/NC, AN3CA/exFOXA1 and AN3CA/exFOXA1 + siAR cells by three independent experiments were subjected to transwell invasion analysis (Left). The mean ± SD number of invased cells of three independent experiments was showed in the right panel. (Magnification, 200×). *$p < 0.05$, **$p < 0.01$, ***$p < 0.001$, and NS $p > 0.05$.

6.3.10 ONCOGENIC ROLE OF FOXA1 IN A TUMOR XENOGRAFT MODEL

Tumors generated by subcutaneous implantation of MFE-296 cells were used to evaluate the effect of FOXA1 on proliferation in a mouse tumor xenograft model. We measured tumor volumes in xenografted mice over a 6-week period following injection of untransfected MFE-296 (MFE-296), stably transfected with shFOXA1 (MFE-296/shFOXA1) or NC (MFE-296/NC). These measurements indicated that tumors in the MFE-296/shFOXA1 group grew significantly slower than those in the MFE-296/NC group and the MFE-296 group (Figure 7A). Six weeks after injection, tumors were removed from the mice (Figure 7C). The final mean weight and volume of tumors in the MFE-296/shFOXA1 group were significantly lower than those in the MFE-296/NC group ($p < 0.05$, Figure 7B and 7C). Tumor tissues were then embedded in paraffin, stained with hematoxylin and eosin (H&E), and immunohistochemically stained with antibodies against FOXA1, AR, Notch1, Hes1, Ki67, or PCNA. Lower FOXA1 expression in the MFE-296/shFOXA1 group also led to reduced staining for AR, indicating that FOXA1 also affected AR expression in vivo, in accordance with the results in vitro. As expected, the MFE-296/shFOXA1 group had significantly lower levels of Notch1 and Hes1 (Figure 7D), thus verifying the role of FOXA1 as a positive regulator of the Notch pathway in vivo. Furthermore, to determine the proliferative ability of MFE-296 cells, we performed immunohistochemical staining of Ki67 and PCNA, which are expressed as proliferation indices. The observed lower expression of Ki67 and PCNA in the MFE-296/shFOXA1 group was consistent with the smaller tumor volumes in the mouse tumor xenograft model (Figure 7D).

6.4 DISCUSSION

Over the past decade, FOXA1 expression has been examined in several human cancers, and oncogenic and tumor-suppressive roles have been proposed for FOXA1 depending on the cancer type and, in some cases, the subtype. In acute myelocytic leukemia, esophageal squamous cell carcinomas, lung adenocarcinomas, thyroid carcinoma, prostate cancer, and AR-positive molecular apocrine breast cancer [12,23-25], FOXA1 acts as an oncogene. However, in hepatocellular carcinoma, pancreatic, and estrogen receptor (ER)-positive breast cancer, FOXA1 has been reported to have a tumor-suppressive function [26-28]. On one hand, FOXA1 acts as a tumor oncogene. In oesophageal squamous cell carcinoma, FOXA1 expression is correlated with lymph node metastases in immunohistochemical specimens and FOXA1 expression inhibition decreases cellular invasion and migration [29]. Also, FOXA1 is over-expressed in aggressive thyroid cancers (ATC) and involved in cell cycle progression via down-regulation of p27^{Kip1} in an ATC cell line [30]. On the other hand, FOXA1 has been reported to act as a tumor suppressor. It has been reported that FOXA1 positively regulates miRNA-122, which is correlated with favourable prognosis in human hepatocellular carcinoma [26]. In addition, FOXA1 acts as an important antagonist of the epithelial-to-mesenchymal transition (EMT) in pancreatic ductal adenocarcinoma through its positive regulation of E-cadherin and maintenance of the epithelial phenotype [27]. It is critical to note that the role of FOXA1, as a tumor oncogene or a tumor suppressor gene, has been reported to vary in prostate and breast cancers depending on multiple cancer subtypes and states of hormone dependence or independence [11,12,28].

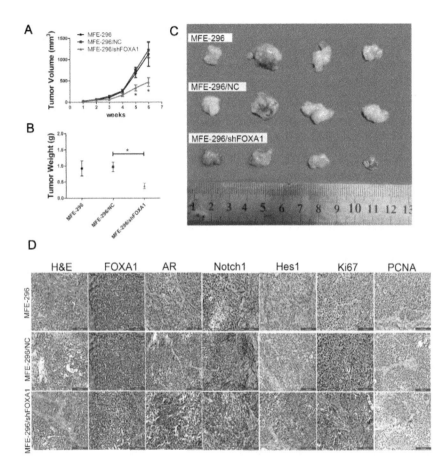

FIGURE 7: Tumorigenicity assay in nude mice. A: The growth rates of tumors formed from untransfected MFE-296 cells (MFE-296) and MFE-296 cells transfected with NC (MFE-296/NC) or shFOXA1 (MFE-296/shFOXA1). After injection, tumor volumes were calculated every seven days. B and C: Six weeks after injection of MFE-296, MFE-296-NC, and MFE-296-shFOXA1 cells, tumors were removed, and the tumor weights and volumes were determined. Arithmetic means and SD are shown. D: Staining with hematoxylin and eosin (H&E) or immunohistochemical staining for FOXA1, AR, Notch1, Hes1, Ki67, and PCNA in mouse tumor tissues (immunohistochemical staining, 200×). *p<0.05 compared with the NC group.

A previous study has addressed the expression and function of FOXA1 in EC; immunohistochemical analysis by Abe et al. indicated that FOXA1 was negatively associated with lymph node status in EC immunohisto-chemical specimens in Japanese, and FOXA1 repressed proliferation and migration in one type of EC cells (Ishikawa) [31]. However, our study found that the FOXA1 level in ECs was significantly higher than that in atypical hyperplasia and normal tissues ($p < 0.05$) in immunohistochemi-cal specimens and that FOXA1 promoted tumor cell proliferation in EC, which differs from the previous results. The difference might be attrib-uted to the immunohistochemical samples in different countries used. Alternatively, the cancer subtype may affect the results: the function of FOXA1 as a tumor suppressor in the Abe et al. study was investigated in the Ishikawa cell line, which is ER-positive [32], whereas we used MFE-296 (high levels of FOXA1 and AR) and AN3CA (low levels of FOXA1 and AR), which are both ER-negative cell lines [33,34]. This idea consists with breast cancer studies that have shown that FOXA1 functions as a tumor suppressor in ER-positive breast cancer cells (MCF-7) [28] but as a tumor activator in ER-negative breast cancer cells (MDA-MB-453) [12]. Furthermore, this idea of the effects of forkhead family members depend-ing on ER expression is also consistent with the study that have shown the Forkhead box class o 3a transcription factor (FoxO3a) has inhibitory effects on motility and invasiveness of ER-positive breast cancer cells but inducing effects on motility and invasiveness of ER-negative breast cancer cells [35]. More comprehensive studies covering several EC cell lines in different cancer subtypes will be necessary to define the role of FOXA1 in EC development.

Most researches on hormone receptors in EC have focused on ER and progesterone receptor (PR). However, the expression of AR in the human normal endometrium and its disorders is not well understood. Though higher serum androgen levels have been certified to exist in the utero-ovarian vein blood samples from women with EC [36], the details of AR expression and its actions in EC are a topic of dispute. Longer CAG re-peats in AR promote carcinogenesis of uterine endometrial cells [37]. An-drogens and AR may be involved in endometrial cell proliferation by regu-lating the expression of insulin growth factor I (IGF-I) in the uterus [38]. Our results suggest that AR expression is significantly higher in EC than in normal endometrium and that AR activated by FOXA1 might promote the Notch pathway, which may be another mechanism involving AR in EC.

Most FOXA1 studies have focused on its role as a pioneer factor that binds to DNA packaged in chromatin and opens the chromatin for binding of additional transcription factors including AR [39,40]. According to our results from qRT-PCR and western blotting, FOXA1 regulates AR target genes by up-regulation of AR expression. Interestingly, our co-immuno-precipitation results (Figure 4A and 4B) showed that FOXA1 interacted with AR at the protein level. Apart from that, our ChIP-PCR results suggested that FOXA1 and AR were directly bound to the same regions upstream of MYC (Figure 4C and 4D). Based on the above results, we suggest that FOXA1 may also directly regulate AR target genes (at least MYC) by binding to AR in EC. Our results regarding an interaction between AR and FOXA1 may be related to the finding that the AR and FOXA1 binding sites are adjacent on multiple promoters of AR target genes in prostatic cells [9,41]. Thus, FOXA1 may regulate the AR target genes through at least two means: AR over-expression or physical interaction with AR in order to induce easy AR accessibility to binding to its target genes. MYC is an immediate early response gene downstream from AR pathway and is tightly regulated through AR cis-regulatory elements identified within its proximal promoters and distal enhancer regions [19], which is consistent with our ChIP-PCR results (Figure 4C and 4D). Interestingly, we showed that FOXA1 and AR more evidently bound to the MYC enhancer regions as compared to MYC promoter regions. These results could be attributed to other co-regulators involved in this binding process. Since TCF7L2, a protein mediating DNA looping for long-distance interactions of distal enhancers and proximal promoters, physically interacts with FOXA1 and AR and mediates the transcription of MYC in breast cancer [19], future investigation will be needed to clarify which co-regulators are involved in FOXA1/AR binding to the enhancer regions upstream of MYC in EC cells.

Although the underlying mechanisms governing the FOXA1-AR correlation in tumor progression are not fully understood, a pathway analysis showed that 187 FOXA1/AR dual target genes were involved in the cellular growth/proliferation pathway in liver cancer [22]. The Notch pathway is implicated in the development of various cancers, and the Notch pathway blockade appears to affect cell proliferation in multiple types of cancers. Notch pathway inhibition in breast cancer cells induces cell cycle arrest and apoptosis [42]. Similarly, downregulation of Notch1 contributes to cell growth inhibition in pancreatic cancer [43]. Our results suggest that downregulation of AR attenuated FOXA1-induced upregulation of the Notch pathway in EC cells. These findings indicate that FOXA1 might promote AR-mediated transcription and ultimately activate the Notch pathway. Here, we describe, for the first time, the association between FOXA1 expression and the Notch pathway in cancer.

The specific mechanism of cell proliferation in EC reported so far has been limited, although several classical transcription factors related to proliferation have been identified, including cyclin D1, p53, IGFBP-1, PTEN, and p27[Kip1][44-48]. In this study, we suggest that FOXA1 promotes cell proliferation in EC by interaction with AR, possibly via the Notch pathway, which may be a newly identified regulatory mechanism of cell proliferation in EC.

We further investigated the effects of FOXA1 and AR on migration and invasion of EC cells, and found that neutralization of AR activity did not inhibit FOXA1-enhanced cancer cell migration or invasion. These observations indicate that the promoting effect of FOXA1 on migration and invasion is not dependent on AR. Our findings in migration and invasion assays are consistent with our findings in immunohistochemical staining, which showed that higher expression of FOXA1 but not AR is found in tumors that displayed a greater depth of myometrial invasion. These results suggest that AR is not the only downstream target of FOXA1 in EC. Future studies will be necessary to define which transcription factors or pathways are involved in FOXA1-enhanced cell migration and invasion in EC.

The traditional endocrine treatment (mainly targeting ER and PR) is ineffective in most ER-negative and PR-negative ECs, and even in some ER-positive and PR-positive ECs [49]. In our investigation, 9 of the15 ER-negative EC cases (60.0%) and 41 of the 61 ER-positive EC cases (67.2%) were AR positive, and the majority of ECs were also FOXA1 positive (Table 1). Thus, AR and FOXA1 might be alternative targets in ECs insensitive to traditional endocrine treatment or could be targets for adjuvant treatment following surgery and traditional endocrine treatment.

There has been speculation about the use of anti-androgens for the treatment of ECs [50]; this hypothesis warrants clinical investigation in light of our findings.

6.5 CONCLUSIONS

In summary, our results suggest a new mechanism for the development of EC, in which FOXA1 promotes tumor cell proliferation through AR and activates the Notch pathway by influencing AR expression. The newly identified FOXA1-AR interaction will help further elucidate the molecular mechanisms underlying EC progression and suggests that FOXA1 and AR are potential targets for EC treatment.

REFERENCES

1. Bokhman JV: Two pathogenetic types of endometrial carcinoma. Gynecol Oncol 1983, 15:10-17.
2. Kandoth C, Schultz N, Cherniack AD, Akbani R, Liu Y, Shen H, Robertson AG, Pashtan I, Shen R, Benz CC, Yau C, Laird PW, Ding L, Zhang W, Mills GB, Kucherlapati R, Mardis ER, Levine DA, Cancer Genome Atlas Research Network: Integrated genomic characterization of endometrial carcinoma. Nature 2013, 497:67-73.
3. Gadducci A, Cosio S, Genazzani AR: Old and new perspectives in the pharmacological treatment of advanced or recurrent endometrial cancer: Hormonal therapy, chemotherapy and molecularly targeted therapies. Crit Rev Oncol Hematol 2006, 58:242-256.
4. Ito K, Suzuki T, Akahira J, Moriya T, Kaneko C, Utsunomiya H, Yaegashi N, Okamura K, Sasano H: Expression of androgen receptor and 5α-reductases in the human normal endometrium and it's disorders. Int J Cancer 2002, 99:652-657.
5. Horie K, Takakura K, Imai K, Liao S, Mori T: Immunohistochemical localization of androgen receptor in the human endometrium, decidua, placenta and pathological conditions of the endometrium. Hum Reprod 1992, 7:1461-1466.
6. Hu YC, Wang PH, Yeh S, Wang RS, Xie C, Xu Q, Zhou X, Chao HT, Tsai MY, Chang C: Subfertility and defective folliculogenesis in female mice lacking androgen receptor. Proc Natl Acad Sci USA 2004, 101:11209-11214.
7. Carlsson P, Mahlapuu M: Forkhead transcription factors: key players in development and metabolism. Dev Biol 2002, 250:1-23.
8. Kaestner KH: The FoxA factors in organogenesis and differentiation. Curr Opin Genet Dev 2010, 20:527-532.

9. Lupien M, Eeckhoute J, Meyer CA, Wang Q, Zhang Y, Li W, Carroll JS, Liu XS, Brown M: FoxA1 translates epigenetic signatures into enhancer-driven lineage-specific transcription. Cell 2008, 132:958-970.

10. Gao N, Zhang J, Rao MA, Case TC, Mirosevich J, Wang Y, Jin R, Gupta A, Rennie PS, Matusik RJ: The role of hepatocyte nuclear factor-3 alpha (Forkhead Box A1) and androgen receptor in transcriptional regulation of prostatic genes. Mol Endocrinol 2003, 17:1484-1507.

11. Sahu B, Laakso M, Ovaska K, Mirtti T, Lundin J, Rannikko A, Sankila A, Turunen JP, Lundin M, Konsti J, Vesterinen T, Nordling S, Kallioniemi O, Hautaniemi S, Jänne OA: Dual role of FoxA1 in androgen receptor binding to chromatin, androgen signalling and prostate cancer. EMBO J 2011, 30:3962-3976.

12. Robinson JL, Macarthur S, Ross-Innes CS, Tilley WD, Neal DE, Mills IG, Carroll JS: Androgen receptor driven transcription in molecular apocrine breast cancer is mediated by FoxA1. EMBO J 2011, 30:3019-3027.

13. Axelson H: Notch signaling and cancer: emerging complexity. Semin Cancer Biol 2004, 14:317-319.

14. Qiao L, Wong BC: Role of Notch signaling in colorectal cancer. Carcinogenesis 2009, 30:1979-1986.

15. Rose SL: Notch signaling pathway in ovarian cancer. Int J Gynecol Cancer 2009, 19:564-566.

16. De Strooper B, Annaert W, Cupers P, Saftig P, Craessaerts K, Mumm JS, Schroeter EH, Schrijvers V, Wolfe MS, Ray WJ, Goate A, Kopan R: A presenilin-1-dependent gamma-secretase-like protease mediates release of Notch intracellular domain. Nature 1999, 398:518-522.

17. Wei Y, Zhang Z, Liao H, Wu L, Wu X, Zhou D, Xi X, Zhu Y, Feng Y: Nuclear estrogen receptor-mediated Notch signaling and GPR30-mediated PI3K/AKT signaling in the regulation of endometrial cancer cell proliferation. Oncol Rep 2012, 27:504-510.

18. Bao W, Wang HH, Tian FJ, He XY, Qiu MT, Wang JY, Zhang HJ, Wang LH, Wan XP: A TrkB-STAT3-miR-204-5p regulatory circuitry controls proliferation and invasion of endometrial carcinoma cells. Mol Cancer 2013, 12:155.

19. Ni M, Chen Y, Fei T, Li D, Lim E, Liu XS, Brown M: Amplitude modulation of androgen signaling by c-MYC. Genes Dev 2013, 27:734-748.

20. Yazawa T, Kawabe S, Kanno M, Mizutani T, Imamichi Y, Ju Y, Matsumura T, Yamazaki Y, Usami Y, Kuribayashi M, Shimada M, Kitano T, Umezawa A, Miyamoto K: Androgen/androgen receptor pathway regulates expression of the genes for cyclooxygenase-2 and amphiregulin in periovulatory granulosa cells. Mol Cell Endocrinol 2013, 369:42-51.

21. Yoshida K, He PJ, Yamauchi N, Hashimoto S, Hattori MA: Up-regulation of circadian clock gene Period 2 in the prostate mesenchymal cells during flutamide-induced apoptosis. Mol Cell Biochem 2010, 335:37-45.

22. Li Z, Tuteja G, Schug J, Kaestner KH: Foxa1 and Foxa2 are essential for sexual dimorphism in liver cancer. Cell 2012, 148:72-83.

23. Dai R, Yan D, Li J, Chen S, Liu Y, Chen R, Duan C, Wei M, Li H, He T: Activation of PKR/eIF2α signaling cascade is associated with dihydrotestosterone-induced cell

cycle arrest and apoptosis in human liver cells. J Cell Biochem 2012, 113:1800-1808.

24. Lin L, Miller CT, Contreras JI, Prescott MS, Dagenais SL, Wu R, Yee J, Orringer MB, Misek DE, Hanash SM, Glover TW, Beer DG: The hepatocyte nuclear factor 3 alpha gene, HNF3alpha (FOXA1), on chromosome band 14q13 is amplified and overexpressed in esophageal and lung adenocarcinomas. Cancer Res 2002, 62:5273-5279.

25. Jain RK, Mehta RJ, Nakshatri H, Idrees MT, Badve SS: High-level expression of forkhead-box protein A1 in metastatic prostate cancer. Histopathology 2011, 58:766-772.

26. Coulouarn C, Factor VM, Andersen JB, Durkin ME, Thorgeirsson SS: Loss of miR-122 expression in liver cancer correlates with suppression of the hepatic phenotype and gain of metastatic properties. Oncogene 2009, 28:3526-3536.

27. Song Y, Washington MK, Crawford HC: Loss of FOXA1/2 is essential for the epithelial-to-mesenchymal transition in pancreatic cancer. Cancer Res 2010, 70:2115-2125.

28. Hurtado A, Holmes KA, Ross-Innes CS, Schmidt D, Carroll JS: FOXA1 is a key determinant of estrogen receptor function and endocrine response. Nat Genet 2011, 43:27-33.

29. Sano M, Aoyagi K, Takahashi H, Kawamura T, Mabuchi T, Igaki H, Tachimori Y, Kato H, Ochiai A, Honda H, Nimura Y, Nagino M, Yoshida T, Sasaki H: Forkhead box A1 transcriptional pathway in KRT7-expressing esophageal squamous cell carcinomas with extensive lymph node metastasis. Int J Oncol 2010, 36:321-330.

30. Nucera C, Eeckhoute J, Finn S, Carroll JS, Ligon AH, Priolo C, Fadda G, Toner M, Sheils O, Attard M, Pontecorvi A, Nose V, Loda M, Brown M: FOXA1 is a potential oncogene in anaplastic thyroid carcinoma. Clin Cancer Res 2009, 15:3680-3689.

31. Abe Y, Ijichi N, Ikeda K, Kayano H, Horie-Inoue K, Takeda S, Inoue S: Forkhead box transcription factor, forkhead box A1, shows negative association with lymph node status in endometrial cancer, and represses cell proliferation and migration of endometrial cancer cells. Cancer Sci 2012, 103:806-812.

32. De Marco P, Bartella V, Vivacqua A, Lappano R, Santolla MF, Morcavallo A, Pezzi V, Belfiore A, Maggiolini M: Insulin-like growth factor-I regulates GPER expression and function in cancer cells. Oncogene 2013, 32:678-688.

33. Hackenberg R, Hawighorst T, Hild F, Schulz KD: Establishment of new epithelial carcinoma cell lines by blocking monolayer formation. J Cancer Res Clin Oncol 1997, 123:669-673.

34. Jiang F, Liu T, He Y, Yan Q, Chen X, Wang H, Wan X: MiR-125b promotes proliferation and migration of type II endometrial carcinoma cells through targeting TP53INP1 tumor suppressor in vitro and in vivo. BMC Cancer 2011, 11:425.

35. Sisci D, Maris P, Cesario MG, Anselmo W, Coroniti R, Trombino GE, Romeo F, Ferraro A, Lanzino M, Aquila S, Maggiolini M, Mauro L, Morelli C, Andò S: The estrogen receptor α is the key regulator of the bifunctional role of FoxO3a transcription factor in breast cancer motility and invasiveness. Cell Cycle 2013, 12:3405-3420.

36. Jongen VH, Sluijmer AV, Heineman MJ: The postmenopausal ovary as an androgen-producing gland; hypothesis on the etiology of endometrial cancer. Maturitas 2002, 43:77-85.

37. Sasaki M, Sakuragi N, Dahiya R: The CAG repeats in exon 1 of the androgen receptor gene are significantly longer in endometrial cancer patients. Biochem Biophys Res Commun 2003, 305:1105-1108.

38. Sahlin L, Norstedt G, Eriksson H: Androgen regulation of the insulin-like growth factor-I and the estrogen receptor in rat uterus and liver. J Steroid Biochem Mol Biol 1994, 51:57-66.

39. Cirillo LA, Zaret KS: An early developmental transcription factor complex that is more stable on nucleosome core particles than on free DNA. Mol Cell 1999, 4:961-969.

40. Laganière J, Deblois G, Lefebvre C, Bataille AR, Robert F, Giguère V: From the Cover: location analysis of estrogen receptor alpha target promoters reveals that FOXA1 defines a domain of the estrogen response. Proc Natl Acad Sci U S A 2005, 102:11651-11656.

41. Jia L, Landan G, Pomerantz M, Jaschek R, Herman P, Reich D, Yan C, Khalid O, Kantoff P, Oh W, Manak JR, Berman BP, Henderson BE, Frenkel B, Haiman CA, Freedman M, Tanay A, Coetzee GA: Functional enhancers at the gene-poor 8q24 cancer-linked locus. PLoS Genet 2009, 5:e1000597.

42. Zang S, Ji Ch QX, Dong X, Ma D, Ye J, Ma R, Dai J, Guo D: A study on Notch signaling in human breast cancer. Neoplasma 2007, 54:304-310.

43. Wang Z, Zhang Y, Li Y, Banerjee S, Liao J, Sarkar FH: Down-regulation of Notch-1 contributes to cell growth inhibition and apoptosis in pancreatic cancer cells. Mol Cancer Ther 2006, 5:483-493.

44. Zorn KK, Bonome T, Gangi L, Chandramouli GV, Awtrey CS, Gardner GJ, Barrett JC, Boyd J, Birrer MJ: Gene expression profiles of serous, endometrioid, and clear cell subtypes of ovarian and endometrial cancer. Clin Cancer Res 2005, 11:6422-6430.

45. Catalano S, Giordano C, Rizza P, Gu G, Barone I, Bonofiglio D, Giordano F, Malivindi R, Gaccione D, Lanzino M, De Amicis F, Andò S: Evidence that leptin through STAT and CREB signaling enhances cyclin D1 expression and promotes human endometrial cancer proliferation. J Cell Physiol 2009, 218:490-500.

46. Rutanen EM, Nyman T, Lehtovirta P, Ammälä M, Pekonen F: Suppressed expression of insulin-like growth factor binding protein-1 mRNA in the endometrium: a molecular mechanism associating endometrial cancer with its risk factors. Int J Cancer 1994, 59:307-312.

47. Matsushima-Nishiu M, Unoki M, Ono K, Tsunoda T, Minaguchi T, Kuramoto H, Nishida M, Satoh T, Tanaka T, Nakamura Y: Growth and gene expression profile analyses of endometrial cancer cells expressing exogenous PTEN. Cancer Res 2001, 61:3741-3749.

48. Huang KT, Pavlides SC, Lecanda J, Blank SV, Mittal KR, Gold LI: Estrogen and progesterone regulate p27kip1 levels via the ubiquitin-proteasome system: pathogenic and therapeutic implications for endometrial cancer. PLoS One 2012, 7:e46072.

49. Thigpen T, Brady MF, Homesley HD, Soper JT, Bell J: Tamoxifen in the treatment of advanced or recurrent endometrial carcinoma: a Gynecologic Oncology Group study. J Clin Oncol 2001, 19:364-367.

50. Day JM, Purohit A, Tutill HJ, Foster PA, Woo LW, Potter BV, Reed MJ: The development of steroid sulfatase inhibitors for hormone-dependent cancer therapy. Ann N Y Acad Sci 2009, 1155:80-87.

There are several supplemental files that are not available in this version of the article. To view this additional information, please use the citation information cited on the first page of this chapter.

CHAPTER 7

PKC α REGULATES NETRIN-1/UNC5B-MEDIATED SURVIVAL PATHWAY IN BLADDER CANCER

JIAO LIU, CHUI-ZE KONG, DA-XIN GONG, ZHE ZHANG, AND YU-YAN ZHU

7.1 BACKGROUND

Bladder cancer (BC) is one of the most deadly urological malignant tumors and also the 2nd most common urologic cancer [1]. In the US, BC is the ninth most common cause of cancer-related mortality, and is the fourth most common cancer in men. Most bladder cancers are initially non-invasive and up to 15% will progress to muscle-invasive carcinoma. Although treatment of bladder cancer has been improved greatly, the mortality of this disease is still increasing [2].

As the central hub of a variety of signal transduction process, PKC involves in cell information transmission, secretion, cell differentiation and proliferation. What's more, it participates in apoptosis and differentiation of tumor cells. PKC α is one subtype of classic protein kinase C,

This chapter was originally published under the Creative Commons Attribution License. Liu J, Kong C-Z, Gong D-X, Zhang Z, and Zhu Y-Y. PKC α Regulates Netrin-1/UNC5B-Mediated Survival Pathway in Bladder Cancer. BMC Cancer 14,93 (2014). doi:10.1186/1471-2407-14-93.

which is closely related to recurrence of bladder cancer [3]. PKC α can promote proliferation, migration and survival of cancer cells through the downstream signal transduction pathways ERK1/2 and NF-κB [4]. Recent research shows that activation, overexpression of PKC α as well as suppressing or depletion of PKC α can regulate the proliferation of cancer cells [5-7]. Thus it can be seen that PKC α is closely related to the biological behaviour of bladder cancer.

As a kind of proto-oncogene, Netrin-1 is the axon guidance factor that attracts the most attention in the family of dependence receptor [8]. Researches show that netrin-1 can activate PKC α after combination with its receptor, which may cause phosphorylation to promote cancer cell proliferation, and then restrain cell proliferation [9]. In recent years, netrin-1 has been found effective in inhibiting apoptosis in lung cancer, advanced neuroblastoma, breast cancer and prostate cancer [10-13]. UNC5B is one of the dependence receptors of netrin-1. Researches show that UNC5B is the downstream gene of p53, down-regulation of UNC5B using small interfering RNA Can significantly inhibit apoptosis, thus concludes that UNC5B plays a role of inducing apoptosis, and it is a kind of tumor suppressor genes [14]. According to reports in the literature, up-regulation of netrin-1transcripts can antagonize apoptosis induced by UNC5B [15]. Since PKC α, netrin-1and UNC5B play a significant role in the process of tumor treatment. Therefore, study the mechanisms of action of PKC alpha regulates netrin-1/UNC5B-mediated survival pathway is of great significance.

In this study, we detect the expression of netrin-1/UNC5B in the bladder cancer tissues as well as in the bladder cancer cell line on both the RNA and protein levels, we found that netrin-1/UNC5B was closely related to the activation of PKC alpha state. Furthermore, netrin-1/UNC5B was closely associated with bladder cancer malignant pathological biological behavior. Therefore, we need to validate that PKC α inhibits bladder cancer cell apoptosis by regulating signaling pathway of netrin-1/UNC5B.

7.2 METHODS

7.2.1 PATIENTS AND SPECIMENS

One hundred and twenty bladder cancer tissues were collected by the surgical resection in the First Affiliated Hospital of China Medical University from 2008 to 2012. Bladder cancer tissues and paired adjacent normal bladder tissues were collected. None of patients underwent chemotherapy, radiotherapy or adjuvant treatment before surgery. Patients' consent for the research use of tumor tissue was obtained, and the research protocol was approved by Ethical Committee at China Medical University. We followed up all patients for the survival time by consulting their case documents and telephoning.

7.2.2 CELL CULTURE, TREATMENT OF CELLS WITH DRUGS AND SIRNA

Human BC cell lines SV, 5637, T24 and BIU-87 were purchased from Cell Bank of Shanghai Institutes for Biological Sciences, Chinese Academy of Sciences. They were maintained in RPMI 1640, or DMEM medium supplemented with 10% fetal bovine serum (FBS). Cells were incubated at 37°C in 5% CO_2.

For PMA treatment, cells were treated at the concentration of 100 nmol/L for 24 hours. For calphostin C treatment, cells were treated by using 100 nmol/L PMA for 4 hours first, then 50 nmol/L calphostin C for 24 hours. For siRNA transfection, Lipofectamine (Invitrogen) was used. PKC siRNA sequences was as follows: forward, 5' GUG CCA UGA AUU UGU UAC UTT 3', reverse, 5' AGU AAC AAA UUC AUG GCA CTT 3'.

7.2.3 REAL-TIME PCR

Total cellular RNA was extracted from cells using the RNeasy Plus Mini Kit (Qiagen). First strand of cDNA was synthesized by using PrimeScript RT reagent kit (Takara). Quantitative real-time polymerase chain reaction (QPCR) was done using SYBR Green PCR Master Mix (Applied Biosystems) in a total volume of 20 μl on a 7900 Real-Time PCR System (Applied Biosystems): 50°C for 2 min, 95°C for 5 min, 45 cycles of 95°C for 40 s, 60°C for 30 s. The sequences of the primer pairs are: UNC5B forward, 5' CAG GGC AAG TTC TAC GAG AT 3', reverse, 5' TGG TCC AGC AGG ATG TGA 3', netrin-1 forward, 5' GTC AAT GCG GCC TTC GG 3', reverse, 5' CTG CTC GTT CTG CTT GGT GAT 3', β-actin forward, 5' TTA GTT GCG TTA CAC CCT TTC 3', reverse, 5' ACC TTC ACC GTT CCA GTT T 3', β-actin was used as the reference gene. Relative gene expression levels were represented as $\Delta CT = CT$ gene $-$ CT reference; fold change of gene expression was computed by the $2-\Delta\Delta CT$ method [16]. Experiments were repeated in triplicate.

7.2.4 WESTERN BLOTTING

Total protein from cells was extracted in lysis buffer (Pierce) and quantified using the Bradford method. Total protein was separated by SDS-PAGE (12%). After transferring to polyvinylidene fluoride (PVDF) membrane (Millipore, Billerica, MA), the membranes were incubated overnight at 4°C with antibodies against UNC5B/netrin-1 (1:1000, Abcam Inc. USA), GAPDH (1:500, Santa Cruz Biotechnology). After incubation with peroxidase-coupled anti-mouse/rabbit IgG (Santa Cruz Biotechnology) at 37°C for 2 h, bound proteins were visualized using ECL (Pierce) and detected using BioImaging Systems (UVP Inc., Upland, CA). The relative protein levels were calculated based on GAPDH protein as a loading control.

7.2.5 IMMUNOHISTOCHEMISTRY AND EVALUATION

Sections were deparaffinized in xylene, hydrated in graded alcohols. After antigen retrieval, sections were incubated in an aqueous solution of 3% hydrogen peroxide followed by incubation with 5% non-fat milk, which served as a blocking agent for nonspecific binding. Slides were incubated with UNC5B & netrin-1 rabbit polyclonal antibody with an optimal dilution of 1:100 overnight at 4°C. Biotinylated goat anti-rabbit serum IgG was used as a secondary antibody. After washing, the sections were incubated with streptavidin–biotin conjugated with horseradish peroxidase at room temperature for 10 min, and the peroxidase reaction was developed with 3, 3'-diaminobenzidine tetrahydrochloride. All the slides were evaluated by 2 pathologists. Five views were examined per slide; 100 cells were observed per view at ×400 magnification. Nucleus and/or cytoplasmic immune-staining in tumor cells were considered positively. Positive reactions were scored for both intensity of staining and percentage of positive cells. Intensity grades were 0 (no staining), 1 (weak, light yellow), 2(moderate, yellowish brown), to 3 (intense, brown) and the percentage of positive tumor cells were scored as 0 (negative), 1 (1–50%), 2 (51–75%), 3 ($\geq$76%). Scores of each sample were multiplied to give final scores of 0–9, and the tumors were finally determined as negative: score 0; low expression: $0 < \text{score} \leq 4$; or high expression: $\text{score} > 4$.

7.2.6 CELL PROLIFERATION AND INVASION ASSAYS

Cell Counting Kit-8 (Dojindo) was employed to determine the number of viable BIU cells. Experiments were performed according the manufacturer's protocol. Invasion ability was examined by wound healing assay. In brief, cells were seeded at a density of 1.0×10^6 cells/well in 6-well culture plates. After they grown into confluence, scratch was performed using a pipette tip, cells were washed with PBS and cultured in the FBF-free medium for 24 hours and photographed.

7.2.7 CELL CYCLE BY FLOW CYTOMETRY

Cells with different treatment were harvested, fixed in 1% paraformaldehyde, washed with phosphate-buffered saline (PBS), and stained in 5 mg/ml propidium iodide in PBS supplemented with RNase A (Roche, Indianapolis, IN) for 30 minutes at room temperature. Data were collected using BD systems.

7.2.8 IMMUNOFLUORESCENCE

Cells were washed with PBS, fixed with 4% formaldehyde, permeabilized with 0.2% Triton X-100 at 37°C, and incubated in 5% BSA. Then cells were incubated with rabbit anti-human netrin-1 & UNC5B antibody (1:100) and mouse anti-human PKC α antibody (1:50) overnight at 4°C. Then fluorescently labeled goat anti-rabbit IgG (1:200) were added at 37°C for 1 h. Nucleus was stained with DAPI. Cells was then observed using fluorescence microscope.

7.2.9 STATISTICAL ANALYSIS

SPSS 13.0(SPSS Inc, Chicago, IL) was used for statistical analysis. The χ^2 test was used to evaluate the association between the expression of netrin-1 & UNC5B and clinicopathologic variables. Kaplan-Meier method and log-rank test were used for survival analysis. The t-test was used to analyze the difference for western blot data. p values < 0.05 was considered significant.

7.3 RESULTS

7.3.1 EXPRESSION OF NETRIN-1 AND UNC5B IN BLADDER CANCER TISSUES AND ASSOCIATION BETWEEN THEIR EXPRESSIONS & CLINICOPATHOLOGIC PARAMETERS

Quantitative real-time PCR (RT-PCR) and western blot analysis were used to evaluate netrin-1 & UNC5B expression in 120 BC tissues and 40 normal bladder epithelial tissues. It showed that the increased netrin-1 expression and decreased UNC5B expression could be detected in BC samples compared with the normal bladder samples (P<0.05). The mRNA expression of netrin-1 was found to be increased, while that of UNC5B decreased in the BC tissues as compared with the normal bladder epithelial tissues. The protein expression of netrin-1 and UNC5B showed the same trend as that of mRNA expression, and the optical density of all tumor (T) & normal (N) tissues were measured and expressed graphically (Figure 1). Differences of mRNA expression and protein levels in different T stage (T1, T2, T3 & T4) and histological grade (G1, G2 & G3) were significant (P<0.05) (Figure 2).

The expression of netrin-1 protein in BC and normal adjacent tissues was located in both cytoplasm and nucleus, while UNC5B protein appeared to be located only in cytoplasm (Figure 3). Elevated expression of netrin-1 and down-regulated level of UNC5B was observed in T4 tumors compared with normal adjacent tissues (P<0.01). To explore the relationship of netrin-1 over-expression and UNC5B down-regulation in a large cohort of BCs, we examined the correlation between the immunostaining of netrin-1 & UNC5B and clinic-pathological features including age, gender, tumor size, tumor grade, etc. There was a statistically significant positive correlation between UNC5B & netrin-1 expression and high grade, aggressive stage and metastasis (Tables 1 and 2), the expression of UNC5B was finally determined T/N<0.5 as low expression & T/N>=0.5 as normal expression and the expression of netrin-1 was considered T/N>2 as high expression & T/N<=2 as normal expression.

During follow-up period, 70.0% (21 of 30) of tumors with high netrin-1 expression developed metastasis compared with 5.6% (5 of 90) of tumors with low netrin-1 expression, (P<0.01). Meanwhile, 43.2% (19 of 44) of tumors with low UNC5B expression showed metastasis, compared with only 9.2% (7 of 76) of tumors with high UNC5B expression having metastasis (P<0.01). Therefore, high expression of netrin-1 and low expression of UNC5B were positively associated with metastasis of BC. Kaplan-Meier plots and log-rank tests showed that patients with high netrin-1 expression and low UNC5B expression in their tumor tissues had statistically significant shorter survival rate compared with those with low netrin-1 expression and high UNC5B expression (P<0.01).

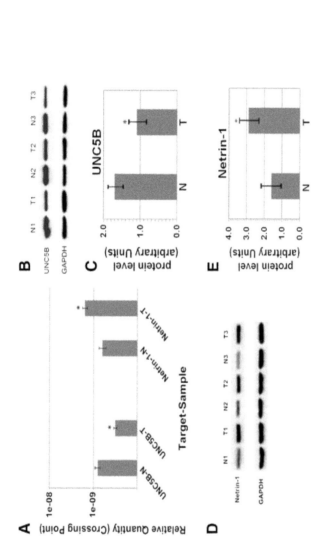

FIGURE 1: Expression by RT-PCR and western blot in BC tissues (T) and normal bladder tissues (N). (A) The average netrin-1 & UNC5B expression±SD for all studied tumors (120 cases) and their corresponding normal tissues (40 cases) by RT-PCR, bar graphs describe significant UNC5B down-regulation and netrin-1 up-regulation in T in comparison with N (P<0.05), β-actin as an internal control. (B) UNC5B protein was detected by western blot in 40 pair tissues and band intensities indicate UNC5B expression conspicuously lower in T comparing with N. GAPDH was used as a loading control. (C) The ratio between the optical densities of UNC5B & GAPDH in the same tissue was calculated and expressed graphically. Significant differences of UNC5B expression between T & N were analyzed statistically and UNC5B expression was obviously greater in N (P<0.05). (D) Netrin-1 protein was detected by western blot and increased in T. (E) The relative protein expression of netrin-1, GAPDH as an internal control (P<0.05).

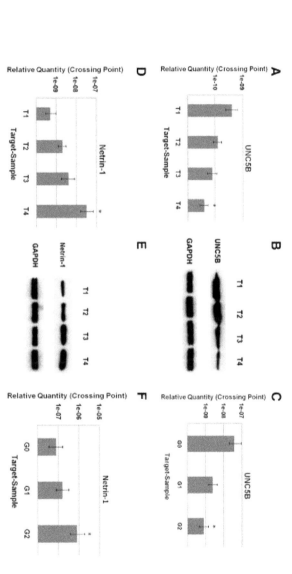

FIGURE 2: Expressions by RT-PCR and western blot in different stages and grades. (A) UNC5B expression was different in BC specimens with different clinical pathological stages (T1, T2, T3 & T4) of ten independent experiments. UNC5B expression in stage T4 was much less than in T1–T3 stages (P<0.05). *P<0.05 compared with stage T1 BC tissues (paired t-test). (B) Different expressions of UNC5B in pT stage were detected by western blot and UNC5B expression in stage T4 was sharply less than in T1–T3 stages (P<0.05). (C) UNC5B mRNA expression was determined by RT-PCR in different grades (G1, G2 & G3). Differences at G1 & G3 were significant. *P<0.05 compared with G1 BC tissues (paired t-test). (D) Netrin-1 expression was different in BC specimens with different pT stage at mRNA level by RT-PCR. Stage T4 was more than in T1–T3 stages (P<0.05). (E) Western blot was used to detect netrin-1 expression in different stages. The band intensities indicate netrin-1 expression more in stage T4. (F) Netrin-1 expression was determined by RT-PCR in different grades (G1, G2 & G3), and G3 was the highest (P<0.05). *P<0.05compared with G1 BC tissues (paired t-test). All data are representative of 3 individual experiments.

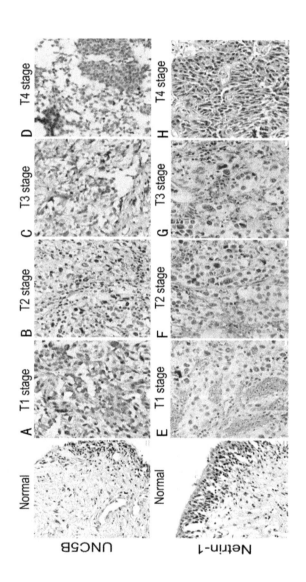

FIGURE 3: Representative images from immunohistochemical staining in different histological stages UNC5B was localized at cytoplasm and netrin-1 was mainly in cell nucleus and partly in cell cytoplasm of tumor tissues with granular brown staining. Almost all T1 tumor tissues showed strong UNC5B staining intensity. UNC5B expression in stage T4 was significantly less than in stage T1–T3 specimens (Figure 3A, B, C & D) (P= 0.021). While netrin-1 showed strong expression in stage T4 than other stages (T3, T2 & T1) and was associated significantly with pT stage (Figure 3E, F, G & H) (P= 0.013). Representative tissue sections with different immunointensity of UNC5B protein in A T1, B T2, C T3 & D T4. E Low level of netrin-1 expression in T1, H High level of netrin-1 expression in T4, F & G represent T2 &T3. Magnification ×400. *P<0.05 compared with stage normal bladder tissues (paired t-test).

However, there was no significant association between tumor recurrence and intense & feeble netrin-1 expression; recurrence curve analysis also indicated that the difference was not statistically significant with high & low UNC5B expression (P>0.01). Moreover, we found that patients with high netrin-1 expression and low UNC5B expression had statistically significant higher metastasis rate compared with those with low netrin-1 expression and high UNC5B expression (P<0.01; Figure 4). Log-rank analysis also showed that the expression of netrin-1 & UNC5B (P<0.01) were significant predictors of the metastasis of BC and had statistically significant independent association with poor prognosis of the patients.

TABLE 1: Relationship between the expression of UNC5B and clinicopathologic factors in BC patients

Factors	Group	UNC5B expression		X^2	P value
		High(%)	Low(%)		
Gender	Male	52(43.3%)	25(20.8%)	0.029	0.162
	female	29(24.2%)	14(11.7%)		
Age at surgery	<55	29(24.2%)	18(15.0%)	2.152	0.247
	≥55	44(36.6%)	29(24.2%)		
Histologic grade	G1	18(15.0%))	5(4.2%)	7.537	0.032
	G2	49(40.8%)	15(12.5%)		
	G3	17(14.2%)	16(13.3%)		
PT Stage	T1	29(24.2%)	6(5.0%)	19.564	0.014
	T2	23(19.2%)	12(10.0%)		
	T3	13(10.8%)	12(10.0%)		
	T4	6(5.0%)	19(15.8%)		
Tumor size(mm)	<10	16(13.3%)	5(4.2%)	2.011	0.436
	10-30	32(26.7%)	24(20.0%)		
	>30	18(15.0%)	25(20.8%)		
Multiplicity	Unifocal	28(23.3%)	24(20.0%)	1.593	0.312
	Multifocal	39(32.5%)	29(24.2%)		
Recurrence	Yes	39(32.5%)	43(35.8%)	31.573	0.220
	No	14(11.7%)	24(20.0%)		
Metastasis	Yes	7(5.8%)	19(15.8%)	13.253	0.001
	No	69(57.5%)	25(20.9%)		

TABLE 2: Relationship between the expression of netrin-1 and clinicopathologic factors in BC patients

Factors	Group	Netrin-1 expression		X^2	P value
		High(%)	Low(%)		
Gender	Male	49(40.9%)	28(23.3%)	0.037	0.213
	female	28(23.3%)	15(12.5%)		
Age at surgery	<55	24(20.0%)	23(19.2%)	1.984	0.198
	≥55	25(20.8%)	48(40.0%)		
Histologic grade	G1	3(2.5%)	20(16.7%)	7.632	0.024
	G2	38(31.6%)	26(21.7%)		
	G3	29(24.2%)	4(3.3%)		
PT Stage	T1	7(5.8%)	28(23.3%)	21.135	0.031
	T2	11(9.2%)	24(20.0%)		
	T3	16(13.4%)	9(7.5%)		
	T4	21(17.5%)	4(3.3%)		
Tumor size(mm)	<10	5(4.2%)	16(13.3%)	1.794	0.153
	10-30	32(26.7%)	24(20.0%)		
	>30	24(20.0%)	19(15.8%)		
Multiplicity	Unifocal	26(21.7%)	26(21.7%)	1.782	0.264
	Multifocal	32(26.6%)	36(30.0%)		
Recurrence	Yes	52(43.3%)	30(25.0%)	32.236	0.658
	No	24(20.0%)	14(11.7%)		
Metastasis	Yes	21(17.5%)	5(4.2%)	10.896	0.002
	No	9(7.5%)	85(70.8%)		

7.3.2 NETRIN-1 & UNC5B EXPRESSION AND LOCATION IN BC CELL LINES

Quantitative real-time PCR and western blot analysis were used to evaluate the expression of netrin-1 & UNC5B in human bladder cell lines SV, BIU-87, 5637 & T24, and immunofluorescence was used to detect of netrin-1 & UNC5B expression and localization. The results showed that highly invasive BC T24 cells had stronger netrin-1 expression than the superficial BC BIU-87 & 5637 cells and normal SV cells which had lowest expression. Opposite trend was observed regarding UNC5B expression (Figure

5), It was further confirmed that the expression of netrin-1 & UNC5B was positively correlated with BC grade. Quantitative real-time PCR and western blot analysis were also used to evaluate netrin-1 & UNC5B's expression after PMA (PKC α agonist) and calphostin C (PKC α inhibitor) treatment. The results showed that netrin-1 expression were significantly inhibited by calphostin C and enhanced by PMA (treat for 24 h), while UNC5B showed the opposite trend (Figure 6). Immunofluorescence results showed that UNC5B was expressed in BC cell cytoplasm in all these four cell lines, while netrin-1 was found mainly located in cell nucleus and partly in cell cytoplasm (Figure 7).

7.3.3 BC CELLS TREATED WITH PKC α AGONIST & INHIBITOR & SIRNA

PKC α inhibitor and agonist, PMA & calphostin C, were applied to treat BC BIU cells. We found that cell proliferative and invasive activities were significantly increased after PMA treatment, but decreased by calphostin C treatment (Figure 8). Moreover, the FCS showed that cell cycle was accelerated by PMA treatment (S phase 22.33% for 24 h, 36.41% for 48 h; G2/M phase 23.39% for 24 h, 34.42% for 48 h) and blocked by calphostin C treatment at both S phase and G2/M phase (S phase 8.39% for 24 h, 4.92% for 48 h; G2/M phase 10.55% for 24 h, 7.46% for 48 h) compared with BIU cell without drugs (S phase 13.14%; G2/M phase 16.72%) (Figure 9); the cells mainly concentrated in G1/G2 phase (almost the same percentage at 57.19%) from mitotic completion to DNA replication.

Migration of bladder cancer cells by wound healing. Cells were seeded at 1.0×10^6 cells/well in 6-well plates. After grown to confluence, the cell monolayer in each well was scraped with a pipette tip to create a scratch. Cells were washed by PBS three times and cultured in the FBS-free medium. Cells were photographed after 24 h and the scratch area was measured using Image software (Figure 10). PKC siRNAs were transfected into bladder cancer cells T24 & BIU-87 transiently. Real-time PCR showed that, netrin-1 expression was elevated after transfection with PKC siRNA, while UNC5B expression was decreased (Figure 11). The immunofluorescence confirmed the co-localization of PKC α and UNC5B, suggesting that the presence of their endogenous binding (Figure 12).

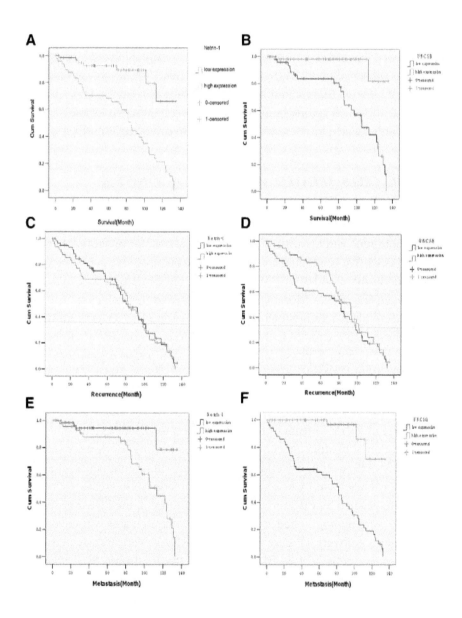

FIGURE 4: Survival, recurrence and metastasis curve analysis by the Kaplan-Meier method. (A) Patients with intense netrin-1 expression had significantly shorter median survival time (76.624 months) than those with weak netrin-1 expression (117.981 months) through log-rank univariate analysis (n=120, P<0.01). (B) Patients with weak UNC5B expression had significantly shorter median survival time (96.881 months) than those with intense UNC5B expression (128.939 months) (P<0.01). (C) Recurrence curve analysis indicated that the difference was not statistically significant with intense netrin-1 expression (74.463 months) and lower expression (79.505 months). (P>0.01) (D) The difference was not statistically significant in more (84.47 months) and less (69.225 months) UNC5B expression by recurrence curve analysis. (P>0.01) (E) Patients with intense netrin-1 expression had significantly shorter median metastasis time (100.836 months) than those with weak netrin-1 expression (124.946 months). (P<0.01) (F) Patients with weak UNC5B expression had significantly shorter median metastasis time (71.243 months) than those with intense UNC5B expression (125.957 months). (P<0.01)

7.4 DISCUSSION AND CONCLUSIONS

Protein Kinase C (PKC), as the hub of a variety of signal transduction process, is not only involved in cell communication, secretion, cell differentiation & proliferation, but more importantly involved in tumor cell apoptosis and differentiation. PKC α is a classical Protein of Kinase C isoforms. Our and others' research have shown that PKC α of high activation status is closely related to activation and apoptosis of bladder cancer recurrence [3]. UNC5B is abnormally expressed and associated with a highly malignant, chemotherapy-related and poor prognosis in colon cancer. It was reported that netrin-1 binding to its receptor can activate PKC α and lead to tumor cell proliferation, but it did not clarify PKC α and netrin-1/UNC5B's regulatory mechanisms. To this end, we explored the mechanism of PKC α with netrin-1/UNC5B in bladder cancer. Our work shows that, PKC α, netrin-1 & UNC5B is closely related to the degree of malignancy and progress in bladder cancer and found PKC α promoted the survival of bladder cancer cell potentially through netrin-1/UNC5B signaling pathway. Thus, PKC α has an important influence on netrin-1/UNC5B signaling pathway & bladder cancer's occurrence and development.

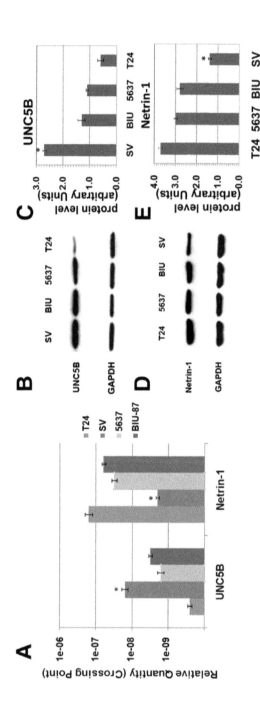

FIGURE 5: Expressions by RT-PCR and western blot in bladder cell lines. (A) Netrin-1 & UNC5B mRNA expression in four bladder cell lines (T24, SV, 5637, and BIU-87) was assayed using quantitative real-time PCR. To normalize Netrin-1 & UNC5B expression, β-actin was used as internal control. Each expression level was shown as mean + SD. UNC5B expression was significantly greater in SV cells than in 5637, T24 or BIU-87, while netrin-1 was lower in SV than in BC cell lines. *P<0.05 compared with BC cell lines including 5637, T24 and BIU-87 (paired t-test). (B) UNC5B &(D) Netrin-1 expression at protein level in different cell lines (SV, BIU, 5637 and T24) were analyzed by western blots with 50 μg protein extracts. GAPDH was selected to be endogenous control. The differences in UNC5B expression between 5637, T24 and BIU-87 were significant. SV cells represent bladder normal cell lines of which UNC5B expression is the most, the others are BC cell lines, and netrin-1 expresses the least in SV. (C) & (E) The ratio between the optical density of UNC5B & netrin-1 and GAPDH of the same tissue was calculated and expressed graphically (P<0.05). *P<0.05 compared with BC cell lines (paired t-test).

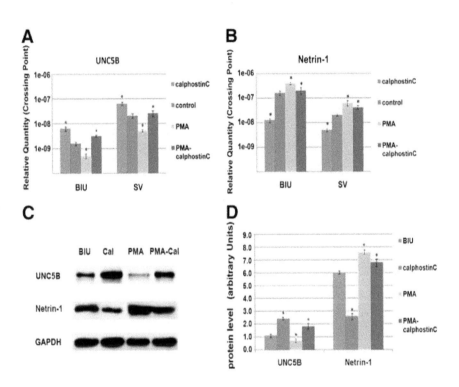

FIGURE 6: Expressions by RT-PCR and western blot in bladder cells with drugs. (A) & (B) Quantitative real-time PCR (RT-PCR) were used to evaluate netrin-1 & UNC5B's expression in treatment of cells (BIU & SV) with drugs- PMA (PKC α agonist) and calphostin C (PKC α inhibitor). The results showed that the cells were significantly inhibited by calphostin C (24 h) in netrin-1's expression and promoted by PMA (24 h), and cells were first promoted, and then inhibited by PMA (4 h) then calphostin C (24 h). Whlie UNC5B was opposite. (C) Confirmation of netrin-1 down-regulation with calphostin C and up-regulation with PMA by western blot. (D) The relative protein expression of netrin-1 & UNC5B in treatment of cells with drugs, GAPDH as an internal control (P<0.05). *P<0.05 compared with BC cell line BIU (paired t-test).

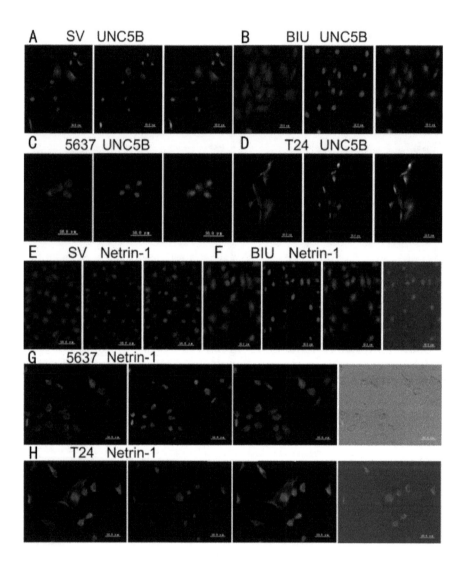

FIGURE 7: Detection of UNC5B expression by immunefluorescence in four cell lines (SV, BIU, 5637 & T24) (A & E) SV, (B & F) BIU, (C & G) 5637, (D & H) T24. (A, B, C & D) The expression of UNC5B located in cytoplasm. The others were cell nucleolus or transmission light figures. From these figures we can see cells' form clearly. UNC5B location was almost the same in SV, 5637, T24 & BIU-87 bladder cancer cells lines. (E, F, G & H) The localization in four cell lines (SV, BIU, 5637 & T24) of netrin-1 was found by immunofluorescence mainly in cell nucleus and partly in cell cytoplasm.

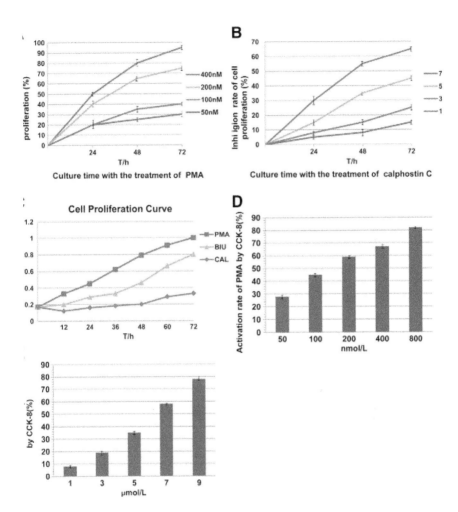

FIGURE 8: Proliferation and invasion ability of BIU were inhibited by calphostin C and promoted by PMA. (A) CCK-8 assay was used to examine BIU cell proliferation. cell proliferation was inhibited on dose-dependent correlation with the increasing concentration (1, 3, 5, 7 μmol/L) of calphostin C at 24 h, 48 h, 72 h. (B) Cell proliferation was promoted by PMA on correlation with the increasing concentration (50, 100, 200, 400 nmol/L) at 24 h, 48 h, 72 h. (C) According A450 of BIU at different times, cell proliferation curve was drawn. Compared with BIU, cells with calphostin C were inhibited and cells with PMA were promoted. (D) & (E) According the optical density value (A) of each well measured, BIU cell growth inhibition/activation ratio was calculated as (1 − A450 of experimental well/A450 of blank control well) × 100%, with different concentration of PMA (50, 100, 200, 400 nmol/L) and calphostin C (1, 3, 5, 7 μmol/L), we calculated the activation/inhibition ratio. Each data are representative of 3 individual experiments. IC50 of calphostin C =7.4 μmol/L, and IC50 of PMA=24 nmol/L.

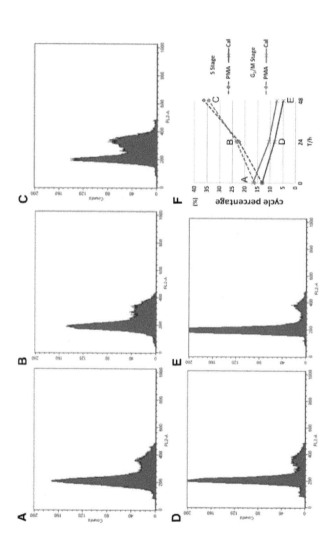

FIGURE 9: Detection of cell cycle by flow cytometry. (A) Flow cytometry was used to examine BIU cell cycle. G1 / G2, S, G2/M phase occupied 57.19%, 13.14%, 16.72%, respectively. (B) & (C) With the increasing treatment time (24 h, 48 h) of PMA, the number of BIU cells increased gradually in S phase and G2/M phase, and S phase was 22.33% for 24 h, 36.41% for 48 h; G2/M phase was 23.39% for 24 h, 34.42% for 48 h. (D) & (E) With the increasing treatment time (24 h, 48 h) of calphostin C, the number of BIU cells decreased noticeably in S phase and G2/M phase, and S phase was 8.39% for 24 h, 4.92% for 48 h; G2/M phase was 10.55% for 24 h, 7.46% for 48 h. The data are representative of 3 individual experiments. (F) Summary of cell cycle distribution in panel A-E.

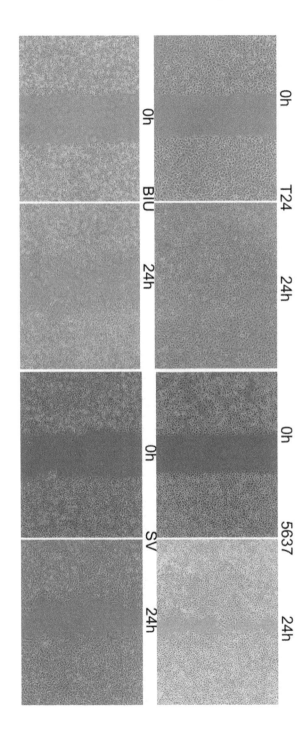

FIGURE 10: Migration and invasion assays were used by wound healing in BC cells (SV, 5637, BIU, T24). Comparing scratches width in order to verify the invasion capability of the prostate carcinoma cells. Results showed that the most invasive cells were T24 and all BC cells were invasive.

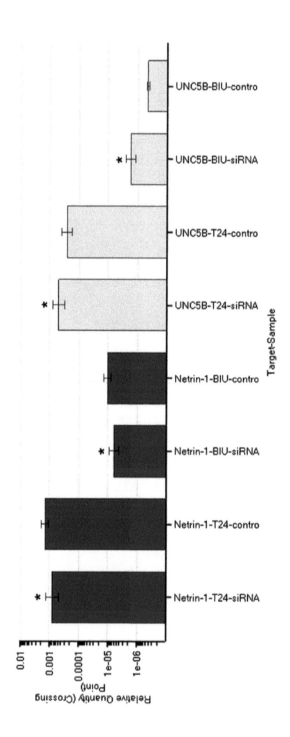

FIGURE 11: PKC siRNAs were transfected into bladder cancer cells T24 & BIU-87 transiently. Real-time PCR results found after transfection PKC siRNA, netrin-1 was inhibited and the expression was decreased, while UNC5B was activated and the expression was elevated.

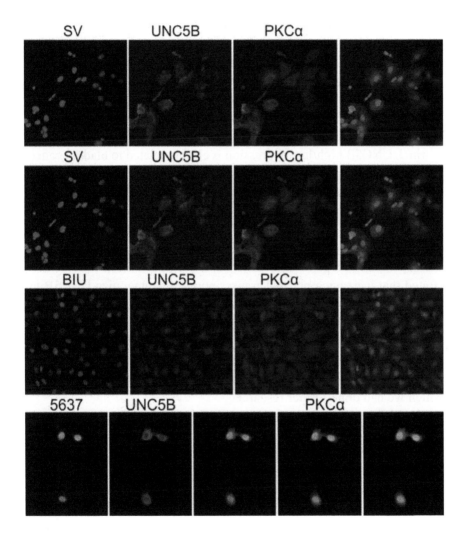

FIGURE 12: The confocal immunofluorescence displayed that the existence of co-localization expression of PKC α and UNC5B. Results showed that PKC α and UNC5B were observed and a co-localization expression of PKC α and UNC5B could be observed. Nuclear staining was performed with DAPI in panels.

The expressions of netrin-1/UNC5B were detected in bladder cancer tissues & adjacent tissues and the relevance and the relationship with clinic pathological parameters was analyzed. The results showed that UNC5B had higher expression in adjacent tissues than bladder cancer tissues and it had higher expression in the low-level cancer tissues than in high-level ones, but netrin-1 in the opposite. According to immuno-histochemical results, it showed UNC5B expression in the cytoplasm and netrin-1 existing in the cytoplasm and nucleus; netrin-1's expression gradually increased from the bladder mucosa-transitional cell carcinoma and high-grade cancer evolution, while UNC5B is gradually reduced; netrin-1/UNC5B high/low expression is closely related to bladder cancer clinical grading, staging & metastasis; and Pearson correlation analysis showed that netrin-1 and UNC5B are negatively correlated. Netrin-1/ UNC5B's expression is proved to exist in kidney cancer and prostate cancer [13,17], and found that netrin-1 inhibits apoptosis in lung cancer, with advanced neuroblastoma, breast cancer [10-12]; UNC5B is one of the dependent receptors of netrin-1, and previous studies had demon-strated that increasing netrin-1 transcription can antagonize UNC5B in-duced apoptosis [15], which is consistent with the results of this study. Previously we have confirmed that PKC α is closely related to bladder cancer cell's apoptosis & recurrence [3], and that netrin-1's binding to its receptor UNC5B can cause PKC α phosphorylation and promote cancer cell proliferation [9], but it had not been confirmed in bladder cancer, for which we had done further research.

From the cellular level, it revealed netrin-1/UNC5B's expression & location in bladder carcinoma. Four kinds of bladder cancer cell line T24, BIU-87, 5637 and SV malignancy has been clearly stated in previous stud-ies: BIU-87, 5637, T24 are all bladder carcinoma cells, and their degree of malignancy increased in turn, and SV-HUC-1 is normal urothelial line [18]. We detected netrin-1/UNC5B expression in bladder cancer cell line from RNA and protein levels by Real-time PCR & Western-blot ion, UN-C5B' expression was the highest in normal bladder cell line (SV), and the expression was the lowest in the most malignant cells of T24, netrin-1 was the opposite. Immunofluorescence results showed that UNC5B was in bladder cancer pulp while expressions of netrin-1 existed in the cytoplasm and the nucleus. Netrin-1/UNC5B's expression in cells and tissues shows consistent trend, and are related with the degree of malignancy of bladder cancer cell lines. PKC α has been shown to be involved in tumor cell apoptosis and differentiation. The high expression of PKC α in bladder cancer cells was found to promote cancer cell proliferation, and inhibit apoptosis and differentiation [3].

When bladder cancer cell was given PKC inhibitors and activators, and detected changes of netrin-1/UNC5B expression and bladder cell cycle, proliferation and apoptosis; it can be further confirmed that netrin-1/UNC5B are closely related with PKC α activation. When bladder cancer cell BIU-87 was given PKC inhibitors (calphoatin C) and activators (PMA), Real-time PCR & Western-blot showed that netrin-1 was inhibited after inhibitor treatment, while UNC5B was activated; netrin-1 was activated after PMA treatment, while UNC5B was suppressed. When CCK-8 and flow cytometry detection was carried out after drug treatment on bladder cancer cycle, proliferation and apoptosis. CCK-8 was found in best status by calphoatin C or PMA for 48 hours, and the inhibition rate & the activation rate increased with the increasing concentration, and at the same time it can be drawn that calphostin C of IC50 = 7.4 μmol/L, PMA's IC50 = 24 nmol/L. Flow cytometry showed S and G2/M were inhibited or activated after calphoatin C or PMA treatment BIU in better condition after 48 hours. These results could confirm that netrin-1/UNC5B was closely associated with PKC α activation, and PKC α activation or inhibition might affect the proliferation and survival of cancer cells [4,6,7].

After transiently transfecting PKC siRNA into the bladder cancer T24 and BIU-87 cells, it clarified PKC α's regulatory mechanismson on netrin-1/UNC5B; Real-time PCR test results showed that netrin-1 was inhibited after PKC siRNA transfection, with its expression decreased, while UNC5B increased. Immunofluorescence results revealed the presence of co-localization of PKC α with UNC5B expression. So we speculate that there may be endogenous binding.

From the above results, we can conclude that: PKC α can promote bladder cancer cell proliferation through the regulation of netrin-1/UNC5B. On this basis, we can intervene any stage in which PKC α and netrin-1/UNC5B affect, so as to control the proliferation of bladder cancer, and provide adequate theoretical basis for bladder cancer's diagnosis and treatment.

REFERENCES

1. Liou LS: Urothelial cancer biomarkers for detection and surveillance. Urology 2006, 67(3 Suppl 1):25-33.
2. Smaldone MC, Jacobs BL, Smaldone AM, Hrebinko RL Jr: Long-term results of selective partial cystectomy for invasive urothelial bladder carcinoma. Urology 2008, 72(3):613-616.
3. Kong CZ, Zhu YY, Liu DH, Yu M, Li S, Li ZL, Sun ZX, Liu GF: Role of protein kinase C-alpha in superficial bladder carcinoma recurrence. Urology 2005, 65(6):1228-1232.
4. Wu B, Zhou H, Hu L, Mu Y, Wu Y: Involvement of PKC α activation in TF/VIIa/ PAR2-induced proliferation, migration, and survival of colon cancer cell SW620. Tumour Biol 2013, 34(2):837-846.
5. Kyuno D, Kojima T, Yamaguchi H, Ito T, Kimura Y, Imamura M, Takasawa A, Murata M, Tanaka S, Hirata K, Sawada N: Protein kinase Cα inhibitor protects against down regulation of claudin-1 during epithelial-mesenchymal transition of pancreatic cancer. Carcinogenesis 2013, 34(6):1232-1243.
6. Lee SK, Shehzad A, Jung JC, Sonn JK, Lee JT, Park JW, Lee YS: Protein kinase Cα protects against multidrug resistance in human colon cancer cells. Mol Cells 2012, 34(1):61-69.
7. Gwak J, Lee JH, Chung YH, Song GY, Oh S: Small molecule-based promotion of PKC α-mediated β-catenin degradation suppresses the proliferation of CRT-positive cancer cells. PLoS One 2012, 7(10):e46697.
8. Ko SY, Dass CR, Nurgali K: Netrin-1 in the developing enteric nervous system and colorectal cancer. Trends Mol Med 2012, 18(9):544-554.
9. Bartoe JL, McKenna WL, Quan TK, Stafford BK, Moore JA, Xia J, Takamiya K, Huganir RL, Hinck L: Protein interacting with C-kinase1/protein kinase Calpha-mediated endocytosis converts netrin-1-mediated repulsion to attraction. J Neurosci 2006, 26(12):3192-3205.
10. Delloye-Bourgeois C, Brambilla E, Coissieux MM, Guenebeaud C, Pedeux R, Firlej V, Cabon F, Brambilla C, Mehlen P, Bernet A: Interference with netrin-1 and tumor cell death in Non–small cell lung cancer. J Natl Cancer Inst 2009, 101(4):237-247.
11. Delloye-Bourgeois C, Fitamant J, Paradisi A, Cappellen D, Douc-Rasy S, Raquin MA, Stupack D, Nakagawara A, Rousseau R, Combaret V, Puisieux A, Valteau-Couanet D, Bénard J, Bernet A, Mehlen P: Netrin-1 acts as a survival factor for aggressive neuroblastoma. J Exp Med 2009, 206(4):833-847.
12. Fitamant J, Guenebeaud C, Coissieux MM, Guix C, Treilleux I, Scoazec JY, Bachelot T, Bernet A, Mehlen P: Netrin-1 expression confers a selective advantage for tumor cell survival in metastatic breast cancer. PNAS 2008, 105(12):4850-4855.
13. Kong CZ, Liu J, Liu L, Zhang Z, Guo KF: Interactional expression of netrin-1 and its dependence receptor UNC5B in prostate carcinoma. Tumour Biol 2013, 34(5):2765-2772.
14. He K, Jang SW, Joshi J, Yoo MH, Ye K: Akt-phosphorylated PIKE-A inhibits UN-C5B- induced apoptosis in cancer cell lines in a p53-dependent manner. Mol Biol Cell 2011, 22(11):1943-1954.

15. Paradisi A, Maisse C, Bernet A, Coissieux MM, Maccarrone M, Scoazec JY, Mehlen P: NF-kappaB regulates netrin-1 expression and affects the conditional tumor suppressive activity of the netrin-1 receptors. Gastroenterology 2008, 135(4):1248-1257.

16. Thomsen R, Sølvsten CA, Linnet TE, Blechingberg J, Nielsen AL: Analysis of qPCR data by converting exponentially related Ct values into linearly related X0 values. J Bioinform Comput Biol 2010, 8(5):885-900.

17. Zhan B, Kong C, Guo K, Zhang Z: PKC α is involved in the progression of kidney carcinoma through regulating netrin-1/UNC5B signaling pathway. Tumour Biol 2013, 34(3):1759-1766.

18. Gai JW, Wahafu W, Hsieh YC, Liu M, Zhang L, Li SW, Zhang B, He Q, Guo H, Jin J: Inhibition of presenilins attenuates proliferation and invasion in bladder cancer cells through multiple pathways. Urol Oncol 2013. *In press*

CHAPTER 8

SIGNALING PATHWAYS IN THE DEVELOPMENT OF INFANTILE HEMANGIOMA

YI JI, SIYUAN CHEN, KAI LI, LI LI, CHANG XU, AND BO XIANG

8.1 BACKGROUND

Infantile hemangioma (IH) is a common disorder in infancy, with an estimated prevalence of 5 to 10%. If left untreated, these tumors are characterized by a rapid growth phase during the first year of life, followed by slow involution, which may continue until the age of 10–12 years (Figure 1) [1,2]. However, some IHs will leave residual changes, such as telangiectasias, fibro-fatty tissue, scars, excessive atrophic skin and pigment changes. In 10% of cases, IHs grow dramatically and destroy tissue, impair function or even threaten life [3]. The standard treatment options for IH include corticosteroids or surgical excision, and the options in life- or sight-threatening cases include treatment with vincristine, interferon or cyclophosphamide. Unfortunately, none of these therapeutic modalities are ideal due to restrictions or potential serious side effects [4-7]. β-blockers have recently been introduced as a safe and effective treatment for IH [8-11].

This chapter was originally published under the Creative Commons Attribution License. Ji Y, Chen S, Li K, Li L, Xu C, and Xiang B. Signaling Pathways in the Development of Infantile Hemangioma. Journal of Hematology & Oncology 7,13 (2014). doi:10.1186/1756-8722-7-13.

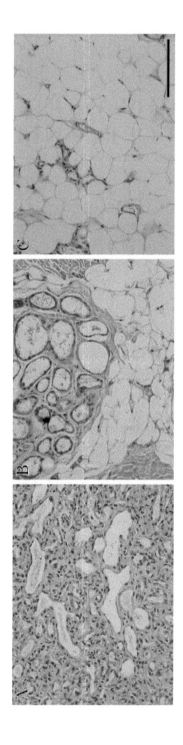

FIGURE 1: Hematoxylin and eosin (H&E) stained sections of proliferating, involuting and involuted phases of IH. The proliferating phase is characterized by densely packed tumor cells that form immature vessels (A). In the involuting phase, disorganized vasculature consists of flat endothelium and pericytes (B). The tumor is replaced by fat and/or connective tissues in the involuted phase (C). Scale bar=100 μm.

However, their use is not without risk, and not all tumors respond to these drugs [12,13]. These issues have spurred extensive research to clarify the signaling pathways implicated in hemangioma neovascularization in the hope that a greater understanding of its molecular pathogenesis will reveal new strategies to tackle IH.

TABLE 1: Cellular components isolated from IH

Cell type	Abbreviation	Cell marker	Characteristics
Hemangioma-derived endothelial cell	HemEC	CD31/PECAM-1, vWF, E-selectin, VEGFR-2, Tie-2 and VE-cadherin	Immature endothelial cells; Clonal expansion; Increased proliferation, migration, tumor formation and survival ability.
Hemangioma-derived endothelial progenitor cell	HemPEC	CD133*, VEGFR-2, CD34, CD31, CD146, VE-cadherin and vWF	Immature endothelial cells; Increased adhesion, migration and proliferation in the presence of endostatin or VEGF.
Hemangioma-derived mesenchymal stem cell	Hem-MSC	SH2(CD105), SH3, SH4, CD90, CD29, α-SMA and CD133	Multilineage differentiation: adipogenic, osteoblastic and myoblastic differentiation
Hemangioma-derived stem cell	HemSC	CD90, CD133, VEGFR-1, VEGFR-2, neuroplin-1 and CD146	Multilineage differentiation: ECs, neuronal cells, adipocytes, osteocytes and chondrocytes; Form hemangioma-like Glut-1+ blood vessels in nude mice.
Hemangioma-derived pericyte	Hem-pericyte	PDGFR-β, neural glial antigen-2, desmin, calponin, smooth muscle 22α, smooth muscle α-actin, α-SMA, smooth muscle myosin heavy chain and CD90	Increased proliferation ability; Reduced contractility; Diminished ability to stabilize blood vessels in IH.

CD133, a pentaspan membrane protein, is used as a stem cell biomarker for the isolation of progenitor/stem-like cells from IH tissues. CD133 is also responsible for self-renewal, tumorigenesis, metabolism, differentiation, autophagy, apoptosis and regeneration [23]. However, little is known about its biological functions in the development of IH.

The initial histochemical work of Mulliken and Glowacki [14], examining endothelial cell (EC) morphology, shed light on the cellular components of IH. In the past decade, hemangioma-derived progenitor/stem cells (HemSCs), mesenchymal stem cells (Hem-MSCs), endothelial progenitor cells (HemEPCs), ECs (HemECs) and perivascular cells (Hem-pericytes), all of which comprise the IH, have been isolated (Table 1) [15-18]. In general, CD133 was used as a stem cell biomarker for the isolation of HemSCs from IH tissues. HemEPCs were purified from HemSCs based on expression of the EC marker CD31. In contrast, Hem-MSCs didn't express CD31 or CD34. In IH tissues, CD133 expression was found to be located in both perivascular region and endothelium [19]. Therefore, HemSCs may contain both of Hem-MSCs and HemEPCs. Studies from different groups have demonstrated that HemSCs have the ability to self-renew and can differentiate into endothelium, adipocytes and pericytes in vitro [15,20]. When implanted subcutaneously into nude mice, HemSCs can produce human glucose transporter-1 (GLUT-1) positive microvessels at 7–14 days [15,20-22].

We now recognize that IH may be not only a disorder of angiogenesis (i.e., the sprouting of new vessels from existing ones) but also – at least in part—a disorder of vasculogenesis (i.e., the de novo formation of new blood vessels from stem cells) [20,24,25]. Improved knowledge of the signaling pathways that regulate angiogenesis and vasculogenesis has led to the identification of several possible therapeutic targets that have driven the development of molecularly targeted therapies. Because many of the signaling pathways are implicated in the pathogenesis of various tumor types, insight gained from these studies will enable the development of target-specific drugs, not only for IH but also for malignant vascular tumors. This review will highlight the most important of these findings. Although the signaling pathways involved in the development of IH are described separately below, there are numerous interactions among them, indirectly reflecting the complexity of IH pathogenesis.

8.1.1 VEGF/VEGFR PATHWAY

The human vascular endothelial growth factor (VEGF) family consists of VEGF-A, VEGF-B, VEGF-C, VEGF-D and placental growth factor (PIGF). These growth factors play pivotal roles in embryonic development and angiogenesis-dependent disease [26]. Many reports have confirmed that excessive VEGF expression in IH tissue parallels the proliferating phase of its growth. Conversely, in the involuting phase, VEGF expression rapidly decreases, and many angiogenesis inhibitors become prominent [21,27,28].

The functions of the different VEGF family members are determined by their receptor specificity. Two receptors for VEGF are members of the tyrosine-kinase family and conserved in ECs. These VEGF receptors (VEGFR) are VEGFR-Flt-1 (VEGFR-1) and VEGFR-Flk-1/KDR (VEGFR-2). VEGFR-1 and VEGFR-2 are located on ECs, bone-marrow derived hematopoietic cells and tumor cells, etc. [26,29]. The expression of these receptors is low in normal tissues and only upregulated during the development of those pathological states when neovascularization occurs [30]. Another receptor, VEGFR-3, is primarily expressed in lymph nodes and tumor blood vessels [31,32]. Neuropilin-1 and neuropilin-2 were discovered as coreceptors that that enhanced the binding and effectiveness of the VEGF stimulation of their receptors [33].

8.1.2 UPREGULATED AUTOCRINE VEGF-A/VEGFR-2 LOOP IN HEMEC

One of the most intensely studied factors involved in angiogenesis and vasculogenesis is VEGF-A. VEGFR-2 is known to mediate the majority of the downstream angiogenic effects of VEGF-A, including microvascular permeability, EC proliferation, migration and survival [34]. Upon the activation of VEGFR-2 in ECs, three major secondary messenger pathways trigger multiple downstream signals that promote angiogenesis. These pathways are the following: the mitogen-activated protein/ERK kinase (MEK)/extracellular signal-regulated kinase (ERK) cascade, the phosphatidylinositol-3 kinase (PI3K)/serine-threonine protein kinase/Akt cascade and the phospholipase C-γ/intracellular Ca^{2+}/(protein kinase C (PKC) cascade [35,36]. The genetic deletion of VEGF-A or its primary signaling receptor VEGFR-2 results in early embryonic lethality, associated with a near-complete block of hematopoietic and vascular development [37].

High-VEGFR-2 cells are well documented to exhibit a higher capacity of self-renewal and superior growth in vitro and in vivo compared with a low-VEGFR-2 cell population [33,38]. HemECs demonstrate the phenotype of a constitutively active autocrine VEGF-A/VEGFR-2 loop (Figure 2), which renders the cells more sensitive to paracrine/external stimulation by VEGF-A and results in the increased proliferation and migration of cells and tumor formation [39,40]. These characteristics likely result from the genetic instability of HemECs as somatic missense mutations in the kinase insert of the VEGFR-2 gene have been found in IHs [41]. In addition, imbalances in gene expression have been reported in mesenchymal compartments compared to normal tissues, suggesting a possible reciprocal interaction between HemECs and the surrounding cells [42-44]. Alternatively, HemECs may originate from progenitor/stem-like cells, which are known to display robust proliferative and clonogenic capabilities and to express high levels of VEGF-A [15,21]. The expression level of VEGF is also increased in HemECs, although this increase is not dramatic [40]. Finally, COSMC was reported to be overexpressed in proliferating IHs, with an association with the enhanced VEGF-mediated phosphorylation of VEGFR-2 and its downstream signaling [45].

The abnormal activation of VEGFR-2 on the cell surface may also be beneficial to the survival of HemECs as VEGF-A plays a critical role in protecting ECs against apoptotic cell death [46,47]. In addition, this inhibition of EC apoptosis can improve angiogenesis and vasculogenesis in patients with ischemia [30]. We recently indicated that maintaining Bcl-2 expression via VEGF-A/VEGFR-2 signaling in primary HemECs blocked the cells from apoptotic death in the absence of external VEGF-A. Moreover, the inactivation of PI3K/Akt suppressed the VEGF-A/VEGFR-2-mediated anti-apoptotic effect and unleashed the inhibitory effect of VEGF-A/VEGFR-2 signaling over the reduction of Bcl-2 expression, thereby amplifying the activation of the caspase cascade [48]. These findings suggest that HemECs may be able to adapt to the abnormal physical environment of the tumor by undergoing a form of reprogramming that involves an increase in apoptosis resistance and by up-regulating a VEGF autocrine survival feedback loop to sustain these effects and stabilize the aberrant phenotype.

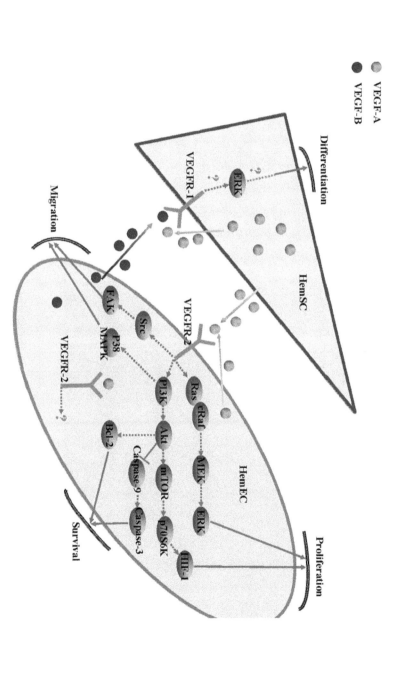

FIGURE 2: The VEGF signaling pathway in HemECs and HemSCs. Upon ligand binding, VEGF receptors dimerize, leading to the phosphorylation of different tyrosine residues. Phosphorylation in turn elicits differential downstream signaling events.

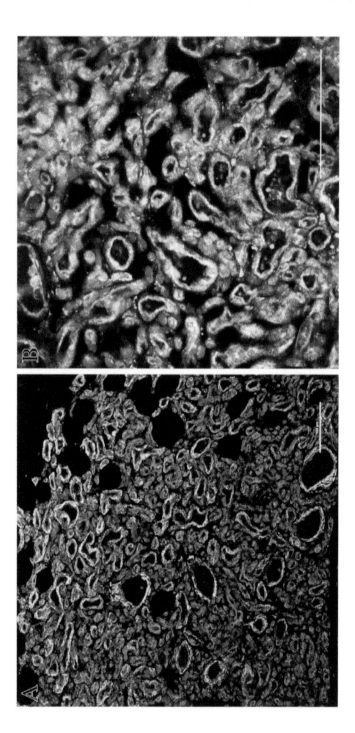

FIGURE 3: Double immunofluorescence staining of IH tissues. (A), Proliferating phase IH tumor section stained for endothelial maker CD31 (red), smooth muscle marker α-SMA and nuclei (laser fluorescent confocal microscopy). (B), Proliferating phase IH tumor section stained for CD31, VEGF-A and nuclei (fluorescent microscopy). The nuclei are stained with DAPI. Scale bars are 100 μm.

Moreover, recent research efforts revealed that pericytes, in addition to producing VEGF-A that acts in a paracrine fashion (Figure 3), can stimulate the autocrine expression of VEGF-A by tumor ECs, both of which could lead to a general suppression of EC apoptosis [49]. Interestingly, HemSCs and Hem-pericytes also secrete high levels of the angiogenic VEGF-A [21,42]. Thus, various combinations of strategies, including the development of novel potent tyrosine kinase inhibitors against VEGFR-2 and potential to abrogate its downstream pathways, can be investigated to achieve synergistic effects on HemEC apoptosis and therefore on hemangioma regression.

8.1.3 DISCREPANCY OF VEGFR-1 SIGNALING IN HEMSCS AND HEMECS

In contrast to other VEGFR genes, VEGFR-1 expresses two types of mRNA, one for a full-length receptor and another for a soluble short protein known as soluble VEGFR-1 (sFlt-1). The binding-affinity of VEG-FR-1 for VEGF-A is one order of magnitude higher than that of VEGFR-2, whereas the kinase activity of VEGFR-1 is about ten-fold weaker than that of VEGFR-2. Therefore, VEGFR-1 is considered a negative regulator of angiogenesis and vasculogenesis during development [50]. VEGFR-1$^{-/-}$ mice show an overabundance of blood vessels and overgrowth of immature ECs, similarly to those features observed in IH [51].

A reduction of VEGFR-1 expression has been implicated in the proliferation of infantile HemECs and tissue [40,52]. The mechanism for this low expression in HemECs was shown to be the sequestration of β1-inergrin in a multiprotein complex composed of tumor endothelial marker-8 (TEM8) and VEGFR-2, which inhibits nuclear factor in activated T cells (NFAT)-mediated VEGFR-1 transcription [40]. However, VEGFR-1 is relatively over-expressed in HemSCs [53]. The involvement of VEGFR-1 in the progression of IH could involve at least one mechanism: the activation of HemSC function, with a subsequent increase in vasculogenesis. VEGF-A, either endogenous or exogenous, significantly induces VEGFR-1-mediated ERK1/2 phosphorylation in HemSCs and promotes the differentiation of HemSCs to HemECs (Figure 2). Moreover, VEGF-B, which is the specific ligand for VEGFR-1, is highly expressed in HemECs and induces similar effects [53]. These results clearly indicate that not only the paracrine function of VEGF-B from HemECs but also the persistent autocrine signaling through the VEGF-A/VEGFR-1 loop in HemSCs contributes to enhanced IH vasculogenesis in general.

8.1.4 NOTCH PATHWAY

The Notch pathway is a conserved ligand-receptor signaling mechanism that modulates cell fate and differentiation. The interaction of Notch receptors (Notch 1 to 4) with their ligands (Delta-like 1, -3, -4, Jagged-1 and -2) leads to the cleavage of the transmembrane Notch receptor, giving rise to the Notch intracellular domain (NICD) that migrates into the nucleus. In the nucleus, the NICD associates with a transcription factor, recombination signal binding protein for immunoglobulin kappa J (RBP-Jk), and activates transcription from the RBP-Jk DNA binding site. The NICD-RBP-Jk complex upregulates the expression of primary target genes of Notch signaling, such as hairy and enhancer of split (HES) and HES-related protein (HERP/HEY) family of transcription factors [54,55].

8.1.5 NOTCH EXPRESSION IN IH

Although the expression levels of the Notch components are likely dynamic during development, making transient expression difficult to detect, current data suggest that many known Notch components, mainly two ligands (Delta-like-4 and Jagged-1), three receptors (North-1, -3 and -4) and four effectors (HES-1, HEY1, HEY2 and HEYL) are involved in the pathogenesis of IH. Both Jagged-1 and Notch-4 are increased in proliferating IHs. All transcript levels of Notch-1, Notch-3, Notch-4, Jagged-1 and Delta-like-4 (Dll4) were higher in the IH than in the placenta (a commonly used tissue for comparisons). Conversely, Notch-2 is strongly decreased in both proliferating and involuting IHs [44,56].

8.1.6 NOTCH SIGNALING TRIGGERS CELL-CELL INTERACTIONS IN IH

Notch signaling is initiated when the extracellular domain of the receptor engages ligands found on neighboring cells that are in close proximity to one another. Thus, Notch signaling depends on cell-contact-dependent interactions. In many cases, the cell that presents the ligand is a cell that does not have Notch signaling present, thus distinguishing two neighboring cells into one with ligand with little Notch signaling and one with receptor and high Notch signaling [57]. In a study by Wu et al. [44], the investigators demonstrated that HemSCs have distinct Notch expression patterns from HemECs. In HemSCs, where Notch3 is strongly expressed, HES1, HEY1, and HEYL were expressed at levels 10 to 100 times to that of HEY-2. In HemECs, however, Notch-1, Notch-4 and Jagged-1 have higher expression levels. HEY-2 proteins were often found to be expressed in HemECs. However, HEY-2 was not uniformly present in all ECs, suggesting that only a subset of IH ECs express the Notch target [56]. These data suggest the possibility that the Notch pathway might also contribute to establishing two distinct subpopulations at different steps of angiogenesis in IH, such as ECs versus smooth muscle cells (SMCs)/pericytes, arteries versus veins and large vessels versus capillaries [54,58,59]. We highlight the concept that ligand-receptor interactions in Notch signaling depend on contact between two cells, which may be two different cell types. Notch ligands involved in IH angiogenesis may be presented by HemECs, pericytes or HemSCs. Interestingly, research by Boscolol et al. [43] revealed that endothelial-derived Jagged-1 can induce HemSCs to acquire a pericyte-like phenotype, which is a crucial step in the vasculogenesis of IH. Disruption of the juxtacrine interaction between endothelial Jagged-1 and Notch receptors on HemSCs inhibited blood vessel formation in IH murine models. However, mice homozygous for a null mutation of several components of the Notch pathway, including Notch-1 and Jagged-1, resulted in embryonic lethality with vascular remodeling defects. Vasculogenesis proceeded normally in these mutants, whereas the next step, angiogenesis, was disrupted [60,61]. These data suggest that the upregulated Jagged-1 expression in the IH endothelium may provide a unique effect to control the vascular development of IH.

8.1.7 IS THERE A SPECIFIC RELATIONSHIP BETWEEN VEGF AND NOTCH PATHWAYS IN IH?

In vivo and in vitro studies have revealed several ways in which the VEGF and Notch pathways interact. Particularly, VEGF increases Dll4 expression [62,63]. Dll4 is strongly expressed by the ECs of sprouting angiogenic vessels, which commonly respond to VEGF signals. There is evidence that the blockade of VEGF in tumors results in a rapid decrease of Dll4 expression in tumor ECs [64]. Interestingly, although HemECs had higher VEGF-A levels and increased activation of VEGFR-2 compared with normal ECs [40], the Dll4 levels in HemECs were lower than those found in normal ECs [44]. These data argue against the idea that VEGF interacts with Notch signaling in IH. However, several lines of evidence indicate otherwise. For example, the disruption of Dll4 or endothelium-specific loss of Notch1 increases the superficial plexus vascular density and causes an excess of angiogenic sprouts. This loss of Notch signaling is associated with an increase in VEGFR-2 activity [57]. Other studies have suggested that reduced Notch activity resulted in reduced VEGFR-1 expression and increased VEGFR-2 expression in cultured ECs [65,66]. In addition, Hellstrom et al. [67] demonstrated that Dll4-Notch signaling within the endothelial cell population serves to suppress the tip-cell phenotype. The retinal vascular abnormalities in Dll4+/- mice and after long-term treatment with γ-secretase inhibitors might also result from changes in the pattern of VEGF-A expression [67]. In contrast to Dll4, Jagged-1 is proangiogenic protein that functions by downregulating Dll4-Notch signaling. Jagged-1 also counteracts Dll4-Notch signaling interactions between stalk ECs, which helps to sustain elevated VEGF receptor expression in the newly formed and therefore immature vascular plexus at the angiogenic front [68]. By analogy to studies of VEGF signaling in HemECs, Notch components may be novel regulators for VEGF signaling in IH (Figure 4).

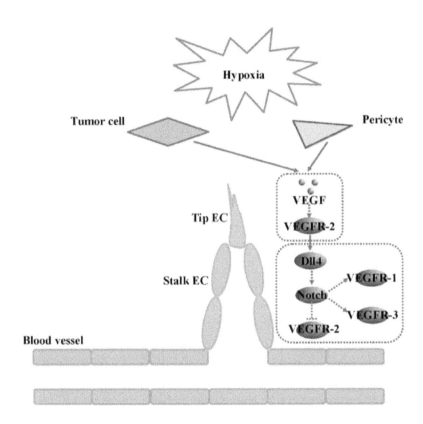

FIGURE 4: Tip/stalk cell specification during spouting angiogenesis and vascular development. Angiogenic sprouts emerge from the newly formed vessels in response to pro-angiogenic cues, such as hypoxia-induced VEGF. VEGF stimulus, acting via VEGFR-2, increases the expression of Dll4 on endothelial cells, which in turn activates Notch receptors on adjacent endothelial cells. Furthermore, VEGFRs are regulated by Notch signaling, providing an additional feedback loop between the two pathways: activated Notch receptors on ECs can positively regulate the expression of VEGFR-1 and VEGFR-3 in those cells. In contrast, Notch activation leads to the reduction of VEGFR-2 expression in cell culture and a concomitant decrease in the proangiogenic response to exogenous VEGF. Both of these effects would likely lead to a lower migratory or proliferative response in connector cells that exhibit Notch activation.

Until now, the involvement of Notch in IH development has remained poorly understood, and many issues still need to be addressed. How does the Notch pathway play a role in the interaction between HemECs and SMCs/pericytes? How are the different roles that the Notch pathway plays, such as arteriovenous patterning, tip cell differentiation and vessel wall formation, integrated during vascular development in IH? How do some of the key downstream Notch target genes affect IH vessels in the presence of high VEGF levels? And finally, does the expression and/or activity of VEGF components in IH depend on the nature of Notch signaling or vice versa? These questions should be addressed by future research efforts.

8.1.8 β–ADRENERGIC SIGNALING

The β-adrenergic receptors (β-ARs), a family of G-protein-coupled receptors that are activated by β-adrenergic agonists (e.g., epinephrine or norepinephrine), can initiate a series of signaling cascades, thereby leading to multiple, cell-specific responses (Figure 5). The ligation of β-ARs by β-adrenergic agonists triggers a G-protein coupled signaling cascade that stimulates cyclic AMP (cAMP) synthesis. This secondary messenger, cAMP, regulates many cellular functions through its effectors, such as cAMP-dependent protein kinase (PKA) and EPAC (exchange proteins directly activated by cAMP) [69-71]. Preclinical studies have demonstrated that β-adrenergic signaling can regulate multiple fundamental biological processes underlying the progression and metastasis of tumors, including the promotion of inflammation [72-74], angiogenesis [75-78], migration [79], invasion [80,81] and resistance to programmed cell death [82-85]. Some evidence suggests that the stimulation of β-adrenergic signaling can also inhibit DNA damage repair and the cellular immune response [86,87] and promote surgery-induced metastasis [88,89]. These findings have led to the hypothesis that commonly prescribed β-blockers may favorably impact cancer progression and metastasis in patients [90].

In the six years since June 2008 when Leaute-Labreze et al. [11] first described their serendipitous observation of the anti-proliferative effect of propranolol on severe IHs, many articles regarding β-blocker therapy for IHs have been published [8,10,91]. However, despite the apparent widespread use of β-blockers, their mechanism of action in IHs has not yet been elucidated. Agonists and antagonists of β-ARs are known to act antithetically via the same intracellular pathways [92]. Given that the expression of all three β-ARs has been demonstrated in IH tumors [93-96], does β-adrenergic signaling play a role in the pathogenesis of IH? This hypothesis was immediately and, to some degree, indirectly testable by Mayer et al. [97], who found that intrauterine exposure to β2-sympathomimetic hexoprenaline can increase the occurrence of IH in preterm infants, suggesting a role for β-AR stimulation in the initiation of IH. Furthermore, we recently demonstrated that the activation of β-ARs resulted in increased HemEC proliferation and upregulation of the ERK signaling cascade. VEGFR-2-mediated ERK signaling was also upregulated upon β-AR activation to mediate the proliferation of HemECs [96]. These findings unveil a functional connection between the β-ARs and IH development. However, confirmatory studies in animal models of IH and mechanistic studies are needed to clearly define the role of β-adrenergic signaling in the growth and involution of IH [98].

8.1.9 TIE-2/ANGIOPOIETIN SIGNALING

Tie-2 and the angiopoietins (Ang), another receptor-ligand system involved in physiological and pathogenic angiogenesis, have also been reported to be associated with the development of IHs. Tie-2 tyrosine kinase receptor is expressed specifically on vascular ECs and on a certain subtype of macrophages implicated in angiogenesis. Ang-1 and Ang-2 have been identified as bona fide ligands of the Tie-2 receptor. Ang-1, which is mainly expressed by pericytes, is a critical player in vessel maturation and mediates the migration, adhesion and survival of ECs. Only tetrameric or higher multimeric forms of Ang-1 activate Tie-2, whereas oligomeric Ang-2 is a weak context-dependent agonist of Tie-2 and may even antagonize the receptor [99]. Ang-1-mediated Tie-2 activation stimulates a number of intracellular signaling pathways, such as the PI3K/Akt pathway, which promotes EC survival and nitric oxide (NO) synthesis by the activation of the mitogen-activated protein kinase (MAPK) pathway [100,101]. The deletion of Ang-1 between E10.5 and E12.5 results in an enlargement of

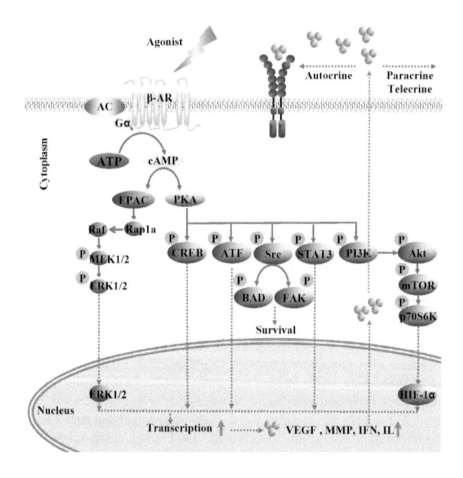

FIGURE 5: β-adrenergic signaling modulates multiple cellular processes in tumor progression and metastasis. The ligation of β-ARs by epinephrine or norepinephrine triggers a G-protein coupled signaling cascade that stimulates cAMP synthesis. cAMP activates the PKA protein, which can mediate multiple signal pathways via the phosphorylation of various downstream signal proteins. In another major pathway, the cAMP activation of EPAC leads to the Rap1A-mediated activation of Raf/MAPK signaling pathways and downstream effects on diverse cellular processes.

vessel diameter, mainly in the capillaries [102,103]. Its phenotypes were comparable with those of IH, i.e., increased numbers of EC and overly covered by pericytes [25].

Using laser capture microdissection, Calicchio et al. [52] found Ang-2 was significantly increased in the IH endothelium compared with the placental vessels. In contrast, Ang-1 was decreased in proliferating IH relative to the placenta. Yu et al. [104] demonstrated that Tie-2 was specifically increased in HemECs and that this increase corresponds to enhanced cellular responses to the Tie-2 agonist Ang-1. Consistent with these findings, Boscolo et al. [42] revealed that hemangioma-derived pericytes exhibited low levels of Ang-1, resulting in a diminished ability to stabilize blood vessels in IH.

8.1.10 HIF–α–MEDIATED PATHWAY

Hypoxia is one of the most powerful inducers of angiogenesis and vasculogenesis. During tumorigenesis, when tumor cells outgrow the limiting diffusion distance to nearby blood vessels and become hypoxic, the balance between pro-angiogenic and anti-angiogenic molecules is tipped towards pro-angiogenic molecules. This angiogenic switch provokes the expression of a variety of angiogenic factors by tumor cells and stromal cells, including VEGF-A, stromal cell-derived factor-1α (SDF-1α), fibroblast growth factor (FGF), platelet-derived growth factors (PDGFs), lysophosphatidic acid (LPA) and Ang [105].

Although the initiating mechanism during the pathogenesis of IH has yet to be discovered, there is evidence that tissue hypoxia may contribute to their explosive growth. The initial clinical description of the promontory mark of IH as an 'area of low blood flow' suggests that tissue ischemia, a powerful stimulus for neovascularization, may be involved [106]. The hypothesis that ischemia/hypoxia plays a crucial role is also supported by the clinical observation of a blanched area of skin in the position of the future hemangioma. This region may be an area of local ischemia in the skin, caused by some unknown events, that creates a hypoxic environment and thus triggers growth factor expression.

In keeping with the observations described above, Ritter et al. [107] proposed a mechanism for myeloid cell-facilitated IH growth involving the hypoxia-induced expression of several growth factors (e.g., insulin-like growth factor-2) that drive endothelial proliferation. Kleinman et al. [108] demonstrated the presence of hypoxia-induced mediators of progenitor/stem cell trafficking in proliferating IH specimens and revealed that the combination of hypoxia and estradiol results in a synergistic effect on the upregulation of matrix metalloproteinase (MMP-9) in ECs in vitro, a key factor in endothelial progenitor cells (EPCs). The transcription factor hypoxia inducible factor (HIF-1α) was also stabilized in proliferating hemangioma specimens. Subsequent investigations revealed that HemECs show significantly a higher expression of HIF-1α than normal ECs. This upregulated HIF-α is a major contributor to the elevated VEGF levels produced in HemECs, and the decreased expression of HIF-α reduces the proliferation of these cells [39]. Moreover, the benefit observed during IH treatment by propranolol has been suggested to also be primarily due to the reduction of HIF-1α expression [109]. This suggestion has been confirmed by Chim et al. [110], who demonstrated that propranolol exerts its suppressive effects on HemECs through the HIF-1α-VEGF-A angiogenesis axis, the effects of which were mediated through the PI3/Akt and p38/MAPK pathways. Altogether, these findings indicate a direct and causative association between HIF-α signaling and the development of IH.

An additional possible effect of HIF-1α signaling in the pathogenesis of IH is mediation of EC autophagy. In their recent study, Chen et al. [111] revealed that a short exposure to hypoxia can induce HIF-α/BNIP3-dependent autophagy, which may promote EC survival growth. In contrast, if the hypoxia stress is prolonged, the autophagy activation may in turn become 5'-AMP-activated protein kinase (AMPK)/mammalian target of rapamin (mTOR) dependent and therefore cause programmed EC death. However, the evidence from this study was weakened by not being performed in IH-derived cells or in IH animal models.

8.1.11 PI3K/AKT/MTOR SIGNALING

PI3K generates 3-phosphorylated inositol lipids, causing the activation of downstream signaling, resulting in the activation of protein kinase B

(PKB; also called c-Akt), which regulates, among others, mTOR, glycogen synthase kinase-3β and Forkhead box O transcription factor activity. Downstream targets of mTOR include p70 ribosomal protein 6S kinase (S6K). The overactivation of the PI3K/Akt/mTOR pathway, a signaling pathway that plays a key role in cellular growth and survival, has been implicated in various tumor pathogeneses, and as such, the inhibition of the PI3K/Akt/mTOR pathway is of therapeutic interest [112-114]. Medici and Olsen [39] found that HemECs had constitutively active PI3K/Akt/mTOR/p70S6K and tested their hypothesis that these cells could be sensitive to mTOR inhibitors (e.g., rapamycin). Finally, they demonstrated that the treatment of HemECs with rapamycin results in a significant decrease in HIF-1 and VEGF-A levels and in reduced proliferation. Strikingly, in vivo and in vitro studies further demonstrated that rapamycin can reduce the self-renewal capacity of the HemSCs, diminish the differentiation potential and inhibit the vasculogenic activity of these cells in vivo [115]. These preclinical data provide us with a pharmacological basis for the potential use of rapamycin in β-blocker-resistant IHs. Nonetheless, the mechanism that accounts for the effects of rapamycin in IH is far from clear, and a growing list of side effects make it doubtful that rapamycin would ultimately be beneficial in pediatric patients [112].

8.1.12 PDGF-B/PDGFR–β SIGNALING

The first evidence for a possible regulatory role of PDGF-B/PDGF receptor-β (PDGFR-β) signaling in IHs was provided by Walter et al. [116]. These researchers established a genetic linkage with chromosome 5q in three familial hemangiomas. The region, 5q31-33, contains three candidate genes involved in blood vessel growth. These genes were fibroblast growth factor receptor-4 (FGFR4), PDGFR-β and VEGFR-3 [116]. Subsequently, a study examining global gene expression changes between the IH growth phases by the genome-wide transcriptional profiling of blood vessels showed a reduction in PDGFR-β expression during the involutive phase [52]. These findings provide the possibility that PDGF-B/PDGFR-β signaling may play a role in IH pathogenesis.

The endothelium is a critical source of PDGF-B for PDGF-β-positive mural cell recruitment, as demonstrated by the endothelium-specific ablation of PDGF-B, which leads to pericyte deficiency [117]. The blockade of pericyte recruitment by abolishing PDGF-B/PDGFR-β signaling causes a lack of basement membrane matrix deposition and concomitantly increased vessel widths [118]. In addition, the ectopic expression of PDGF-B by tumor cells results in the increased recruitment of mural cells to blood vessels on the establishment of subcutaneous tumors [119,120]. Unfortunately, despite the tight physical and functional association between ECs and pericytes, there is a paucity of information about the signals exchanged between the two cell types in IHs. Reassuringly, data from a separate study demonstrated that PDGF/PDGF-R-β signaling may act as an intrinsic negative regulator of IH involution. In this study, Roach and colleagues [121] found that PDGF is elevated during the proliferating phase and may inhibit adipocyte differentiation. The exposure of HemSCs to exogenous PDGF results in an activation of autocrine PDGF/PDGF-R-β signaling, thereby inhibiting IH involution. These findings highlight the involvement of PDGF/PDGF-R-β signaling in the development of IHs. Moreover, hemangioma-derived pericytes also express PDGFR-β, although its effect has not been elucidated in IH pathogenesis [42]. Thus, the possibility of targeting HemSC and Hem-pericyte function, for example, via their PDGF receptors, to gain enhanced efficacy of antiangiogenic treatment regiments is supported by the reports of beneficial effects of combining PDGFR inhibitors with antiangiogenic drugs or regimens [118,122,123].

8.1.13 CONCLUSION AND FUTURE CHALLENGES

In conclusion, the findings summarized above demonstrate that signaling pathways involved in the development of IH are increasingly being clarified, underscoring their significant relevance to understanding IH pathogenesis. However, similar to malignant tumors, there is extensive crosstalk between individual signaling pathways in IH. This crosstalk is generally due to two factors. First, multiple pathways often control a common process. Second, many signaling outcomes impact other processes through feedback loops and compensatory responses. Therefore, elucidating the molecular pathogenesis of IH presents an intriguing challenge. To solve this puzzle, an organized reconstruction of the sequential molecular perturbations during IH neovascularization is required. Such an analysis needs to combine data from different levels, including genetic aberrations, expression alterations and protein modification in a comprehensive set of tissue samples. These issues should highlight the important role that the increased knowledge of the molecular pathways involved in the pathogenesis of IH will have in guiding the development of effective, rationally designed therapeutic strategies. Future research efforts will not only provide us with a pharmacological basis of the therapeutic use of β-blocker in IHs but also a basis for the further investigation of other potential antihemangioma agents.

REFERENCES

1. Mulliken JB, Fishman SJ, Burrows PE: Vascular anomalies. Curr Probl Surg 2000, 37(8):517-584.
2. Drolet BA, Esterly NB, Frieden IJ: Hemangiomas in children. N Engl J Med 1999, 341(3):173-181.
3. Margileth AM, Museles M: Cutaneous hemangiomas in children. Diagnosis and conservative management. JAMA 1965, 194(5):523-526.
4. George ME, Sharma V, Jacobson J, Simon S, Nopper AJ: Adverse effects of systemic glucocorticosteroid therapy in infants with hemangiomas. Arch Dermatol 2004, 140(8):963-969.
5. Goyal R, Watts P, Lane CM, Beck L, Gregory JW: Adrenal suppression and failure to thrive after steroid injections for periocular hemangioma. Ophthalmology 2004, 111(2):389-395.
6. Neri I, Balestri R, Patrizi A: Hemangiomas: new insight and medical treatment. Dermatol Ther 2012, 25(4):322-334.
7. Chang LC, Haggstrom AN, Drolet BA, Baselga E, Chamlin SL, Garzon MC, Horii KA, Lucky AW, Mancini AJ, Metry DW, et al.: Growth characteristics of infantile hemangiomas: implications for management. Pediatrics 2008, 122(2):360-367.
8. Chan H, McKay C, Adams S, Wargon O: RCT of timolol maleate gel for superficial infantile hemangiomas in 5- to 24-week-olds. Pediatrics 2013, 131(6):e1739-e1747.
9. Malik MA, Menon P, Rao KL, Samujh R: Effect of propranolol vs prednisolone vs propranolol with prednisolone in the management of infantile hemangioma: a randomized controlled study. J Pediatr Surg 2013, 48(12):2453-2459.
10. Hogeling M, Adams S, Wargon O: A randomized controlled trial of propranolol for infantile hemangiomas. Pediatrics 2011, 128(2):e259-e266.
11. Leaute-Labreze C, Dumas DLRE, Hubiche T, Boralevi F, Thambo JB, Taieb A: Propranolol for severe hemangiomas of infancy. N Engl J Med 2008, 358(24):2649-2651.

12. Causse S, Aubert H, Saint-Jean M, Puzenat E, Bursztejn AC, Eschard C, Mahe E, Maruani A, Mazereeuw-Hautier J, Dreyfus I, et al.: Propranolol-resistant infantile haemangiomas. Br J Dermatol 2013, 169(1):125-129.

13. Shehata N, Powell J, Dubois J, Hatami A, Rousseau E, Ondrejchak S, McCuaig C: Late rebound of infantile hemangioma after cessation of oral propranolol. Pediatr Dermatol 2013, 30(5):587-591.

14. Mulliken JB, Glowacki J: Hemangiomas and vascular malformations in infants and children: a classification based on endothelial characteristics. Plast Reconstr Surg 1982, 69(3):412-422.

15. Khan ZA, Boscolo E, Picard A, Psutka S, Melero-Martin JM, Bartch TC, Mulliken JB, Bischoff J: Multipotential stem cells recapitulate human infantile hemangioma in immunodeficient mice. J Clin Invest 2008, 118(7):2592-2599.

16. Yu Y, Fuhr J, Boye E, Gyorffy S, Soker S, Atala A, Mulliken JB, Bischoff J: Mesenchymal stem cells and adipogenesis in hemangioma involution. Stem Cells 2006, 24(6):1605-1612.

17. Yu Y, Flint AF, Mulliken JB, Wu JK, Bischoff J: Endothelial progenitor cells in infantile hemangioma. Blood 2004, 103(4):1373-1375.

18. Dosanjh A, Chang J, Bresnick S, Zhou L, Reinisch J, Longaker M, Karasek M: In vitro characteristics of neonatal hemangioma endothelial cells: similarities and differences between normal neonatal and fetal endothelial cells. J Cutan Pathol 2000, 27(9):441-450.

19. Yuan SM, Chen RL, Shen WM, Chen HN, Zhou XJ: Mesenchymal stem cells in infantile hemangioma reside in the perivascular region. Pediatr Dev Pathol 2012, 15(1):5-12.

20. Xu D, TM O, Shartava A, Fowles TC, Yang J, Fink LM, Ward DC, Mihm MC, Waner M, Ma Y: Isolation, characterization, and in vitro propagation of infantile hemangioma stem cells and an in vivo mouse model. J Hematol Oncol 2011, 4:54.

21. Greenberger S, Boscolo E, Adini I, Mulliken JB, Bischoff J: Corticosteroid suppression of VEGF-A in infantile hemangioma-derived stem cells. N Engl J Med 2010, 362(11):1005-1013.

22. Mai HM, Zheng JW, Wang YA, Yang XJ, Zhou Q, Qin ZP, Li KL: CD133 selected stem cells from proliferating infantile hemangioma and establishment of an in vivo mice model of hemangioma. Chin Med J (Engl) 2013, 126(1):88-94.

23. Li Z: CD133: a stem cell biomarker and beyond. Exp Hematol Oncol 2013, 2(1):17.

24. Greenberger S, Bischoff J: Pathogenesis of infantile haemangioma. Br J Dermatol 2013, 169(1):12-19.

25. Boscolo E, Bischoff J: Vasculogenesis in infantile hemangioma. Angiogenesis 2009, 12(2):197-207.

26. Verheul HM, Pinedo HM: The role of vascular endothelial growth factor (VEGF) in tumor angiogenesis and early clinical development of VEGF-receptor kinase inhibitors. Clin Breast Cancer 2000, 1(Suppl 1):S80-S84.

27. Przewratil P, Sitkiewicz A, Andrzejewska E: Local serum levels of vascular endothelial growth factor in infantile hemangioma: intriguing mechanism of endothelial growth. Cytokine 2010, 49(2):141-147.

28. Zhang L, Lin X, Wang W, Zhuang X, Dong J, Qi Z, Hu Q: Circulating level of vascular endothelial growth factor in differentiating hemangioma from vascular malformation patients. Plast Reconstr Surg 2005, 116(1):200-204.

29. Shibuya M: Vascular endothelial growth factor receptor-1 (VEGFR-1/Flt-1): a dual regulator for angiogenesis. Angiogenesis 2006, 9(4):225-230.
30. Dimmeler S, Zeiher AM: Endothelial cell apoptosis in angiogenesis and vessel regression. Circ Res 2000, 87(6):434-439.
31. Huang HY, Ho CC, Huang PH, Hsu SM: Co-expression of VEGF-C and its receptors, VEGFR-2 and VEGFR-3, in endothelial cells of lymphangioma. Implication in autocrine or paracrine regulation of lymphangioma. Lab Invest 2001, 81(12):1729-1734.
32. Shawber CJ, Funahashi Y, Francisco E, Vorontchikhina M, Kitamura Y, Stowell SA, Borisenko V, Feirt N, Podgrabinska S, Shiraishi K, et al.: Notch alters VEGF responsiveness in human and murine endothelial cells by direct regulation of VEGFR-3 expression. J Clin Invest 2007, 117(11):3369-3382.
33. Hamerlik P, Lathia JD, Rasmussen R, Wu Q, Bartkova J, Lee M, Moudry P, Bartek JJ, Fischer W, Lukas J, et al.: Autocrine VEGF-VEGFR2-Neuropilin-1 signaling promotes glioma stem-like cell viability and tumor growth. J Exp Med 2012, 209(3):507-520.
34. Olsson AK, Dimberg A, Kreuger J, Claesson-Welsh L: VEGF receptor signalling - in control of vascular function. Nat Rev Mol Cell Biol 2006, 7(5):359-371.
35. Gupta K, Kshirsagar S, Li W, Gui L, Ramakrishnan S, Gupta P, Law PY, Hebbel RP: VEGF prevents apoptosis of human microvascular endothelial cells via opposing effects on MAPK/ERK and SAPK/JNK signaling. Exp Cell Res 1999, 247(2):495-504.
36. Gerber HP, McMurtrey A, Kowalski J, Yan M, Keyt BA, Dixit V, Ferrara N: Vascular endothelial growth factor regulates endothelial cell survival through the phosphatidylinositol 3'-kinase/Akt signal transduction pathway. Requirement for Flk-1/KDR activation. J Biol Chem 1998, 273(46):30336-30343.
37. Ferrara N, Gerber HP, LeCouter J: The biology of VEGF and its receptors. Nat Med 2003, 9(6):669-676.
38. Gerber HP, Malik AK, Solar GP, Sherman D, Liang XH, Meng G, Hong K, Marsters JC, Ferrara N: VEGF regulates haematopoietic stem cell survival by an internal autocrine loop mechanism. Nature 2002, 417(6892):954-958.
39. Medici D, Olsen BR: Rapamycin inhibits proliferation of hemangioma endothelial cells by reducing HIF-1-dependent expression of VEGF. PLoS One 2012, 7(8):e42913.
40. Jinnin M, Medici D, Park L, Limaye N, Liu Y, Boscolo E, Bischoff J, Vikkula M, Boye E, Olsen BR: Suppressed NFAT-dependent VEGFR1 expression and constitutive VEGFR2 signaling in infantile hemangioma. Nat Med 2008, 14(11):1236-1246.
41. Walter JW, North PE, Waner M, Mizeracki A, Blei F, Walker JW, Reinisch JF, Marchuk DA: Somatic mutation of vascular endothelial growth factor receptors in juvenile hemangioma. Genes Chromosomes Cancer 2002, 33(3):295-303.
42. Boscolo E, Mulliken JB, Bischoff J: Pericytes from infantile hemangioma display proangiogenic properties and dysregulated angiopoietin-1. Arterioscler Thromb Vasc Biol 2013, 33(3):501-509.
43. Boscolo E, Stewart CL, Greenberger S, Wu JK, Durham JT, Herman IM, Mulliken JB, Kitajewski J, Bischoff J: JAGGED1 signaling regulates hemangioma stem cell-

to-pericyte/vascular smooth muscle cell differentiation. Arterioscler Thromb Vasc Biol 2011, 31(10):2181-2192.

44. Wu JK, Adepoju O, De Silva D, Baribault K, Boscolo E, Bischoff J, Kitajewski J: A switch in Notch gene expression parallels stem cell to endothelial transition in infantile hemangioma. Angiogenesis 2010, 13(1):15-23.

45. Lee JJ, Chen CH, Chen YH, Huang MJ, Huang J, Hung JS, Chen MT, Huang MC: COSMC is overexpressed in proliferating infantile hemangioma and enhances endothelial cell growth via VEGFR2. PLoS One 2013, 8(2):e56211.

46. Lichtenberger BM, Tan PK, Niederleithner H, Ferrara N, Petzelbauer P, Sibilia M: Autocrine VEGF signaling synergizes with EGFR in tumor cells to promote epithelial cancer development. Cell 2010, 140(2):268-279.

47. Lee S, Chen TT, Barber CL, Jordan MC, Murdock J, Desai S, Ferrara N, Nagy A, Roos KP, Iruela-Arispe ML: Autocrine VEGF signaling is required for vascular homeostasis. Cell 2007, 130(4):691-703.

48. Ji Y, Chen S, Li K, Xiao X, Xu T, Zheng S: Upregulated autocrine vascular endothelial growth factor (VEGF)/VEGF receptor-2 loop prevents apoptosis in haemangioma-derived endothelial cells. Br J Dermatol 2014, 170(1):78-86.

49. Franco M, Roswall P, Cortez E, Hanahan D, Pietras K: Pericytes promote endothelial cell survival through induction of autocrine VEGF-A signaling and Bcl-w expression. Blood 2011, 118(10):2906-2917.

50. Kearney JB, Ambler CA, Monaco KA, Johnson N, Rapoport RG, Bautch VL: Vascular endothelial growth factor receptor Flt-1 negatively regulates developmental blood vessel formation by modulating endothelial cell division. Blood 2002, 99(7):2397-2407.

51. Fong GH, Rossant J, Gertsenstein M, Breitman ML: Role of the Flt-1 receptor tyrosine kinase in regulating the assembly of vascular endothelium. Nature 1995, 376(6535):66-70.

52. Calicchio ML, Collins T, Kozakewich HP: Identification of signaling systems in proliferating and involuting phase infantile hemangiomas by genome-wide transcriptional profiling. Am J Pathol 2009, 174(5):1638-1649.

53. Boscolo E, Mulliken JB, Bischoff J: VEGFR-1 mediates endothelial differentiation and formation of blood vessels in a murine model of infantile hemangioma. Am J Pathol 2011, 179(5):2266-2277.

54. Iso T, Hamamori Y, Kedes L: Notch signaling in vascular development. Arterioscler Thromb Vasc Biol 2003, 23(4):543-553.

55. Gridley T: Notch signaling during vascular development. Proc Natl Acad Sci U S A 2001, 98(10):5377-5378.

56. Adepoju O, Wong A, Kitajewski A, Tong K, Boscolo E, Bischoff J, Kitajewski J, Wu JK: Expression of HES and HEY genes in infantile hemangiomas. Vasc Cell 2011, 3:19.

57. Dufraine J, Funahashi Y, Kitajewski J: Notch signaling regulates tumor angiogenesis by diverse mechanisms. Oncogene 2008, 27(38):5132-5137.

58. Lanner F, Sohl M, Farnebo F: Functional arterial and venous fate is determined by graded VEGF signaling and notch status during embryonic stem cell differentiation. Arterioscler Thromb Vasc Biol 2007, 27(3):487-493.

59. Li JL, Harris AL: Notch signaling from tumor cells: a new mechanism of angiogenesis. Cancer Cell 2005, 8(1):1-3.

60. Krebs LT, Xue Y, Norton CR, Shutter JR, Maguire M, Sundberg JP, Gallahan D, Closson V, Kitajewski J, Callahan R, et al.: Notch signaling is essential for vascular morphogenesis in mice. Genes Dev 2000, 14(11):1343-1352.

61. Xue Y, Gao X, Lindsell CE, Norton CR, Chang B, Hicks C, Gendron-Maguire M, Rand EB, Weinmaster G, Gridley T: Embryonic lethality and vascular defects in mice lacking the Notch ligand Jagged1. Hum Mol Genet 1999, 8(5):723-730.

62. Hainaud P, Contreres JO, Villemain A, Liu LX, Plouet J, Tobelem G, Dupuy E: The role of the vascular endothelial growth factor-Delta-like 4 ligand/Notch4-ephrin B2 cascade in tumor vessel remodeling and endothelial cell functions. Cancer Res 2006, 66(17):8501-8510.

63. Patel NS, Li JL, Generali D, Poulsom R, Cranston DW, Harris AL: Up-regulation of delta-like 4 ligand in human tumor vasculature and the role of basal expression in endothelial cell function. Cancer Res 2005, 65(19):8690-8697.

64. Noguera-Troise I, Daly C, Papadopoulos NJ, Coetzee S, Boland P, Gale NW, Lin HC, Yancopoulos GD, Thurston G: Blockade of Dll4 inhibits tumour growth by promoting non-productive angiogenesis. Nature 2006, 444(7122):1032-1037.

65. Williams CK, Li JL, Murga M, Harris AL, Tosato G: Up-regulation of the Notch ligand Delta-like 4 inhibits VEGF-induced endothelial cell function. Blood 2006, 107(3):931-939.

66. Zhang J, Ye J, Ma D, Liu N, Wu H, Yu S, Sun X, Tse W, Ji C: Cross-talk between leukemic and endothelial cells promotes angiogenesis by VEGF activation of the Notch/Dll4 pathway. Carcinogenesis 2013, 34(3):667-677.

67. Hellstrom M, Phng LK, Hofmann JJ, Wallgard E, Coultas L, Lindblom P, Alva J, Nilsson AK, Karlsson L, Gaiano N, et al.: Dll4 signalling through Notch1 regulates formation of tip cells during angiogenesis. Nature 2007, 445(7129):776-780.

68. Benedito R, Roca C, Sorensen I, Adams S, Gossler A, Fruttiger M, Adams RH: The notch ligands Dll4 and Jagged1 have opposing effects on angiogenesis. Cell 2009, 137(6):1124-1135.

69. Zhang X, Odom DT, Koo SH, Conkright MD, Canettieri G, Best J, Chen H, Jenner R, Herbolsheimer E, Jacobsen E, et al.: Genome-wide analysis of cAMP-response element binding protein occupancy, phosphorylation, and target gene activation in human tissues. Proc Natl Acad Sci U S A 2005, 102(12):4459-4464.

70. Luttrell LM, Ferguson SS, Daaka Y, Miller WE, Maudsley S, Della RG, Lin F, Kawakatsu H, Owada K, Luttrell DK, et al.: Beta-arrestin-dependent formation of beta2 adrenergic receptor-Src protein kinase complexes. Science 1999, 283(5402):655-661.

71. Ji Y, Chen S, Xiao X, Zheng S, Li K: beta-blockers: a novel class of antitumor agents. Onco Targets Ther 2012, 5:391-401.

72. Shahzad MM, Arevalo JM, Armaiz-Pena GN, Lu C, Stone RL, Moreno-Smith M, Nishimura M, Lee JW, Jennings NB, Bottsford-Miller J, et al.: Stress effects on FosB- and interleukin-8 (IL8)-driven ovarian cancer growth and metastasis. J Biol Chem 2010, 285(46):35462-35470.

73. Bernabe DG, Tamae AC, Biasoli ER, Oliveira SH: Stress hormones increase cell proliferation and regulates interleukin-6 secretion in human oral squamous cell carcinoma cells. Brain Behav Immun 2011, 25(3):574-583.
74. Cole SW, Arevalo JM, Takahashi R, Sloan EK, Lutgendorf SK, Sood AK, Sheridan JF, Seeman TE: Computational identification of gene-social environment interaction at the human IL6 locus. Proc Natl Acad Sci U S A 2010, 107(12):5681-5686.
75. Chakroborty D, Sarkar C, Basu B, Dasgupta PS, Basu S: Catecholamines regulate tumor angiogenesis. Cancer Res 2009, 69(9):3727-3730.
76. Thaker PH, Han LY, Kamat AA, Arevalo JM, Takahashi R, Lu C, Jennings NB, Armaiz-Pena G, Bankson JA, Ravoori M, et al.: Chronic stress promotes tumor growth and angiogenesis in a mouse model of ovarian carcinoma. Nat Med 2006, 12(8):939-944.
77. Pasquier E, Street J, Pouchy C, Carre M, Gifford AJ, Murray J, Norris MD, Trahair T, Andre N, Kavallaris M: beta-blockers increase response to chemotherapy via direct antitumour and anti-angiogenic mechanisms in neuroblastoma. Br J Cancer 2013, 108(12):2485-2494.
78. Ji Y, Chen S: Comment on 'Beta-blockers increase response to chemotherapy via direct anti-tumour and anti-angiogenic mechanisms in neuroblastoma'. Br J Cancer 2013, 109(7):2022-2023.
79. Entschladen F, Drell TT, Lang K, Joseph J, Zaenker KS: Tumour-cell migration, invasion, and metastasis: navigation by neurotransmitters. Lancet Oncol 2004, 5(4):254-258.
80. Sood AK, Bhatty R, Kamat AA, Landen CN, Han L, Thaker PH, Li Y, Gershenson DM, Lutgendorf S, Cole SW: Stress hormone-mediated invasion of ovarian cancer cells. Clin Cancer Res 2006, 12(2):369-375.
81. Armaiz-Pena GN, Allen JK, Cruz A, Stone RL, Nick AM, Lin YG, Han LY, Mangala LS, Villares GJ, Vivas-Mejia P, et al.: Src activation by beta-adrenoreceptors is a key switch for tumour metastasis. Nat Commun 2013, 4:1403.
82. Sastry KS, Karpova Y, Prokopovich S, Smith AJ, Essau B, Gersappe A, Carson JP, Weber MJ, Register TC, Chen YQ, et al.: Epinephrine protects cancer cells from apoptosis via activation of cAMP-dependent protein kinase and BAD phosphorylation. J Biol Chem 2007, 282(19):14094-14100.
83. Sood AK, Armaiz-Pena GN, Halder J, Nick AM, Stone RL, Hu W, Carroll AR, Spannuth WA, Deavers MT, Allen JK, et al.: Adrenergic modulation of focal adhesion kinase protects human ovarian cancer cells from anoikis. J Clin Invest 2010, 120(5):1515-1523.
84. Hassan S, Karpova Y, Baiz D, Yancey D, Pullikuth A, Flores A, Register T, Cline JM, D'Agostino RJ, Danial N, et al.: Behavioral stress accelerates prostate cancer development in mice. J Clin Invest 2013, 123(2):874-886.
85. Ji Y, Li K, Xiao X, Zheng S, Xu T, Chen S: Effects of propranolol on the proliferation and apoptosis of hemangioma-derived endothelial cells. J Pediatr Surg 2012, 47(12):2216-2223.
86. Hara MR, Kovacs JJ, Whalen EJ, Rajagopal S, Strachan RT, Grant W, Towers AJ, Williams B, Lam CM, Xiao K, et al.: A stress response pathway regulates DNA damage through beta2-adrenoreceptors and beta-arrestin-1. Nature 2011, 477(7364):349-353.

87. Glaser R, Kiecolt-Glaser JK: Stress-induced immune dysfunction: implications for health. Nat Rev Immunol 2005, 5(3):243-251.

88. Goldfarb Y, Sorski L, Benish M, Levi B, Melamed R, Ben-Eliyahu S: Improving postoperative immune status and resistance to cancer metastasis: a combined perioperative approach of immunostimulation and prevention of excessive surgical stress responses. Ann Surg 2011, 253(4):798-810.

89. Glasner A, Avraham R, Rosenne E, Benish M, Zmora O, Shemer S, Meiboom H, Ben-Eliyahu S: Improving survival rates in two models of spontaneous postoperative metastasis in mice by combined administration of a beta-adrenergic antagonist and a cyclooxygenase-2 inhibitor. J Immunol 2010, 184(5):2449-2457.

90. Ji Y, Chen S: Do antihypertensive medications influence breast cancer risk? JAMA Intern Med 2014, In press

91. Pope E, Chakkittakandiyil A, Lara-Corrales I, Maki E, Weinstein M: Expanding the therapeutic repertoire of infantile haemangiomas: cohort-blinded study of oral nadolol compared with propranolol. Br J Dermatol 2013, 168(1):222-224.

92. Powe DG, Entschladen F: Targeted therapies: using beta-blockers to inhibit breast cancer progression. Nat Rev Clin Oncol 2011, 8(9):511-512.

93. Hadaschik E, Scheiba N, Engstner M, Flux K: High levels of beta2-adrenoceptors are expressed in infantile capillary hemangiomas and may mediate the therapeutic effect of propranolol. J Cutan Pathol 2012, 39(9):881-883.

94. Chisholm KM, Chang KW, Truong MT, Kwok S, West RB, Heerema-McKenney AE: beta-Adrenergic receptor expression in vascular tumors. Mod Pathol 2012, 25(11):1446-1451.

95. Rossler J, Haubold M, Gilsbach R, Juttner E, Schmitt D, Niemeyer CM, Hein L: beta1-Adrenoceptor mRNA level reveals distinctions between infantile hemangioma and vascular malformations. Pediatr Res 2013, 73(4 Pt 1):409-413.

96. Ji Y, Chen S, Li K, Xiao X, Zheng S, Xu T: The role of beta-adrenergic receptor signaling in the proliferation of hemangioma-derived endothelial cells. Cell Div 2013, 8(1):1.

97. Mayer M, Minichmayr A, Klement F, Hroncek K, Wertaschnigg D, Arzt W, Wiesinger-Eidenberger G, Lechner E: Tocolysis with the beta-2-sympathomimetic hexoprenaline increases occurrence of infantile haemangioma in preterm infants. Arch Dis Child Fetal Neonatal Ed 2013, 98(2):F108-F111.

98. Ji Y, Chen S, Li K, Xiao X, Zheng S: Propranolol: a novel anti-hemangioma agent with multiple potential mechanisms of action. Ann Surg 2013, In press

99. Kim KT, Choi HH, Steinmetz MO, Maco B, Kammerer RA, Ahn SY, Kim HZ, Lee GM, Koh GY: Oligomerization and multimerization are critical for angiopoietin-1 to bind and phosphorylate Tie2. J Biol Chem 2005, 280(20):20126-20131.

100. Jones N, Master Z, Jones J, Bouchard D, Gunji Y, Sasaki H, Daly R, Alitalo K, Dumont DJ: Identification of Tek/Tie2 binding partners. Binding to a multifunctional docking site mediates cell survival and migration. J Biol Chem 1999, 274(43):30896-30905.

101. Jones N, Chen SH, Sturk C, Master Z, Tran J, Kerbel RS, Dumont DJ: A unique autophosphorylation site on Tie2/Tek mediates Dok-R phosphotyrosine binding domain binding and function. Mol Cell Biol 2003, 23(8):2658-2668.

102. Jeansson M, Gawlik A, Anderson G, Li C, Kerjaschki D, Henkelman M, Quaggin SE: Angiopoietin-1 is essential in mouse vasculature during development and in response to injury. J Clin Invest 2011, 121(6):2278-2289.

103. Koh GY: Orchestral actions of angiopoietin-1 in vascular regeneration. Trends Mol Med 2013, 19(1):31-39.

104. Yu Y, Varughese J, Brown LF, Mulliken JB, Bischoff J: Increased Tie2 expression, enhanced response to angiopoietin-1, and dysregulated angiopoietin-2 expression in hemangioma-derived endothelial cells. Am J Pathol 2001, 159(6):2271-2280.

105. Boye E, Olsen BR: Signaling mechanisms in infantile hemangioma. Curr Opin Hematol 2009, 16(3):202-208.

106. Chen TS, Eichenfield LF, Friedlander SF: Infantile hemangiomas: an update on pathogenesis and therapy. Pediatrics 2013, 131(1):99-108.

107. Ritter MR, Reinisch J, Friedlander SF, Friedlander M: Myeloid cells in infantile hemangioma. Am J Pathol 2006, 168(2):621-628.

108. Kleinman ME, Greives MR, Churgin SS, Blechman KM, Chang EI, Ceradini DJ, Tepper OM, Gurtner GC: Hypoxia-induced mediators of stem/progenitor cell trafficking are increased in children with hemangioma. Arterioscler Thromb Vasc Biol 2007, 27(12):2664-2670.

109. Storch CH, Hoeger PH: Propranolol for infantile haemangiomas: insights into the molecular mechanisms of action. Br J Dermatol 2010, 163(2):269-274.

110. Chim H, Armijo BS, Miller E, Gliniak C, Serret MA, Gosain AK: Propranolol induces regression of hemangioma cells through HIF-1alpha-mediated inhibition of VEGF-A. Ann Surg 2012, 256(1):146-156.

111. Chen G, Zhang W, Li YP, Ren JG, Xu N, Liu H, Wang FQ, Sun ZJ, Jia J, Zhao YF: Hypoxia-induced autophagy in endothelial cells: a double-edged sword in the progression of infantile haemangioma? Cardiovasc Res 2013, 98(3):437-448.

112. Lamming DW, Ye L, Sabatini DM, Baur JA: Rapalogs and mTOR inhibitors as anti-aging therapeutics. J Clin Invest 2013, 123(3):980-989.

113. Benjamin D, Colombi M, Moroni C, Hall MN: Rapamycin passes the torch: a new generation of mTOR inhibitors. Nat Rev Drug Discov 2011, 10(11):868-880.

114. Slomovitz BM, Coleman RL: The PI3K/AKT/mTOR pathway as a therapeutic target in endometrial cancer. Clin Cancer Res 2012, 18(21):5856-5864.

115. Greenberger S, Yuan S, Walsh LA, Boscolo E, Kang KT, Matthews B, Mulliken JB, Bischoff J: Rapamycin suppresses self-renewal and vasculogenic potential of stem cells isolated from infantile hemangioma. J Invest Dermatol 2011, 131(12):2467-2476.

116. Walter JW, Blei F, Anderson JL, Orlow SJ, Speer MC, Marchuk DA: Genetic mapping of a novel familial form of infantile hemangioma. Am J Med Genet 1999, 82(1):77-83.

117. Bjarnegard M, Enge M, Norlin J, Gustafsdottir S, Fredriksson S, Abramsson A, Takemoto M, Gustafsson E, Fassler R, Betsholtz C: Endothelium-specific ablation of PDGFB leads to pericyte loss and glomerular, cardiac and placental abnormalities. Development 2004, 131(8):1847-1857.

118. Stratman AN, Schwindt AE, Malotte KM, Davis GE: Endothelial-derived PDGF-BB and HB-EGF coordinately regulate pericyte recruitment during vasculogenic tube assembly and stabilization. Blood 2010, 116(22):4720-4730.

119. Sennino B, Falcon BL, McCauley D, Le T, McCauley T, Kurz JC, Haskell A, Epstein DM, McDonald DM: Sequential loss of tumor vessel pericytes and endothelial cells after inhibition of platelet-derived growth factor B by selective aptamer AX102. Cancer Res 2007, 67(15):7358-7367.

120. Pietras K, Hanahan D: A multitargeted, metronomic, and maximum-tolerated dose "chemo-switch" regimen is antiangiogenic, producing objective responses and survival benefit in a mouse model of cancer. J Clin Oncol 2005, 23(5):939-952.

121. Roach EE, Chakrabarti R, Park NI, Keats EC, Yip J, Chan NG, Khan ZA: Intrinsic regulation of hemangioma involution by platelet-derived growth factor. Cell Death Dis 2012, 3:e328.

122. Wnuk M, Hlushchuk R, Tuffin G, Huynh-Do U, Djonov V: The effects of PTK787/ZK222584, an inhibitor of VEGFR and PDGFRbeta pathways, on intussusceptive angiogenesis and glomerular recovery from Thy1.1 nephritis. Am J Pathol 2011, 178(4):1899-1912.

123. Erber R, Thurnher A, Katsen AD, Groth G, Kerger H, Hammes HP, Menger MD, Ullrich A, Vajkoczy P: Combined inhibition of VEGF and PDGF signaling enforces tumor vessel regression by interfering with pericyte-mediated endothelial cell survival mechanisms. Faseb J 2004, 18(2):338-340.

CHAPTER 9

STAG2 IS A CLINICALLY RELEVANT TUMOR SUPPRESSOR IN PANCREATIC DUCTAL ADENOCARCINOMA

LISA EVERS, PEDRO A. PEREZ-MANCERA,
ELIZABETH LENKIEWICZ, NANYUN TANG,
DANIELA AUST, THOMAS KNLISEL, PETRA RЬMMELE,
TARA HOLLEY, MICHELLE KASSNER, MERAJ AZIZ,
RAMESH K. RAMANATHAN, DANIEL D .VON HOFF, HOLLY YIN,
CHRISTIAN PILARSKY, AND MICHAEL T. BARRETT

9.1 BACKGROUND

A genetic hallmark of pancreatic ductal adenocarcinoma (PDA) is the presence of somatic *KRAS* mutations in over 90 to 95% of tumors, the most prevalent being *KRAS*[G12D][1,2]. A fundamental question remains the identification of somatic aberrations arising in the complex genomic landscape of PDA that drive the progression of *KRAS* mutant neoplastic cells in humans in vivo. Furthermore, of significant interest are those selected aberrations that create therapeutic vulnerabilities that can be exploited to advance improved and more personalized care of patients. The *STAG2*

This chapter was originally published under the Creative Commons Attribution License. Evers L, Perez-Mancera PA, Lenkiewicz E, Tang N, Aust D, Knösel T, Rümmele P, Holley T, Kassner M, Aziz M, Ramanathan K, Von Hoff DD, Yin H, Pilarsky C, and Barrett MT. STAG2 is a Clinically Relevant Tumor Suppressor in Pancreatic Ductal Adenocarcinoma. Genome Medicine 6,9 (2014). doi:10.1186/ gm526.

gene encodes a subunit of the cohesion complex, which plays an essential role in the proper division and segregation of chromosomes, a process that is essential for the maintenance of genome stability and cell survival [3,4]. Mutations targeting this class of genes have been studied in model systems, but have been detected in a relatively small number of somatic tumors arising in patients in vivo[5-7]. Recently, somatic aberrations and loss of STAG2 expression have been reported in a subset of tumors and cell lines, including melanomas, sarcomas, and glioblastomas [3]. Notably, truncating mutations in STAG2 have been shown to be one of the most common genetic lesions in bladder carcinoma [8]. Functional analysis has shown that loss of STAG2 leads to chromosome missegregation and aneuploidy in human cell lines and may promote a mutator phenotype [3,4]. The development of a $KRAS^{G12D}$-driven genetically engineered mouse (GEM) model of PDA has provided a powerful resource for the study of the events that accelerate tumorigenesis and drive tumor progression in the pancreas [9]. Strikingly, STAG2 was one of the most frequent and significant insertion targets reported in a transposon-mediated screen of the $KRAS^{G12D}$ GEM model of PDA. However, mutations in STAG2 and clinically relevant variations in its protein expression levels have not been reported to date in human PDA samples [3,8]. Thus, the clinical significance of STAG2 expression and its role as a tumor suppressor gene in human PDA remains to be elucidated.

PDA is a highly lethal cancer that is difficult to molecularly characterize at the biopsy level due to complex genomes and heterogeneous cellularity, as cancer cells represent, on average, only 25% of the cells within the tumor [10]. The presence of admixtures of non-neoplastic cells in patient samples can obscure the detection of somatic aberrations, including mutations, homozygous deletions, and breakpoints, in biopsies of interest. Furthermore, clinical samples frequently contain multiple neoplastic populations that cannot be distinguished by morphology-based methods [11,12]. In order to investigate whether STAG2 is a tumor suppressor in human PDA, we used DNA content-based flow cytometry to sort PDA samples from 50 patients. The genome of each sorted tumor population was then interrogated for somatic mutations and aberrations with oligonucleotide array comparative genomic hybridization (aCGH) and targeted resequencing using our established protocols [13]. In this present study we also sought to confirm the inactivation of *STAG2* in those *KRAS*[G12D] GEM tumors with transposon insertions within the gene to provide further evidence of a tumor suppressor role in PDA. We then used a tissue microarray (TMA) to survey STAG2 protein expression levels in 344 human PDA tumor samples and adjacent tissues. The clinical annotation available for each tumor represented on the TMA allowed the assessment of STAG2 expression relative to overall survival and responses to adjuvant therapy. Finally, given the role of *STAG2* in maintaining genomic stability, we used RNA interference (RNAi)-based cellular assays with PDA cell lines to assess the potential therapeutic consequence of STAG2 expression in response to a panel of 18 currently used therapeutic agents. Our results provide evidence for a clinically relevant tumor suppressor role for *STAG2* in *KRAS* mutant PDA. These highly iterative findings have implications for the development of personalized approaches for patients with PDA.

9.2 METHODS

9.2.1 CLINICAL SAMPLES

PDA samples were obtained under a Western Institutional Review Board protocol (20040832) for a National Institutes of Health-funded bio-specimen repository (NCI P01 grant CA109552) and two American Association for Cancer Research/Stand up to Cancer (SU2C) sponsored clinical trials, 20206-001 and 2026-003. Additional PDA samples were obtained with approved consent of the Ethics Committee of Basel (252/08, 302/09). All patients in this study gave informed consent for collection and use of all the samples, which were collected in liquid nitrogen and stored at -80°C. All tumor samples were histopathologically evaluated prior to genomic analysis. All research conformed to the Helsinki Declaration [14].

9.2.2 FLOW CYTOMETRY

Biopsies were minced in the presence of NST buffer and DAPI according to published protocols [11,15,16]. Nuclei were disaggregated then filtered through a 40 µm mesh prior to flow sorting with an Influx cytometer (Becton-Dickinson, San Jose, CA, USA) with ultraviolet excitation and DAPI emission collected at >450 nm. DNA content and cell cycle were analyzed using the software program MultiCycle (Phoenix Flow Systems, San Diego, CA, USA).

9.2.3 ACGH

DNAs were extracted using QIAGEN micro kits (Valencia, CA, USA). For each hybridization, 100 ng of genomic DNA from each sample and of pooled commercial 46XX reference (Promega, Madison, WI, USA were amplified using the GenomiPhi amplification kit (GE Healthcare, Piscataway, NJ, USA). Subsequently, 1 µg of amplified sample and 1 µg of amplified reference template were digested with DNaseI then labeled with Cy-5 dUTP and Cy-3 dUTP, respectively, using a BioPrime labeling kit (Invitrogen, Carlsbad, CA, USA). All labeling reactions were assessed using a Nanodrop assay (Nanodrop, Wilmington, DE, USA) prior to mixing and hybridization to CGH arrays with either 244,000 or 400,000 oligonucleotide features (Agilent Technologies, Santa Clara, CA, USA). The aCGH data have been deposited in the National Center for Biotechnology Information (NCBI) Gene Expression Omnibus (accession numbers GSE54328 and GSE21660).

9.2.4 RESEQUENCING

For each sequencing reaction 50 ng of DNA was amplified in 10 µl reactions for 35 cycles using MyTaq™ HSMix (Bioline, Taunton, MA, USA). All PCR products were verified by visual inspection on agarose gels. Samples were then purified by column filtration prior to analysis with an Applied Biosystems 3730 capillary sequencer (Life Technologies, Carlsbad,

CA, USA). We surveyed 33 of the 35 exons in *STAG2* using published primers [3]. All sequences were analyzed using Mutation Surveyor v4.0.5 (Softgenetics, State College, PA, USA). *KRAS* mutational status was determined using custom primers designed to flank, amplify and sequence codons 12, 13 and 61 in exons 2 and 3 of the *KRAS* gene (Caris MPI, Irving, TX, USA) [17]. The STAG2 resequencing data have been deposited in the NCBI BankIt (ID 1699484).

9.2.5 CLINICAL PATIENT SAMPLES, IMMUNOHISTOCHEMISTRY, AND TISSUE MICROARRAY ANALYSIS

Tissue microarrays of 459 patients with 344 patients eligible for analysis were prepared from patient samples obtained after appropriate informed consent in Dresden (Institute of Pathology, University Hospital Dresden), Regensburg (Institute of Pathology, University Hospital Regensburg) and Jena (Institute of Pathology, University Hospital Jena). All samples were obtained with approved consent of the ethics committee of the Technischen Universität Dresden. Samples were collected consecutively from patients undergoing routine surgery for pancreatic carcinoma. Histological diagnosis was performed in the individual centers by pathologists trained in the routine work-up of pancreatic cancer specimens. The tumor samples were collected from 1993 to 2010, and most of the patients (68%) did not undergo adjuvant chemotherapy. Those that did undergo adjuvant therapy (32%) were chiefly treated with 5-fluorouracil (5-FU) or gemcitabine-based regimens and received a survival benefit of median 5.5 months (P=0.02). The median survival times of patients after surgery from each center were indistinguishable. Immunohistochemistry was performed on 5 µm sections that were prepared using silanized slides (Menzel Gläser, Braunschweig, Lower Saxony, Germany). Staining was performed manually. In brief, sections were treated in PTM buffer pH6.0 for 45 minutes in the pressure cooker. After blocking of the endogenous peroxidase, the sections were incubated with the STAG2 antibody sc-81852 (1:400 for 30 minutes at room temperature). Antibody binding was detected using the Ultravision LP detection System (Thermo Fisher Scientific, Fremont, CA, USA) and Bright DAB (Medac, Wedel, Germany). Slides were counterstained with hematoxylin. Staining of STAG2 immunohistochemistry was performed by one (CP) and checked by another scientist (DA) with random point controls. Staining intensities were scored as the percentage of stained nuclei independent of staining intensity. Samples with a minimal staining of less than 30% on nuclei were scored as negative.

9.2.6 INSERTIONAL MUTAGENESIS SCREEN

The generation and characterization of the KCTSB13 cohort, and the common insertion sites (CISs) analysis is described in [9].

9.2.7 DETECTION OF STAG2-T2/ONC FUSION MRNA BY RT-PCR IN SLEEPING BEAUTY TUMORS

Total RNA was extracted from snap-frozen pancreatic tumors developed in KCTSB13 mice using the RNeasy Mini kit (QIAGEN), and total RNA (1 μg) was reverse transcribed into cDNA using the High Capacity RNA-to-cDNA kit (Applied Biosystems). RT-PCR was carried out with a nested PCR approach using primers of mouse STAG2 exon 1 and the carp β-actin splice acceptor sequence of the T2/Onc transposon cassette. cDNA was used as a template in a first round of PCR using specific primers corresponding to exon 1 of *STAG2* (5′-GAGGGAACAACATTCATGTG-3′) and the carp β-actin splice acceptor sequence (5′-CATACCGGCTACGTTGCTAA-3′). The product of this reaction was used as a template in a second round of nested PCR using an internal primer in the STAG2 exon 1 (5′-CCCTCGGCTTCTCTCCCCCG-3′) and a second primer in the carp β-actin splice acceptor sequence (5′-ACGTTGC-TAACAACCAGTGC-3′). PCR products were cloned into pCR 2.1-TOPO vector (Invitrogen) and positives clones sequenced.

9.2.8 CELL CULTURE

The 13 pancreatic and 3 control cell lines in this study were obtained directly from ATCC, which performs cell line characterizations using short tandem repeat profiling [18], and passaged for fewer than 6 months after resuscitation. Cells were maintained in RPMI medium (Invitrogen) supplemented with 10% fetal bovine serum (FBS) and penicillin/streptomycin (Invitrogen), and passaged for 3 months for all assays.

9.2.9 WESTERN BLOT ANALYSIS

The 13 PDA cell lines and 3 control cell lines (SK-ES-1, U87-MG, and A375) were lysed in buffer (Roche, Indianapolis, IN) containing protease and phosphatase inhibitors. Total protein (25 µg per sample) from each cell line was resolved on NuPAGE Novex 4-12% Bis-Tris precast gels (Invitrogen) then transferred to PVDF membranes. Antibodies against β-actin (Cell Signaling Technology, Danvers, MA, USA) and STAG2 (Santa Cruz Biotechnology, sc-81852, Santa Cruz, CA, USA) were used at 1:30,000 and 1:100 dilutions, respectively. Amersham ECL Prime Western Blotting Detection Reagent (GE Healthcare) was used to detect antibody binding on a BioSpectrum Imaging System (UVP, Upland, CA, USA).

9.2.10 SMALL INTERFERING RNA

The pancreatic cancer cell lines PANC-1 and Panc 04.03 were reverse transfected with four small interfering RNA (siRNA) sequences (QIAGEN) targeting STAG2, in addition to GFP, UBB, and ACDC control siRNAs (QIAGEN). siRNA (1 µl of 0.667 µM) was printed into each well of barcoded 384-well plates with a solid-white bottom (Corning 8749, Corning, NY, USA) using a Biomek FX Laboratory Automation Workstation (Beckman Coulter, Indianapolis, IN, USA). A transfection reagent, SilentFect lipid (Bio-Rad, Hercules, CA, USA), was used to introduce the siRNA sequences into the cells. A mixture of SilentFect and serum-free RPMI medium was added to the plates (20 µl per well) using a BioTek µFill Dispenser (BioTek, Winooski, VT, USA). Plates were then incubated for 30 minutes at room temperature to allow the formation of transfection reagent-nucleic acid complexes. Cells were trypsinized, quantified and resuspended in 10% FBS-RPMI assay medium and dispensed into the plates

(20 μl per well) containing the siRNA using a μFill Dispenser (1,000 cells per well for PANC-1, 2,000 cells per well for Panc 04.03). Cells were incubated at 37°C for 24 hours before treatment with varying concentrations of each of 18 drugs (ranging from 0.6 nM to 100 μM) currently in use in our clinical trials, or vehicle alone in medium with 5% FBS by dispensing a 10 μl volume per well. The plates were further incubated at 37°C for 5 days before cellular viability was measured using CellTiter-Glo luminescent reagent (Promega) and an Analyst GT Multimode Microplate Reader (Molecular Devices, Sunnyvale, CA, USA). The final assay volume (per well) contained a 13 nM concentration of each siRNA, 40 nl of SilentFect transfection reagent and 5% FBS (20 μl serum-free medium + 20 μl of medium with 10% FBS + 10 μl of drug solution prepared in medium with 5% FBS).

9.3 RESULTS

Rapid autopsy samples, consisting of patient-matched primary and distant metastatic tissues, have been used to study the clonal evolution of PDA [13,19]. In our studies we have screened multiple examples of these tumors using our flow sorting methods to identify clonal populations of PDA cells for genome analysis. In one of these cases we detected a 42.5 kb homozygous deletion in *STAG2* in a 4.5 N population sorted from the primary pancreatic tissue (Figure 1). The clonal deletion mapped to the 5′ region of the gene and included exon 1 and a series of regulatory regions [20]. Our use of flow sorted samples allowed objective discrimination of homozygous loss with a rigorous threshold of $\log_2$ ratio < -3.0 in our aCGH results. The same homozygous deletion was detected in the 4.5 N aneuploid populations found in the two distinct distant metastatic sites surveyed within the same patient. In contrast, the patient-matched sorted diploid population had a normal aCGH profile with an intact *STAG2* locus. The presence of the same somatic homozygous deletion within a 4.5 N PDA population in each anatomical site suggests that cells that lost *STAG2* arose early in the history of this tumor and contributed to its progression to metastatic disease. Tumor suppressor genes targeted by homozygous deletions are frequently inactivated by alternative mechanisms, including somatic mutations [21,22].

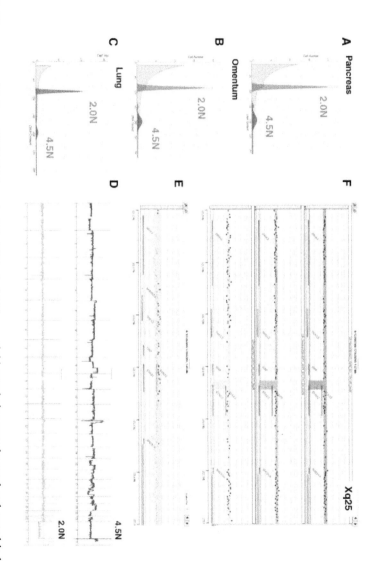

FIGURE 1: Homozygous deletion of STAG2 in metastatic PDA. A 4.5 N aneuploid population was detected and sorted in biopsies from the (A) primary and (B,C) two metastatic sites from rapid autopsy specimen UNMC 12R. (D) aCGH analysis of the sorted diploid (2.0 N; bottom) and the sorted 4.5 N (top) populations. Chromosome Xq25 CGH plots of (E) sorted 2.0 N population and (F) clonal homozygous deletion of *STAG2* in the 4.5 N populations sorted from the pancreas (top panel), omentum (middle panel), and lung (bottom panel). Shaded areas in (E,F) denote ADM2 step gram defined genomic intervals.

Given the role of chromosomal instability and aneuploidy in the development and progression of PDA, we hypothesized that STAG2 would be somatically deficient in a subset of PDA patients.

To identify somatic mutations we sequenced the first 33 of the 35 *STAG2* exons in 50 flow sorted clinical samples. Tissue samples were obtained from three sources. The first was a phase II trial of patients with advanced metastatic disease that had progressed on at least one prior therapy. The second was a phase III trial of patients with resectable PDA. The final source was a tumor bank that included tissues from a series of rapid autopsy samples. Whenever possible we used a patient matched blood sample as a control. However, for those samples of interest without matching normal tissue samples we evaluated the flow cytometry and aCGH profiles of the diploid and aneuploid fractions in each biopsy. In all cases the aneuploid fractions represent pure (>95%) tumor populations as determined by their separation from other peaks in the histograms and the presence of distinct genomic copy number aberrations, including homozygous deletions ($\log_2$ratios < -3.0) and focal amplicons. In contrast, the total diploid fractions from PDA biopsies may contain admixtures of neoplastic and non-neoplastic cells. Thus, for each sample of interest we profiled the total diploid fraction by aCGH. In all cases the genomes of the sorted diploid cells were non-aberrant by copy number analysis. This allowed discrimination of germ line from somatic events for tumor samples of interest, including those tissues without matching blood samples. A non-conserved *STAG2* mutation, E20Q, was found in one additional patient sample (Figure 2). The somatic nature of this mutation was confirmed by sequencing the 2.0 N population sorted from the primary tumor sample. These data provide the first report of *STAG2* somatic aberrations in human PDA. In addition we detected six recurring previously reported polymorphisms, and one novel polymorphism, throughout the gene in multiple patient samples (Additional file 1).

The previously reported transposon-mediated mutational screen of a pancreatic ductal preneoplasia mouse model identified a series of genes whose targeted inactivation cooperated with *KRAS*G[12D] in the development and progression of PDA [9]. Strikingly, *STAG2* was identified as one of the top 20 candidate genes. Transposon insertions within the *STAG2* locus were found in 37/198 (18.7%) tumors that arose in the mouse model.

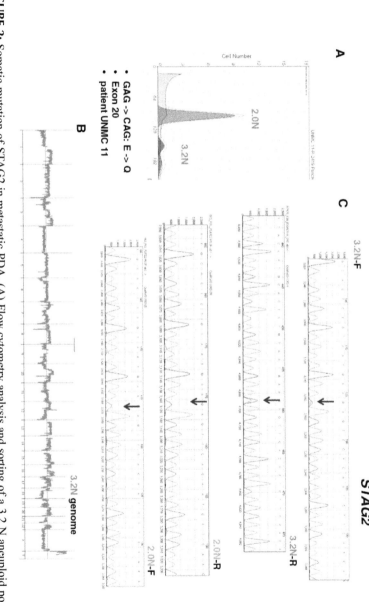

FIGURE 2: Somatic mutation of STAG2 in metastatic PDA. (A) Flow cytometry analysis and sorting of a 3.2 N aneuploid population in primary pancreas tissue from rapid autopsy specimen UNMC 11R. (B) Whole genome aCGH plot of the 3.2 N population. (C) Detection of somatic mutation in forward (3.2 N-F) and reverse (3.2 N-R) sequences in the 3.2 N genome. Mutation was not detected in the matching 2.0 N population sorted from the same tissue.

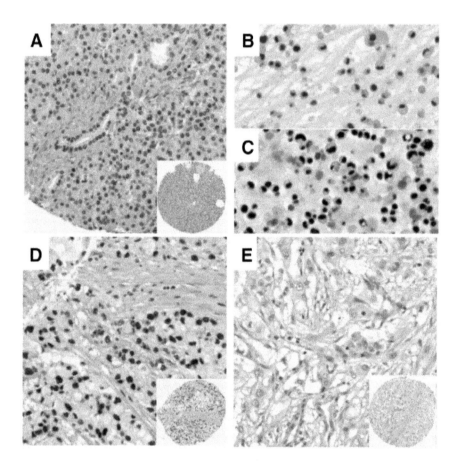

FIGURE 3: Immunohistochemical staining of STAG2 in pancreatic cancer. Near ubiquitous staining in (A) ductal cells of normal tissue, (B) the primary cancer cell line PaCaDD161, and (D) pancreatic cancer. (C) The cell line PaCaDD135 displays a heterogeneous staining pattern, whereas in (E) a minor fraction of pancreatic cancer STAG2 expression is lost. Magnification 200× (insets 100×).

In this present study we confirmed that *STAG2* expression was disrupted in these tumors by isolating chimeric fusion mRNAs that spliced the *STAG2* transcript to the T2/Onc transposon (Additional file 2). The insertions were found in preinvasive lesions and primary tumors, suggesting that early inactivation of the gene could be important in the progression of the disease. Insertions were also found in eight mice with metastasis. This included one animal with sufficient tumor tissue available for resequencing from primary and multiple metastases that confirmed the clonal nature of the *STAG2* insertion. Interestingly, the insertions are found in both males and females, suggesting that X-inactivation or haploinsufficiency could be contributing to the loss of *STAG2* expression in the females.

To further assess the clinical significance of STAG2 expression in human tumors, we screened a TMA containing a collection of 344 specimens obtained from resected German patients (Additional file 3). In normal tissue nearly all ductal cells stained with a high intensity (Figure 3). There was a broad range of signal intensities with a statistically significant loss of STAG2 expression in the tumor tissue (Wilcoxon rank test) and complete absence of STAG2 staining observed in 15 (4.3%) patients (Additional files 4 and 5). In univariate Kaplan-Meier analysis nearly complete STAG2-positive staining (>95% of nuclei positive) was associated with a median survival benefit of 6.41 months (P=0.031) (Figure 4). Interestingly, the survival benefit of adjuvant chemotherapy can only be identified in the group of patients with a STAG2 staining <95% (median survival benefit 7.65 months; P=0.028) (Figure 5). Multivariate cox regression analysis showed that STAG2 is an independent prognostic factor for survival in pancreatic cancer patients (Table 1).

Given the role of STAG2 in the maintenance of genomic stability, we hypothesized that loss of STAG2 would create a synthetic lethal condition to exposure to one or more therapeutic agents. To test this we screened 13 PDA cell lines for expression of STAG2 by western blot analysis (Additional file 6). These included five cell lines previously reported as wild type [3]. All 13 cell lines have been shown to express *STAG2* mRNA; however, 2 of the cell lines had decreased protein levels relative to the other cell lines (Additional file 7) [23]. We selected a cell line (PANC-1) positive for STAG2 protein expression and one cell line (Panc 04.03) with low STAG2 protein expression for siRNA assays.

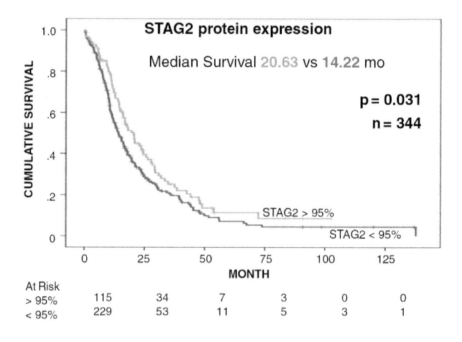

FIGURE 4: STAG2 expression and overall survival in patients with PDA. Tissue microarray (TMA) analysis of STAG2 protein expression in PDA samples from 344 patients. Cumulative survival (y-axis) of patients with intact versus deficient STAG2 expression levels is plotted versus time (x-axis). Mo, months.

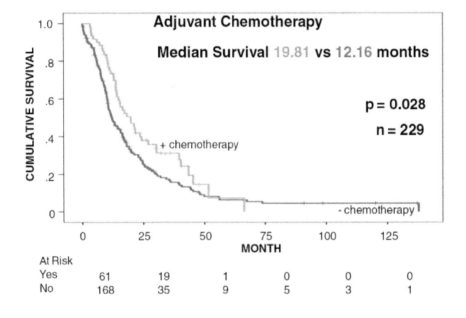

FIGURE 5: STAG2 expression and response to adjuvant therapy. Cumulative survival (y-axis) plotted versus time (x-axis) is compared for the 229 STAG2-deficient patients who received (n=61) or did not receive (n=168) adjuvant chemotherapy. Mo, months.

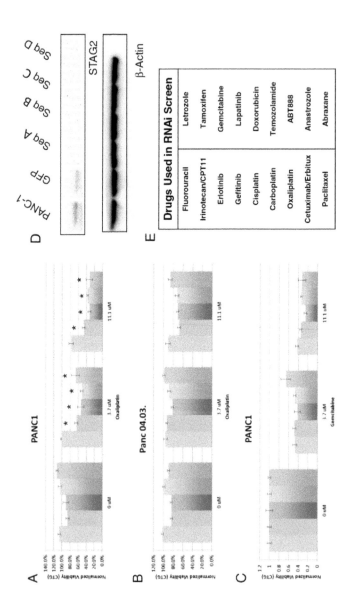

FIGURE 6: Synthetic lethal analysis of STAG2 knockdown and exposure to chemotherapeutic agents in PDA cells. (A) Synthetic lethal response of PANC-1 cells and (B) Panc 04.03 cells in the presence of oxaliplatin. (C) Absence of response of PANC-1 cells in the presence of gemcitabine. Asterisks indicate that all four siRNA sequences at drug concentrations of 3.7 and 11.1 μM have P<0.01 when compared to GFP control. Error bars in (A–C) represent standard deviations. (D) Western blot of STAG2 expression in PANC1 cells in the presence of control (GFP) and four STAG2 siRNAs. (E) Drugs used in siRNA synthetic lethal screen.

Each of these two cell lines was exposed to 8 concentrations of the 18 clinical drugs available from our recently completed clinical trial in the presence or absence of 4 siRNAs targeting *STAG2* (Figure 6). Depletion of STAG2 did not affect the viability of these cell lines in the absence of drug. The STAG2-positive PANC-1 cell line displayed an increased sensitivity to the three platinum-based drugs (cisplatin, carboplatin, oxaliplatin) used in our assays in the presence of *STAG2* targeting siRNAs. For example, each of the four *STAG2* siRNAs significantly increased ($P<0.01$) the sensitivity of PANC-1 cells to oxaliplatin at drug concentrations of 3.7, and 11.1 μM when compared to GFP control. In contrast, there was no difference in the response of the low STAG2 protein-expressing Panc 04.03 cells in the presence of any of the siRNAs. The knockdown of STAG2 in PANC-1 and Panc 04.03 cells was confirmed by western blot analysis for each of the siRNAs.

9.4 DISCUSSION

Somatic mutations in *STAG2* have now been reported in a variety of human cancers [3,8,24]. One notable exception to date has been pancreatic tumors. The initial analysis of *STAG2* in PDA was limited to five established cell lines [3]. These same cell lines were also positive for STAG2 protein expression in our current study (Additional file 6). In addition, recent next generation sequencing-based studies of the genomic landscape of PDA tumors have not detected somatic aberrations that selectively disrupted the *STAG2* locus [1,25]. Thus, to our knowledge, this report represents the first description of a tumor suppressor role for STAG2 in human PDA. A key feature of our study of human tumors is the use of DNA content flow assays to identify and purify distinct populations of tumor cells in each biopsy of interest. These assays enable an unbiased analysis of clinical human PDA tissues regardless of tumor content. For example, in the two confirmed cases of somatic genomic targeting of *STAG2* the tumor cell content was less than 10% of the total cellular content of the biopsy (Figures 1 and 2). Our use of highly purified flow sorted tumor populations and the ability to detect homozygous deletions and mutations in PDA clinical samples, regardless of cellularity and tumor content, provides a robust screen for potential tumor suppressors.

TABLE 1: Multivariate analysis of tissue microarray cohort

Model	Variable	P-value	Hazard ratio (95% CI))
A. Clinical pathology and STAG2	STAG2 (< 95%)	0.017	1.421 (1.064-1.898)
	T stage (1/2)	0.729	1.066 (0.743-1.529)
	N stage (N0)	0.196	0.826 (0.618-1.104)
	Grade (1/2)	0.013	0.711 (0.544-0.93)
	Sex (female)	0.865	1.023 (0.785-1.333)
	Margin involvement (negative)	0.297	0.852 (0.63-1.151)
	M stage (M0)	0.132	1.487 (0.887-2.492)
B. Clinical pathology and STAG2	STAG2 (< 95%)	0.017	1.415 (1.064-1.883)
	T stage (1/2)	0.719	1.068 (0.746-1.531)
	N stage (N0)	0.195	0.825 (0.618-1.103)
	Grade (1/2)	0.013	0.713 (0.546-0.931)
	Margin involvement (negative)	0.3	0.853 (0.631-1.152)
	M stage (M0)	0.126	1.494 (0.894-2.497)
C. Clinical pathology and STAG2	STAG2 (< 95%)	0.016	1.42 (1.068-1.888)
	N stage (N0)	0.198	0.827 (0.619-1.105)
	Grade (1/2)	0.014	0.717 (0.55-0.934)
	Margin involvement (negative)	0.321	0.861 (0.64-1.158)
	M stage (M0)	0.131	1.484 (0.889-2.478)
D. Clinical pathology and STAG2	STAG2 (< 95%)	0.014	1.428 (1.074-1.898)
	N stage (N0)	0.185	0.822 (0.616-1.098)
	Grade (1/2)	0.011	0.709 (0.545-0.924)
	M stage (M0)	0.086	1.556 (0.94-2.574)
E. Clinical pathology and STAG2 (final model)	STAG2 (< 95%)	0.021	1.394 (1.051-1.849)
	Grade (1/2)	0.007	0.697 (0.536-0.906)
	M stage (M0)	0.061	1.615 (0.979-2.665)

CI, confidence interval.

Recent cancer genome studies have identified multiple examples of mutations and aberrations of potential driver genes in relatively small subsets (1 to 5%) of patients with a given tumor type [26-28]. Strikingly, many of these low frequency driver events occur in multiple tumor types, highlighting their potential clinical significance. Notably, a *STAG2* mutation was identified as a driver of clonal evolution from myeloid dysplastic syndrome to secondary acute myeloid leukemia [24]. The relatively low prevalence of *STAG2* somatic aberrations that we report in PDA is consistent with those in many other tumors [21]. The identification of *STAG2* as one of the most significant tumor suppressors of $KRAS^{G12D}$-driven PDA in the GEM model supports an interactive role of these two genes in the development and progression of PDA. In our studies of human PDA we have confirmed the presence of *KRAS* mutations in over 95% of the flow sorted samples we have profiled. These include the two patients with *STAG2* somatic aberrations in the current study, one of whom had a clonal homozygous deletion in the primary tumor and two distant metastatic lesions. Thus, our current results further support the findings from the mouse model that *STAG2* inactivation cooperates with *KRAS* mutation as an early event in the evolution and progression of human PDA.

Our TMA-based expression analysis provides further clinical validation of a tumor suppressor role for *STAG2* in PDA. A recent study reporting *STAG2* mutations in bladder cancer included an immunohistochemistry screen across a broad panel of tumor types [8]. These included 36 PDA tumors, none of which showed loss of STAG2 expression. Tumor suppressor genes that are located on chromosome X are frequently subject to epigenetic regulation and silencing by chromosome X inactivation [29,30]. The reactivation of silent but intact copies of these tumor suppressors with agents such as 5-aza-2'-deoxycytidine and histone deacetylase (HDAC) inhibitors such as trichostatin has been proposed as a strategy to enhance therapeutic responses in solid tumors. The number of tumors (4.3%, 15/344) with a complete absence of protein expression in our panel of primary surgical resection tissues and our finding of somatic genetic lesions in 4% (2/50) of tumors surveyed reflects the overall prevalence of those PDAs with complete STAG2 loss (Additional file 8). However, a striking finding was the effect of even modest decreases of STAG2 protein levels on survival and response to adjuvant therapies (Figures 4 and 5).

The presence of intact but under-expressed copies of this mediator of genome stability provides a potential therapeutic vulnerability similar to that proposed for the X-linked deubiquitinase *USP9X*[9]. Specifically, up-regulation of deficient but intact *STAG2* may provide a therapeutic benefit for patients with PDA, including those who have already progressed to invasive stages of disease. Furthermore, given the synergistic interaction with *KRAS* mutations, an early and ubiquitous genetic event in the development of PDA and the clinical significance of deficient expression, development of agents that increase STAG2 expression may provide a strategy for both treatment and prevention. However, a caveat for this approach is that, unlike *USP9X*, there is a low (approximately 4%) but potentially significant prevalence of genetic lesions in *STAG2* in human PDA.

Current therapeutic options for patients with advanced PDA include a series of chemotherapeutics with a broad range of mechanisms and targets (Figure 6). *STAG2* has been shown to regulate proper chromosome segregation in diploid cells. Loss of function gives rise to chromosome aneuploidy and a mutator phenotype in model systems. Thus, we hypothesized inactivation of *STAG2* could create a synthetic lethal combination with one or more currently used therapeutic agents that target either DNA replication or repair. Consistent with this, our TMA-based analysis of localized surgically resected tumors showed that those PDAs with deficient *STAG2* levels derived the most benefit from standard adjuvant therapy with either gemcitabine or 5-FU. Our finding that RNAi-based silencing of *STAG2* sensitizes PDA cells specifically to platinum-based therapies in vitro suggests that the increased therapeutic benefit seen in patients receiving adjuvant therapies could be further enhanced. However, it remains to be determined if this added benefit may be limited to those relatively rare PDAs with complete loss of STAG2 expression.

9.5 CONCLUSIONS

The clinical significance and translational importance of low frequency mutations and genomic lesions is a challenge to the study of PDA and other solid tumors. Our iterative approach, involving clonal genomic analysis of clinical samples, genetically engineered mouse models, clinical validation of genes of interest, and functional interrogation of candidate therapeutic targets and agents provides a comprehensive and highly translational approach to the study of PDA. This current work validates that *STAG2* behaves as a tumor suppressor gene in human PDA. Given its role in the maintenance of genome stability, the synergy with *KRAS* mutations, and the clinical significance of altered expression, we propose that deficiencies in *STAG2* represent a potential therapeutic vulnerability that can be exploited for improved treatment and possible prevention of PDA.

REFERENCES

1. Biankin AV, Waddell N, Kassahn KS, Gingras MC, Muthuswamy LB, Johns AL, Miller DK, Wilson PJ, Patch AM, Wu J, Chang DK, Cowley MJ, Gardiner BB, Song S, Harliwong I, Idrisoglu S, Nourse C, Nourbakhsh E, Manning S, Wani S, Gongora M, Pajic M, Scarlett CJ, Gill AJ, Pinho AV, Rooman I, Anderson M, Holmes O, Leonard C, Taylor D, et al.: Pancreatic cancer genomes reveal aberrations in axon guidance pathway genes. Nature 2012, 491:399-405.

2. Jones S, Zhang X, Parsons DW, Lin JC, Leary RJ, Angenendt P, Mankoo P, Carter H, Kamiyama H, Jimeno A, Hong SM, Fu B, Lin MT, Calhoun ES, Kamiyama M, Walter K, Nikolskaya T, Nikolsky Y, Hartigan J, Smith DR, Hidalgo M, Leach SD, Klein AP, Jaffee EM, Goggins M, Maitra A, Iacobuzio-Donahue C, Eshleman JR, Kern SE, Hruban RH, et al.: Core signaling pathways in human pancreatic cancers revealed by global genomic analyses. Science 2008, 321:1801-1806.

3. Solomon DA, Kim T, Diaz-Martinez LA, Fair J, Elkahloun AG, Harris BT, Toretsky JA, Rosenberg SA, Shukla N, Ladanyi M, Samuels Y, James CD, Yu H, Kim JS, Waldman T: Mutational inactivation of STAG2 causes aneuploidy in human cancer. Science 2011, 333:1039-1043.

4. Kolodner RD, Cleveland DW, Putnam CD: Cancer. Aneuploidy drives a mutator phenotype in cancer. Science 2011, 333:942-943.

5. Olesen SH, Thykjaer T, Orntoft TF: Mitotic checkpoint genes hBUB1, hBUB1B, hBUB3 and TTK in human bladder cancer, screening for mutations and loss of heterozygosity. Carcinogenesis 2001, 22:813-815.

6. Lengauer C, Kinzler KW, Vogelstein B: Genetic instabilities in human cancers. Nature 1998, 396:643-649.

7. Cahill DP, Lengauer C, Yu J, Riggins GJ, Willson JK, Markowitz SD, Kinzler KW, Vogelstein B: Mutations of mitotic checkpoint genes in human cancers. Nature 1998, 392:300-303.

8. Solomon DA, Kim JS, Bondaruk J, Shariat SF, Wang ZF, Elkahloun AG, Ozawa T, Gerard J, Zhuang D, Zhang S, Navai N, Siefker-Radtke A, Phillips JJ, Robinson BD, Rubin MA, Volkmer B, Hautmann R, Kufer R, Hogendoorn PC, Netto G, Theodorescu D, James CD, Czerniak B, Miettinen M, Waldman T: Frequent truncating mutations of STAG2 in bladder cancer. Nat Genet 2013, 45:1428-1430.

9. Perez-Mancera PA, Rust AG, van der Weyden L, Kristiansen G, Li A, Sarver AL, Silverstein KA, Grutzmann R, Aust D, Rummele P, Knosel T, Herd C, Stemple DL, Kettleborough R, Brosnan JA, Li A, Morgan R, Knight S, Yu J, Stegeman S, Collier LS, ten Hoeve JJ, de Ridder J, Klein AP, Goggins M, Hruban RH, Chang DK, Biankin AV, Grimmond SM, Wessels LF, Australian Pancreatic Cancer Genome Initiative, et al.: The deubiquitinase USP9X suppresses pancreatic ductal adenocarcinoma. Nature 2012, 486:266-270.

10. Seymour AB, Hruban RH, Redston M, Caldas C, Powell SM, Kinzler KW, Yeo CJ, Kern SE: Allelotype of pancreatic adenocarcinoma. Cancer Res 1994, 54:2761-2764.

11. Maley CC, Galipeau PC, Finley JC, Wongsurawat VJ, Li X, Sanchez CA, Paulson TG, Blount PL, Risques RA, Rabinovitch PS, Reid BJ: Genetic clonal diversity predicts progression to esophageal adenocarcinoma. Nat Genet 2006, 38:468-473.

12. Rabinovitch PS, Dziadon S, Brentnall TA, Emond MJ, Crispin DA, Haggitt RC, Bronner MP: Pancolonic chromosomal instability precedes dysplasia and cancer in ulcerative colitis. Cancer Res 1999, 59:5148-5153.

13. Ruiz C, Lenkiewicz E, Evers L, Holley T, Robeson A, Kiefer J, Demeure MJ, Hollingsworth MA, Shen M, Prunkard D, Rabinovitch PS, Zellweger T, Mousses S, Trent JM, Carpten JD, Bubendorf L, Von Hoff D, Barrett MT: Advancing a clinically relevant perspective of the clonal nature of cancer. Proc Natl Acad Sci U S A 2011, 108:12054-12059.

14. WMA Declaration of Helsinki – Ethical Principles for Medical Research Involving Human Subjects World Med J 2013, 59:199-202.

15. Galipeau PC, Li X, Blount PL, Maley CC, Sanchez CA, Odze RD, Ayub K, Rabinovitch PS, Vaughan TL, Reid BJ: NSAIDs modulate CDKN2A, TP53, and DNA content risk for progression to esophageal adenocarcinoma. PLoS Med 2007, 4:e67.

16. Rabinovitch PS, Longton G, Blount PL, Levine DS, Reid BJ: Predictors of progression in Barrett's esophagus III: baseline flow cytometric variables. Am J Gastroenterol 2001, 96:3071-3083.

17. Von Hoff D, Ramanathan R, Evans D, Demeure MJ, Maney T, Wright B, Gatalica Z, McGinniss MJ, Piper VG: Actionable targets in pancreatic cancer detected by immunohistochemistry (IHC), microarray (MA), fluorescent in situ hybridization (FISH) and mutational analysis. J Clin Oncol 2012, 30:Abstract 4013.

18. The International Cell Line Authentication Committee (ICLAC) [http://standards. atcc.org/kwspub/home/the_international_cell_line_authentication_committee-iclac_/]

19. Yachida S, Jones S, Bozic I, Antal T, Leary R, Fu B, Kamiyama M, Hruban RH, Eshleman JR, Nowak MA, Velculescu VE, Kinzler KW, Vogelstein B, Iacobuzio-Donahue CA: Distant metastasis occurs late during the genetic evolution of pancreatic cancer. Nature 2010, 467:1114-1117.

20. Consortium EP, Dunham I, Kundaje A, Aldred SF, Collins PJ, Davis CA, Doyle F, Epstein CB, Frietze S, Harrow J, Kaul R, Khatun J, Lajoie BR, Landt SG, Lee BK, Pauli F, Rosenbloom KR, Sabo P, Safi A, Sanyal A, Shoresh N, Simon JM, Song L, Trinklein ND, Altshuler RC, Birney E, Brown JB, Cheng C, Djebali S, Dong X, et al.: An integrated encyclopedia of DNA elements in the human genome. Nature 2012, 489:57-74.

21. Veeriah S, Taylor BS, Meng S, Fang F, Yilmaz E, Vivanco I, Janakiraman M, Schultz N, Hanrahan AJ, Pao W, Ladanyi M, Sander C, Heguy A, Holland EC, Paty PB, Mischel PS, Liau L, Cloughesy TF, Mellinghoff IK, Solit DB, Chan TA: Somatic mutations of the Parkinson's disease-associated gene PARK2 in glioblastoma and other human malignancies. Nat Genet 2010, 42:77-82.

22. Cox C, Bignell G, Greenman C, Stabenau A, Warren W, Stephens P, Davies H, Watt S, Teague J, Edkins S, Birney E, Easton DF, Wooster R, Futreal PA, Stratton MR: A survey of homozygous deletions in human cancer genomes. Proc Natl Acad Sci U S A 2005, 102:4542-4547.

23. Barretina J, Caponigro G, Stransky N, Venkatesan K, Margolin AA, Kim S, Wilson CJ, Lehar J, Kryukov GV, Sonkin D, Reddy A, Liu M, Murray L, Berger MF, Monahan JE, Morais P, Meltzer J, Korejwa A, Jane-Valbuena J, Mapa FA, Thibault J, Bric-Furlong E, Raman P, Shipway A, Engels IH, Cheng J, Yu GK, Yu J Jr, Aspesi P, de Silva M, et al.: The Cancer Cell Line Encyclopedia enables predictive modelling of anticancer drug sensitivity. Nature 2012, 483:603-607.

24. Walter MJ, Shen D, Ding L, Shao J, Koboldt DC, Chen K, Larson DE, McLellan MD, Dooling D, Abbott R, Fulton R, Magrini V, Schmidt H, Kalicki-Veizer J, O'Laughlin M, Fan X, Grillot M, Witowski S, Heath S, Frater JL, Eades W, Tomasson M, Westervelt P, DiPersio JF, Link DC, Mardis ER, Ley TJ, Wilson RK, Graubert TA: Clonal architecture of secondary acute myeloid leukemia. N Engl J Med 2012, 366:1090-1098.

25. Campbell PJ, Yachida S, Mudie LJ, Stephens PJ, Pleasance ED, Stebbings LA, Morsberger LA, Latimer C, McLaren S, Lin ML, McBride DJ, Varela I, Nik-Zainal SA, Leroy C, Jia M, Menzies A, Butler AP, Teague JW, Griffin CA, Burton J, Swerdlow H, Quail MA, Stratton MR, Iacobuzio-Donahue C, Futreal PA: The patterns and dynamics of genomic instability in metastatic pancreatic cancer. Nature 2010, 467:1109-1113.

26. Guichard C, Amaddeo G, Imbeaud S, Ladeiro Y, Pelletier L, Maad IB, Calderaro J, Bioulac-Sage P, Letexier M, Degos F, Clement B, Balabaud C, Chevet E, Laurent A, Couchy G, Letouze E, Calvo F, Zucman-Rossi J: Integrated analysis of somatic mutations and focal copy-number changes identifies key genes and pathways in hepatocellular carcinoma. Nat Genet 2012, 44:694-698.

27. Nik-Zainal S, Van Loo P, Wedge DC, Alexandrov LB, Greenman CD, Lau KW, Raine K, Jones D, Marshall J, Ramakrishna M, Shlien A, Cooke SL, Hinton J, Menzies A, Stebbings LA, Leroy C, Jia M, Rance R, Mudie LJ, Gamble SJ, Stephens PJ, McLaren S, Tarpey PS, Papaemmanuil E, Davies HR, Varela I, McBride DJ, Bignell GR, Leung K, Butler AP, et al.: The life history of 21 breast cancers. Cell 2012, 149:994-1007.

28. Stephens PJ, Tarpey PS, Davies H, Van Loo P, Greenman C, Wedge DC, Nik-Zainal S, Martin S, Varela I, Bignell GR, Yates LR, Papaemmanuil E, Beare D, Butler A, Cheverton A, Gamble J, Hinton J, Jia M, Jayakumar A, Jones D, Latimer C, Lau KW, McLaren S, McBride DJ, Menzies A, Mudie L, Raine K, Rad R, Chapman MS, Teague J, et al.: The landscape of cancer genes and mutational processes in breast cancer. Nature 2012, 486:400-404.

29. Rivera MN, Kim WJ, Wells J, Driscoll DR, Brannigan BW, Han M, Kim JC, Feinberg AP, Gerald WL, Vargas SO, Chin L, Iafrate AJ, Bell DW, Haber DA: An X

chromosome gene, WTX, is commonly inactivated in Wilms tumor. Science 2007, 315:642-645.

30. Spatz A, Borg C, Feunteun J: X-chromosome genetics and human cancer. Nat Rev Cancer 2004, 4:617-629.

There are several supplemental files that are not available in this version of the article. To view this additional information, please use the citation information cited on the first page of this chapter.

CHAPTER 10

MOLECULAR DETERMINANTS OF CONTEXT-DEPENDENT PROGESTERONE RECEPTOR ACTION IN BREAST CANCER

CHRISTY R. HAGAN AND CAROL A. LANGE

10.1 INTRODUCTION

The mitogenic activity of estrogen is well established, but an under-studied ovarian steroid hormone, progesterone, is emerging as a primary mitogen in the breast, contributing significantly to genetic programming required for mammary stem cell self-renewal, mammary gland development, proliferation, and hyperplasia [1]. The effects of progesterone are triggered after binding of progesterone to its intracellular receptor, the progesterone receptor (PR). The PR exists in two primary isoforms, differing structurally by the inclusion of an N-terminal segment unique to the full-length isoform, PR-B [2] (Figure 1). This region, termed the B-upstream segment, is missing from the shorter isoform, PR-A [3]. The two isoforms are encoded by the same gene (regulated by distinct but tandem upstream promoters) and are most often co-expressed [4]. The PR is a member of the steroid hormone receptor subgroup of ligand-activated transcription factors within the large nuclear receptor superfamily, and is an important

This chapter was originally published under the Creative Commons Attribution License. Hagan CR and Lange CA. Molecular Determinants of Context-Dependent Progesterone Receptor Action in Breast Cancer. BMC Medicine 12,32 (2014). doi:10.1186/1741-7015-12-32.

down-stream effector of estrogen-receptor (ER) signaling; in most cir-cumstances, estrogen is required for robust PR expression. PR binding to DNA, either directly through progesterone response elements or indirectly through tethering interactions with other transcription factors, activates transcriptional profiles associated with mammary gland proliferation and breast cancer [5-9]. Additionally, PR binding interactions with transcrip-tional co-activators and repressors are critical to PR transcription factor function [10].

PRs are highly post-translationally modified, primarily through N-ter-minal phosphorylation (select phosphorylation sites most relevant to breast cancer biology are highlighted in Figure 1), acetylation, SUMOylation, and ubiquitination [9,11-17]. These receptor modifications dramatically alter PR function, receptor localization and turnover, and promoter selec-tivity. The PR can be phosphorylated basally in the absence of the hor-monal ligand, but is potently modified after ligand treatment, in response to local growth factors or in a cell cycle-dependent manner [12,13,15-17] (G. Dressing and C. Lange, unpublished data). Mitogenic protein kinase—such as CDK2, CK2, and MAPK—have been shown to phosphorylate PR and subsequently modify PR action. Therefore, PR can be thought of as a 'mitogenic sensor' in the cell, with PR phosphorylation serving as a read-out of kinase activity. Highly mitogenic environments like cancer, where kinase activities are frequently high, may be a situation where PR is per-sistently phosphorylated in the absence of ligand. Moreover, in this case, mitogenic signals (that is, growth factors) may diminish or replace the need for ligand, thus activating PRs inappropriately.

In addition to receiving direct inputs from protein kinases via phos-phorylation, PR interacts with and activates members of cytoplasmic sig-naling cascades, such c-SRC [18,19]. These rapid signaling actions of PR (previously termed non-genomic actions) are independent of PR's DNA-binding transcriptional activity [19]. However, direct PR interactions with components of kinase cascades and subsequent signaling pathway activa-tion are highly integrated with PR genomic actions. Indeed, kinases that modify PR, as well as other growth factor-activated kinases, have been found in association with DNA-bound (that is, phosphorylated) PRs that function as part of the same transcription complexes that regulate PR-tar-get gene promoters and enhancers [16,20,21]. Increasing knowledge about

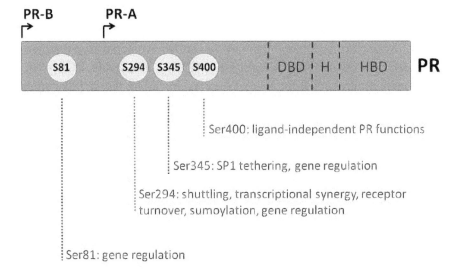

FIGURE 1: Schematic of progesterone receptor structure and select phosphorylation sites. Progesterone receptor (PR) isoforms A and B differ in their inclusion of an N-terminal upstream segment unique to PR-B. Both isoforms contain an identical DNA binding domain (DBD), hinge region (H) and hormone binding domain (HBD). Full-length PR-B contains 14 phosphorylation sites; serines 81, 294, 345 and 400 have known links to PR action and gene expression in breast cancer.

post-translational PR modifications and PR-modifying binding partners suggests that these events (such as phosphorylation, SUMOylation, and so on) are required for context-dependent activation of PR.

Understanding PR action is of great clinical significance in breast cancer, as evidenced by large-scale clinical trials conducted more than 10 years ago that demonstrated that PR actions fuel breast cancer growth. In two independent trials, women whose hormone-replacement therapy (HRT) regimens included estrogen and synthetic progesterone (that is, medroxyprogesterone acetate, norethisterone, or norgestrel/levonorgestrel) had a higher risk of developing breast cancer than women whose regimens included only estrogen and no progestins [22,23]. The results of these trials remain controversial for several reasons, including the fact that study participants were well past the onset of menopause when HRT was initiated. Additionally, although synthetic progestins clearly closely mimic progesterone in vitro, some synthetic progestins (medroxyprogesterone acetate) may alter androgen receptor (AR) [24] or glucocorticoid receptor (GR) [25] signaling, exhibit different half-lives, and are metabolized differently than natural progesterone, and therefore may be associated with different breast cancer risks relative to their naturally occurring counterparts [26,27]. Finally, continuous dosing of progestins as part of HRT may fail to mimic cyclical lifetime exposure to natural ligand in vivo. However, taken together, these landmark clinical studies implicate PR in human breast cancer development and progression, a finding that is well-supported by animal studies [28,29]. It is thus important to fully understand how activated PRs may contribute to early breast cancer progression, perhaps by driving the transition from steroid receptor (SR)-positive tumors with better clinical prognoses to more aggressive, poorer outcome SR-negative and luminal-B-type tumors.

Convincing preclinical and clinical evidence suggests that progestins increase breast cancer risk in part by driving the proliferation of early lesions [28,30-35]. Even so, at least five main sources of confusion remain regarding the role of PR actions in breast cancer (expanded on in Box 1). First, PR action is context dependent—that is, PR action differs in normal versus neoplastic tissue and according to hormone exposure (for example, in the presence versus absence of estrogen), as well as organ site (for example, proliferative in the breast versus inhibitory in the uterus).

Moreover, despite convincing progestin-dependent proliferative responses in murine models [32,36,37], early reports showed that progesterone was anti-proliferative or non-proliferative in human cells [38-40]. However, recent work from the laboratory of C. Brisken [41] has shown that progesterone is proliferative in human breast tissue microstructures isolated from normal human breast specimens. Interestingly, progesterone-dependent proliferation and signaling is preserved only when the tissue architecture remains intact; human tissues (previously dissociated) grown in two- or three-dimensional cultures did not display this proliferative phenotype, suggestive of further context-dependent PR actions. Second, PR isoform-specific activities (PR-A versus PR-B) overlap but can have very disparate activities within a given target tissue and at selected gene promoters; however, despite their distinct activities, the two PR-isoforms are not distinguished clinically. Third, ligand-independent (that is, growth factor- or kinase-dependent) activities of PR are poorly understood. Fourth, the dosing (cyclical versus continuous) and source (natural versus synthetic) of ligand are likely to be key determinants of the kinetics of PR action. Fifth, although anti-progestins showed clinical promise in early clinical trials, their use was limited by liver toxicities (onapristone; [42]) largely attributable to cross-reactivity with other nuclear receptors, such as GR. This review will focus on the molecular determinants of PR's context-dependent actions and their clinical significance. These PR actions are primarily determined by the availability of PR-binding partners and direct modifications to PR that dictate promoter selection.

10.1.1 POST-TRANSLATIONAL MODIFICATIONS AND MOLECULAR INTERACTIONS ALTER PROMOTER SELECTIVITY

Mounting evidence suggests that post-translational modifications of PR are key determinants of promoter selectivity and, in turn, the spectrum of target genes activated in response to ligand binding (reviewed in [43,44]). PR promoter preference is partially dictated by differences in the recruitment of PR and/or its co-activators or co-repressors to specific DNA sequences. In microarray analyses, cells expressing wild-type PR or PRs containing single point-mutations at specific phosphorylation or

SUMOylation sites exhibit dramatic changes in PR-dependent gene expression, specific to precise post-translational modifications. For example, recent analyses from the Lange laboratory revealed that PR phosphorylation on serine 294 favors the subsequent deSUMOylation on PR lysine 388 [45], thereby yielding a hyperactive receptor that regulates a unique gene expression signature found in high ERBB2-expressiong tumors; this unique phospho-PR gene expression signature predicted decreased survival in patients treated with tamoxifen [9]. By contrast, a separate gene expression pattern is observed when PR is phosphorylated on Ser81 by CK2, a kinase commonly overexpressed in breast cancers; this modification is associated with the expression of gene sets involved in interferon and STAT5 signaling (discussed in more detail below) [8]. Therefore, in response to ligand, growth factor-mediated PR phosphorylation (or phosphorylation-dependent alterations of other post-translational modifications such as SUMOylation) dictates the selective expression of specific subsets of target genes and subsequently their transcriptional programs.

Target gene selectivity is achieved not only through differential recruitment of PR [8,16], but also through associated transcriptional co-activators and repressors that are critical to PR function [9,10,46]. For example, pioneer factors are specialized subsets of transcription factors that open defined regions of chromatin, making it accessible for other transcription factors, like SRs (reviewed in [47,48]). These types of factors have been identified for other nuclear receptors, such as ER and AR; however, they have yet to be identified for PR. Preliminary data suggest that FOXA1 and STAT5 may be putative pioneer factors for PR [8,49,50]; differential binding interactions between PR and these factors provide a mechanism for promoter selectivity, perhaps based on PR post-translational modifications (that is, via phosphorylation-specific interactions with pioneer factors).

Emerging evidence suggests that interactions between members of the SR superfamily is an additional regulatory step in determining target-gene specificity. Interactions between ER and AR have been the focus of recent investigations [51,52]. Recent data from the Lanari group demonstrate the existence of functional cross-talk between ER and PR; both receptors are localized together on regulatory regions of PR-target genes, such as CCND1 and MYC, primarily in response to treatment with progestins [53]. Moreover, work recently published from our group suggests a complimentary story whereby ER and PR cooperate to regulate a subset of ER-target genes in response to estrogen, but fully independent of exogenously added progestin. In this case, PR-B appears to act as a scaffolding molecule for increased recruitment of signaling adaptors and protein kinases that phosphorylate ER within ER/PR-containing transcription complexes [54]. Taken together, these studies suggest that context-dependent progesterone/PR action may in part depend on the presence of other steroid hormones and their receptors. Detailed biochemical studies of steroid hormone receptor cross-talk are needed to provide a framework for a better understanding of differential hormone actions in pre- and post-menopausal conditions where endogenous hormone levels dramatically differ, as well as during breast or prostate cancer treatment with hormone-ablation therapies where closely related steroid hormone receptors (PR, GR, AR, ER) may substitute for the blocked activity of another (ER or AR).

10.1.2 PROGESTERONE RECEPTOR PHOSPHORYLATION BY CK2 AS A PARADIGM FOR RECEPTOR MODIFICATION AND REGULATION

Recent data from our laboratory characterizing PR phosphorylation on Ser81 by CK2 exemplifies how the aforementioned modifications and signaling inputs can alter PR function. CK2 is a ubiquitously expressed kinase often up-regulated in many different types of cancer, including breast [55-57]. We and others have shown that CK2 phosphorylates PR on Ser81, a site that is basally phosphorylated; however, Ser81 phosphorylation levels increase markedly in response to ligand (or when cells enter S phase in the absence of ligand) [16,58]. PR phosphorylation at Ser81 is associated with a specific gene expression profile, which is correlated with pathways altered in breast cancer, including genes implicated in mammary stem cell maintenance and renewal [8,16]. Additionally, the PR target genes whose expression require phosphorylation at Ser81 are significantly associated with interferon/inflammation and STAT-signaling datasets, a unique observation for SRs that represents a novel link between steroid hormone action, inflammation, and cancer [8]. A key target gene regulated by Ser81 phosphorylation is STAT5 itself, and notably, JAK/STAT signaling is required for potent activation of PR Ser81-regulated genes, indicating a feed-forward mechanism for gene program activation (Figure 2). STAT5 is present, along with phosphorylated PR, on the regulatory region of *WNT1*, a key Ser81 target gene known to be involved in cancer and stem cell biology. Moreover, an in silico analysis of a publically available PR whole genome chromatin immunoprecipitation dataset reveals that there is significant enrichment of STAT5 consensus sites within PR-bound chromatin regions, indicating that STAT5 may function as a pioneer factor for phosphorylated PR (perhaps specifically when PR Ser81 is phosphorylated). These data suggest that CK2-mediated Ser81 phosphorylation of PR may activate gene expression programs involved in modulating inflammation related to breast cancer development and progression, including mammary stem cell maintenance and self-renewal.

Recent studies have defined a new mechanism by which CK2 and PR interact. Direct interaction between PR and DUSP6, a negative regulator of the MAPK pathway, is required to achieve phosphorylation on PR Ser81 [8]. This regulation occurs independently of DUSP6 phosphatase activity, suggesting that DUSP6 is acting as a scaffold for the interaction between PR and the kinase that phosphorylates Ser81, CK2. Related to this finding, an interaction between DUSP6 and CK2 has previously been identified [59]. Together, this suggests a model whereby DUSP6 binding to CK2 brings the kinase (CK2) in close proximity to its substrate (PR Ser81), allowing for efficient phosphorylation and subsequent selection of target genes within a given (that is, inflammatory, pro-growth, survival) genetic program.

Cumulatively, in this vignette describing one context-dependent scenario of PR action, there exists cross-talk between mitogenic kinases (that is, CK2 phosphorylation of PR Ser81), MAPK pathway components (that is, DUSP6 interaction with PR is required for Ser81 phosphorylation), phosphorylation-dependent gene regulation (that is, Ser81 phosphorylation is required for PR recruitment to specific subsets of PR target genes), and putative phosphorylation-specific interactions with a pioneer factor/co-factor (that is, JAK/STAT-dependence of PR Ser81-regulated gene expression). PR phosphorylation by CK2 on Ser81 is an exemplary case study of how the molecular determinants of PR action differentially determine receptor function in breast cancer models (Figure 2).

Complexities of PR actions

- Tissue-specific effects (breast vs. reproductive tract)

- Actions in normal vs. neoplastic tissues

- Isoform-specific actions (PR-A vs. PR-B)

- Lack of clinical designation between PR isoforms

- Ligand-independent actions

- Timing of hormone delivery (continuous vs. cyclical)

- Source of hormone (synthetic vs. natural progesterone)

- PR actions are both ER-dependent and ER-independent

- Efficacy of early anti-progestins in the clinic

FIGURE 2: Molecular determinants of progesterone receptor action. Co-activators/repressors: interactions between PR and known transcriptional co-activators (for example, SRC1) and co-repressors (for example, NCOR/SMRT) are a key determinant of promoter specificity. Pioneer factors: interactions with predicted PR pioneer factors (for example, STAT5, putatively) lead to chromatin remodeling, allowing for efficient PR recruitment and subsequent target-gene transcription. Different pioneer factors would be predicted to determine differential PR recruitment. Post-translational modifications: phosphorylation (P), acetylation (Ac), ubiquitination (Ub), and SUMOylation (Sumo) primarily on N-terminal serine and lysine residues dictate receptor localization, turnover, subcellular localization, and promoter selectivity. Steroid receptor (SR) interactions: emerging evidence suggests that interactions between members of the steroid receptor superfamily (such as ER and PR) determine PR target-gene specificity. Scaffolding interactions: PR interaction with proteins acting as scaffolds (such as DUSP6) determine receptor post-translational modifications, thereby contributing to promoter selection. Cell cycle: phosphorylation on select PR serine residues and cell cycle-dependent protein complex formation determine receptor function and recruitment of PR to specific target genes.

10.1.3 PROGESTERONE RECEPTOR CLINICAL SIGNIFICANCE IN BREAST CANCER

Luminal breast tumors are characterized by their expression of ER and PR, both of which are good prognostic markers for predicted response to endocrine therapies. Interestingly, analysis of The Cancer Genome Atlas data for the luminal A/B subtype of breast tumors reveals that heterozygous loss of the PR locus occurs in 40% of luminal tumors, while 25% of luminal tumors are also heterozygous for the ER locus. However, these tumors are overwhelmingly ER-positive and largely respond well to ER-targeted therapies [60]. Interestingly, PR and ER copy number is often correlated in individual tumors; tumors with altered copy numbers for ER are likely to have changes in PR copy number. Despite these genomic alterations, both PR and ER mRNA levels are similar in luminal tumors that are diploid versus those that have lost an allele at these loci. Thus, gene copy number may not be a robust measure of the functional (that is, protein) readout for these steroid hormone receptors and should be interpreted with caution. Moreover, complex intra- and inter-tumoral heterogeneity may be reflected in analyses of genomic copy number. Because PR-positive cells release pro-proliferative factors (that is, PR target-gene products) that induce paracrine signaling, a small percentage of PR-positive cells within an individual tumor could have significant effects on tumor stem cell maintenance and/or tumor growth and progression. This is a complex situation that makes PR loci genomic heterozygosity difficult to interpret. Cumulatively, these data underscore the need to gain a much better understanding of PR signaling within the clinical context.

HRT clinical trial data (discussed above) suggest an important role for progestins and PR as drivers (that is, tumor promoters) of breast cancer cell growth. Progesterone-dependent expression of secreted paracrine factors is required for self-renewal of (PR-null) stem cells in the normal mammary gland [32,37] (see below). PR target genes include soluble factors known to modify cancer stem cells (WNT1 and RANKL). However, the role of PR target genes in the maintenance or expansion of cancer progenitor or stem cells is currently unknown. While a minority of normal (non-pregnant) breast epithelial cells contain steroid hormone receptors, the majority of luminal breast cancers express ER and PR (discussed above);

heterogeneous cells within the breast may contain both ER and PR, only ER, or only PR [61]. Interestingly, very few somatic mutations have been identified in ER [62] or PR. With regard to PR, isolated genetic polymorphisms linked to breast and reproductive cancers appear to increase levels of PR-B isoform expression, rather than affect PR transcriptional activity [63-65]. Additionally, the PR-A promoter is more frequently methylated (that is, silenced) relative to the PR-B promoter in advanced endocrine-resistant breast cancers [66]. These data imply that genetic alteration of PR itself is usually not sufficient to promote tumorigenesis. Alternatively, we propose that oncogenic mutations that drive signaling pathways provide the context for heightened ER and PR transcriptional activity. For example, high levels of kinases, such as CK2, CDKs or MAPKs, may induce persistent progesterone-independent phosphorylation of PR-B on serines 81 or 294, respectively, thereby leading to activation of phospho-isoform-specific transcriptional programs shown to be significantly altered in luminal breast cancer [8,9]. Therapeutic strategies that target receptor-modifying protein kinases (that is, anti-CK2, CDK2 or MAPK) and/or their transcriptional co-factors (that is, STATs, AP1, SP1, FOXO1, FOXA1) are likely to be very successful at treating breast cancer and must remain a direction of robust exploration within the SR field.

Historically, clinical testing of anti-progestins has been limited [42,67-70]. The results of a clinical trial released in 1999 showed promise for anti-progestins as front-line breast cancer endocrine therapy [42]. Although patient accrual in this study was small (19 patients), 67% of patients achieved tumor remission when treated with onapristone, a PR type I antagonist that blocks PR binding to DNA, as front-line endocrine therapy for locally advanced or primary breast cancer [42]. Liver function test abnormalities were seen early in this trial, and for that reason new patient accrual was stopped. These liver-associated effects were likely due to inhibition of GR, a closely related SR. The clinical efficacy of lonaprisan, a type III PR antagonist that promotes PR repression through the recruitment of transcriptional co-repressors (while maintaining DNA binding), was measured in a phase II study as second-line therapy for PR-positive breast cancer [70]. The results from this trial were disappointing, and the trial was terminated before full patient accrual. Although a small percentage (14%) of patients achieved stable disease, no patients achieved complete or partial responses. This trial likely failed for a number of reasons, including lack of patient classification, patients having previous exposure to endocrine therapies, and a lack of mechanistic understanding of PR inhibitor action and isoform specificity. Notably, clinically used anti-progestins that target the ligand-binding domain of PR may fail to block ligand-independent actions of PR (discussed above).

Renewed optimism for the use of anti-progestins to prevent or inhibit breast cancer growth is provided by more recent preclinical studies of anti-progestins in murine mammary tumor models. In a dramatic example, treatment of nulliparous *Brca1/Trp53*-deficient mice with mifepristone, a PR antagonist, completely inhibited the formation of mammary gland tumors normally observed in virgin mice [71], perhaps via modulation of the stem cell compartment [30,32]. Newer, highly selective anti-progestins, which are currently in development by several pharmaceutical companies, may increase the clinical utility of anti-progestins in breast cancer prevention and treatment and is an area of renewed research interest. Notably, many patients that relapse while on tamoxifen therapy retain expression of PR, underscoring the clinical significance of considering PRs as potentially acting independently of ER in the context of breast cancer progression during estrogen ablation (that is, PR expression is most often used clinically as a measure of ER function) [72,73]. Based on our current understanding of ligand-dependent and ligand-independent (kinase-induced) PR actions, classification of patients based on gene-expression profiling could better identify the subpopulation of patients that would respond well to selective anti-progestins. In addition, cross-talk between ER and PR (or AR), and growth-factor signaling pathways (discussed above) is a likely confounding component of development to endocrine-resistant disease, and should therefore be considered (for example, via the use of pathway-specific gene biomarkers) when selecting anti-progestins as potentially beneficial front-line or second-line therapy [74-76].

As mentioned above (and in Box 1), the clinical significance of PR isoforms is likely vastly under-appreciated. In mammary tissue, PR exists as two primary isoforms, PR-A and PR-B. Although PR-B is required for mammary gland development and PR-A for uterine development, these isoforms are most often co-expressed in the same tissues, typically at a ratio of 1:1. Single isoform expression in tissues is rare [77-79]. Interestingly, in pre-neoplastic lesions and samples from patients with breast cancer, this balanced A:B ratio is often altered, frequently due to apparent loss of PR-B [78,80]. Cumulative data from the Lange laboratory has revealed that this imbalance may be explained by phosphorylation-dependent turnover of transcriptionally active PR-B receptors relative to more stable and less active PR-A receptors. PR-B but not PR-A undergoes extensive crosstalk with mitogenic protein kinases [8,16,45,81,82]. Thus, PR-B is heavily phosphorylated in response to ligand or via the action of growth factors, and although this isoform-specific phosphorylation (on PR-B Ser294) is linked to high transcriptional activity, it is also coupled to rapid ubiquitin-dependent turnover of the receptor; regulated PR-B turnover is tightly linked to transcriptional activity (that is, stable non-degradable mutants of PR are poor transcriptional activators) [83,84]. Of note, this phosphorylation event (PR-B Ser294) has been detected in a subset of human tumors [9]. Therefore, loss of PR-B, as measured by protein levels in clinical immunohistochemistry tests or western blotting may actually reflect high PR-B transcriptional activity coupled with rapid protein turnover; peak PR target-gene expression (mRNA) is coincident with nearly undetectable PR protein in experimental models [85]. Mouse models (mammary gland) predominantly express PR-A prior to pregnancy. In humans, normal mammary gland function may rely upon balanced expression of the two PR isoforms. Unfortunately, current immunohistochemistry clinical testing for PR in breast cancer samples does not differentiate between PR-A and PR-B isoforms. Because an imbalance between the two isoforms appears to be linked to cancerous phenotypes, clinical isoform distinction may have great diagnostic potential and should be considered as part of routine luminal cancer work-up.

Emerging data linking progesterone regulation to the expansion of the mammary stem cell compartment highlight the role that PR and progesterone may play in early events in breast cancer. Recent seminal work in murine models has shown that progesterone can induce the rapid expansion of mammary stem cells, a population of SR-negative (that is, ER- and PR-negative) cells located in the basal epithelial compartment of the mammary gland [32,37]. Because these cells are PR negative, this expansion likely occurs through the production of paracrine factors secreted by neighboring or nearby PR-positive luminal epithelial cells. Progesterone-dependent expansion of the mammary stem cell population is mediated by key PR-target genes, including *RANKL* and *WNT4*[32,37]. Brisken and colleagues have shown that progesterone-dependent control of RANKL expression in human tissues is dependent on intact breast tissue microstructure, and have confirmed that RANKL is required for progesterone-induced proliferation [41]; estrogen is a permissive hormone (for PR expression) in this context. Interestingly, PR-dependent RANKL expression requires STAT5A [50]. This observation is similar to what has been published for PR regulation of WNTs [8], highlighting an emerging role for co-ordinate STAT5/PR regulation of select subsets of PR-target genes related to proliferation and stem cell self-renewal (see above). Moreover, a PR-positive subpopulation of mammary gland progenitor cells has been recently discovered [61], challenging the current dogma that mammary gland precursors are strictly SR-negative. These exciting findings suggest that this long-lived population of cells, one that is exquisitely sensitive to mutagenic events, can expand in response to progesterone in both a paracrine and autocrine fashion [36]. Notably, these PR-positive mammary stem cells are devoid of ER protein or mRNA expression, further underscoring the need for understanding PR action as independent of ER in this context.

10.2 CONCLUSIONS

Recent clinical and preclinical studies clearly demonstrate the significance of fully understanding the determinants of context-dependent PR action. They not only challenge the current clinical diagnostic paradigm in which PR is only used as a marker of ER transcriptional activity, but also support a renewed interest in understanding PR as a driver of breast tumor progression and thus a potentially very useful target for improved breast cancer therapy [1,86]. In this review, we have highlighted the concept that gene-expression analyses linked to PR actions suggest different transcriptional programs are activated in response to specific post-translational modifications (phosphorylation events) and protein-protein interactions. Although these unique PR gene signatures highlight functional differences between modified PRs and their components, the overlap between these (predominantly proliferative) programs supports a strong role for PR in early tumor progression toward more aggressive cancer phenotypes, and in some cases, even highlights a phospho-PR gene signature associated with poor response to endocrine treatment [9]. Therefore, gene signatures that define PR action will likely provide a useful paired diagnostic for clinically applied selective anti-progestins. We conclude that PR function is highly dependent on the molecular context, which is defined by such factors as protein kinase activity (as a major input to receptor post-translational modifications), co-factor availability, and the presence of progesterone and other steroid hormone levels and receptors (Figure 2). Future therapeutic approaches should consider targeting receptor-modifying activities in place of or in conjunction with anti-hormone therapies. With progesterone emerging as the primary mitogen in the adult breast (wherein estrogen is permissive for PR expression), understanding PR function and identifying or targeting modifiers of PR action are of critical importance to advancing the treatment of breast cancer.

BOX 1: COMPLEXITIES OF PROGESTERONE RECEPTOR ACTIONS.

- Tissue-specific effects (breast vs. reproductive tract)
- Actions in normal vs. neoplastic tissues
- Isoform-specific actions (PR-A vs. PR-B)
- Lack of clinical designation between PR isoforms
- Ligand-independent actions
- Timing of hormone delivery (continuous vs. cyclical)
- Source of hormone (synthetic vs. natural progesterone)
- PR actions are both ER-dependent and ER-independent
- Efficacy of early anti-progestins in the clinic

REFERENCES

1. Brisken C: Progesterone signalling in breast cancer: a neglected hormone coming into the limelight. Nat Rev Cancer 2013, 13:385-396.
2. Kraus WL, Montano MM, Katzenellenbogen BS: Cloning of the rat progesterone receptor gene 5′-region and identification of two functionally distinct promoters. Mol Endocrinol 1993, 7:1603-1616.
3. Hill KK, Roemer SC, Churchill ME, Edwards DP: Structural and functional analysis of domains of the progesterone receptor. Mol Cell Endocrinol 2012, 348:418-429.
4. Kastner P, Krust A, Turcotte B, Stropp U, Tora L, Gronemeyer H, Chambon P: Two distinct estrogen-regulated promoters generate transcripts encoding the two functionally different human progesterone receptor forms A and B. Embo J 1990, 9:1603-1614.
5. Owen GI, Richer JK, Tung L, Takimoto G, Horwitz KB: Progesterone regulates transcription of the p21(WAF1) cyclin-dependent kinase inhibitor gene through Sp1 and CBP/p300. J Biol Chem 1998, 273:10696-10701.
6. Stoecklin E, Wissler M, Schaetzle D, Pfitzner E, Groner B: Interactions in the transcriptional regulation exerted by Stat5 and by members of the steroid hormone receptor family. J Steroid Biochem Mol Biol 1999, 69:195-204.
7. Cicatiello L, Addeo R, Sasso A, Altucci L, Petrizzi VB, Borgo R, Cancemi M, Caporali S, Caristi S, Scafoglio C, et al.: Estrogens and progesterone promote persistent CCND1 gene activation during G1 by inducing transcriptional derepression via c-Jun/c-Fos/estrogen receptor (progesterone receptor) complex assembly to a distal regulatory element and recruitment of cyclin D1 to its own gene promoter. Mol Cell Biol 2004, 24:7260-7274.
8. Hagan CR, Knutson TP, Lange CA: A common docking domain in progesterone receptor-B links DUSP6 and CK2 signaling to proliferative transcriptional programs in breast cancer cells. Nucleic Acids Res 2013, 41:8962-8942.
9. Knutson TP, Daniel AR, Fan D, Silverstein KA, Covington KR, Fuqua SA, Lange CA: Phosphorylated and sumoylation-deficient progesterone receptors drive proliferative gene signatures during breast cancer progression. Breast Cancer Res 2012, 14:R95.
10. McKenna NJ, Lanz RB, O'Malley BW: Nuclear receptor coregulators: cellular and molecular biology. Endocr Rev 1999, 20:321-344.
11. Daniel AR, Faivre EJ, Lange CA: Phosphorylation-dependent antagonism of sumoylation derepresses progesterone receptor action in breast cancer cells. Mol Endocrinol 2007, 21:2890-2906.
12. Lange CA, Shen T, Horwitz KB: Phosphorylation of human progesterone receptors at serine-294 by mitogen-activated protein kinase signals their degradation by the 26S proteasome. Proc Natl Acad Sci U S A 2000, 97:1032-1037.
13. Weigel NL, Bai W, Zhang Y, Beck CA, Edwards DP, Poletti A: Phosphorylation and progesterone receptor function. J Steroid Biochem Mol Biol 1995, 53:509-514.
14. Daniel AR, Gaviglio AL, Czaplicki LM, Hillard CJ, Housa D, Lange CA: The progesterone receptor hinge region regulates the kinetics of transcriptional responses

through acetylation, phosphorylation, and nuclear retention. Mol Endocrinol 2011, 24:2126-2138.

15. Pierson-Mullany LK, Lange CA: Phosphorylation of progesterone receptor serine 400 mediates ligand-independent transcriptional activity in response to activation of cyclin-dependent protein kinase 2. Mol Cell Biol 2004, 24:10542-10557.

16. Hagan CR, Regan TM, Dressing GE, Lange CA: CK2-dependent phosphorylation of progesterone receptors (PR) on Ser81 regulates PR-B isoform-specific target gene expression in breast cancer cells CK2. Mol Cell Biol 2011, 31:2439-2452.

17. Faivre EJ, Daniel AR, Hillard CJ, Lange CA: Progesterone receptor rapid signaling mediates serine 345 phosphorylation and tethering to specificity protein 1 transcription factors. Mol Endocrinol 2008, 22:823-837.

18. Ballare C, Uhrig M, Bechtold T, Sancho E, Di Domenico M, Migliaccio A, Auricchio F, Beato M: Two domains of the progesterone receptor interact with the estrogen receptor and are required for progesterone activation of the c-Src/Erk pathway in mammalian cells. Mol Cell Biol 2003, 23:1994-2008.

19. Boonyaratanakornkit V, Scott MP, Ribon V, Sherman L, Anderson SM, Maller JL, Miller WT, Edwards DP: Progesterone receptor contains a proline-rich motif that directly interacts with SH3 domains and activates c-Src family tyrosine kinases. Mol Cell 2001, 8:269-280.

20. Diaz Flaque MC, Vicario R, Proietti CJ, Izzo F, Schillaci R, Elizalde PV: Progestin drives breast cancer growth by inducing p21(CIP1) expression through the assembly of a transcriptional complex among Stat3, progesterone receptor and ErbB-2. Steroids 2013, 78:559-567.

21. Narayanan R, Adigun AA, Edwards DP, Weigel NL: Cyclin-dependent kinase activity is required for progesterone receptor function: novel role for cyclin A/Cdk2 as a progesterone receptor coactivator. Mol Cell Biol 2005, 25:264-277.

22. Beral V: Breast cancer and hormone-replacement therapy in the Million Women Study. Lancet 2003, 362:419-427.

23. Chlebowski RT, Anderson GL, Gass M, Lane DS, Aragaki AK, Kuller LH, Manson JE, Stefanick ML, Ockene J, Sarto GE, et al.: Estrogen plus progestin and breast cancer incidence and mortality in postmenopausal women. JAMA 2010, 304:1684-1692.

24. Birrell SN, Butler LM, Harris JM, Buchanan G, Tilley WD: Disruption of androgen receptor signaling by synthetic progestins may increase risk of developing breast cancer. FASEB J 2007, 21:2285-2293.

25. Courtin A, Communal L, Vilasco M, Cimino D, Mourra N, de Bortoli M, Taverna D, Faussat AM, Chaouat M, Forgez P, et al.: Glucocorticoid receptor activity discriminates between progesterone and medroxyprogesterone acetate effects in breast cells. Breast Cancer Res Treat 2012, 131:49-63.

26. Fournier A, Berrino F, Riboli E, Avenel V, Clavel-Chapelon F: Breast cancer risk in relation to different types of hormone replacement therapy in the E3N-EPIC cohort. Int J Cancer 2005, 114:448-454.

27. Lyytinen H, Pukkala E, Ylikorkala O: Breast cancer risk in postmenopausal women using estradiol-progestogen therapy. Obstet Gynecol 2009, 113:65-73.

28. Lanari C, Lamb CA, Fabris VT, Helguero LA, Soldati R, Bottino MC, Giulianelli S, Cerliani JP, Wargon V, Molinolo A: The MPA mouse breast cancer model: evidence

for a role of progesterone receptors in breast cancer. Endocr Relat Cancer 2009, 16:333-350.

29. Lanari C, Molinolo AA: Progesterone receptors-animal models and cell signalling in breast cancer. Diverse activation pathways for the progesterone receptor: possible implications for breast biology and cancer. Breast Cancer Res 2002, 4:240-243.

30. Horwitz KB, Dye WW, Harrell JC, Kabos P, Sartorius CA: Rare steroid receptor-negative basal-like tumorigenic cells in luminal subtype human breast cancer xenografts. Proc Natl Acad Sci U S A 2008, 105:5774-5779.

31. Horwitz KB, Sartorius CA: Progestins in hormone replacement therapies reactivate cancer stem cells in women with preexisting breast cancers: a hypothesis. J Clin Endocrinol Metab 2008, 93:3295-3298.

32. Joshi PA, Jackson HW, Beristain AG, Di Grappa MA, Mote PA, Clarke CL, Stingl J, Waterhouse PD, Khokha R: Progesterone induces adult mammary stem cell expansion. Nature 2010, 465:803-807.

33. Santen RJ: Risk of breast cancer with progestins: critical assessment of current data. Steroids 2003, 68:953-964.

34. Hofseth LJ, Raafat AM, Osuch JR, Pathak DR, Slomski CA, Haslam SZ: Hormone replacement therapy with estrogen or estrogen plus medroxyprogesterone acetate is associated with increased epithelial proliferation in the normal postmenopausal breast. J Clin Endocrinol Metab 1999, 84:4559-4565.

35. Santen RJ: Menopausal hormone therapy and breast cancer. J Steroid Biochem Mol Biol 2013, 2013:2013.

36. Beleut M, Rajaram RD, Caikovski M, Ayyanan A, Germano D, Choi Y, Schneider P, Brisken C: Two distinct mechanisms underlie progesterone-induced proliferation in the mammary gland. Proc Natl Acad Sci U S A 2010, 107:2989-2994.

37. Asselin-Labat ML, Shackleton M, Stingl J, Vaillant F, Forrest NC, Eaves CJ, Visvader JE, Lindeman GJ: Steroid hormone receptor status of mouse mammary stem cells. J Natl Cancer Inst 2006, 98:1011-1014.

38. Groshong SD, Owen GI, Grimison B, Schauer IE, Todd MC, Langan TA, Sclafani RA, Lange CA, Horwitz KB: Biphasic regulation of breast cancer cell growth by progesterone: role of the cyclin-dependent kinase inhibitors, p21 and p27(Kip1). Mol Endocrinol 1997, 11:1593-1607.

39. Clarke RB, Howell A, Anderson E: Estrogen sensitivity of normal human breast tissue in vivo and implanted into athymic nude mice: analysis of the relationship between estrogen-induced proliferation and progesterone receptor expression. Breast Cancer Res Treat 1997, 45:121-133.

40. Communal L, Vilasco M, Hugon-Rodin J, Courtin A, Mourra N, Lahlou N, Dumont S, Chaouat M, Forgez P, Gompel A: Ulipristal acetate does not impact human normal breast tissue. Hum Reprod 2012, 27:2785-2798.

41. Tanos T, Sflomos G, Echeverria PC, Ayyanan A, Gutierrez M, Delaloye JF, Raffoul W, Fiche M, Dougall W, Schneider P, et al.: Progesterone/RANKL is a major regulatory axis in the human breast. Sci Transl Med 2013, 5:182ra155.

42. Robertson JF, Willsher PC, Winterbottom L, Blamey RW, Thorpe S: Onapristone, a progesterone receptor antagonist, as first-line therapy in primary breast cancer. Eur J Cancer 1999, 35:214-218.

43. Hagan CR, Daniel AR, Dressing GE, Lange CA: Role of phosphorylation in progesterone receptor signaling and specificity. Mol Cell Endocrinol 2012, 357:43-49.

44. Dressing GE, Hagan CR, Knutson TP, Daniel AR, Lange CA: Progesterone receptors act as sensors for mitogenic protein kinases in breast cancer models. Endocr Relat Cancer 2009, 16:351-361.

45. Daniel AR, Faivre EJ, Lange CA: Phosphorylation-dependent antagonism of sumoylation derepresses progesterone receptor action in breast cancer cells. Mol Endocrinol 2007, 21:2890-2906.

46. Daniel AR, Lange CA: Protein kinases mediate ligand-independent derepression of sumoylated progesterone receptors in breast cancer cells. Proc Natl Acad Sci U S A 2009, 106:14287-14292.

47. Jozwik KM, Carroll JS: Pioneer factors in hormone-dependent cancers. Nat Rev Cancer 2012, 12:381-385.

48. Magnani L, Eeckhoute J, Lupien M: Pioneer factors: directing transcriptional regulators within the chromatin environment. Trends Genet 2011, 27:465-474.

49. Clarke CL, Graham JD: Non-overlapping progesterone receptor cistromes contribute to cell-specific transcriptional outcomes. PLoS One 2012, 7:e35859.

50. Obr AE, Grimm SL, Bishop KA, Pike JW, Lydon JP, Edwards DP: Progesterone receptor and Stat5 signaling crosstalk through RANKL in mammary epithelial cells. Mol Endocrinol 2013, 27:1808-1824.

51. Peters AA, Buchanan G, Ricciardelli C, Bianco-Miotto T, Centenera MM, Harris JM, Jindal S, Segara D, Jia L, Moore NL, et al.: Androgen receptor inhibits estrogen receptor-alpha activity and is prognostic in breast cancer. Cancer Res 2009, 69:6131-6140.

52. Need EF, Selth LA, Harris TJ, Birrell SN, Tilley WD, Buchanan G: Research resource: interplay between the genomic and transcriptional networks of androgen receptor and estrogen receptor alpha in luminal breast cancer cells. Mol Endocrinol 2012, 26:1941-1952.

53. Giulianelli S, Vaque JP, Soldati R, Wargon V, Vanzulli SI, Martins R, Zeitlin E, Molinolo AA, Helguero LA, Lamb CA, et al.: Estrogen receptor alpha mediates progestin-induced mammary tumor growth by interacting with progesterone receptors at the cyclin D1/MYC promoters. Cancer Res 2012, 72:2416-2427.

54. Daniel AR, Gaviglio AL, Knutson TP, Ostrander JH, D'Assoro AB, Ravindranathan P, Peng Y, Raj GV, Yee D, Lange CA: Progesterone receptor-B enhances estrogen responsiveness of breast cancer cells via scaffolding PELP1- and estrogen receptor-containing transcription complexes. Oncogene 2014. doi:10.1038/onc.2013.579. [Epub ahead of print]

55. Tawfic S, Yu S, Wang H, Faust R, Davis A, Ahmed K: Protein kinase CK2 signal in neoplasia. Histol Histopathol 2001, 16:573-582.

56. Guerra B, Issinger OG: Protein kinase CK2 in human diseases. Curr Med Chem 2008, 15:1870-1886.

57. Meggio F, Pinna LA: One-thousand-and-one substrates of protein kinase CK2? FASEB J 2003, 17:349-368.

58. Zhang Y, Beck CA, Poletti A, Edwards DP, Weigel NL: Identification of phosphorylation sites unique to the B form of human progesterone receptor. In vitro phosphorylation by casein kinase II. J Biol Chem 1994, 269:31034-31040.

59. Castelli M, Camps M, Gillieron C, Leroy D, Arkinstall S, Rommel C, Nichols A: MAP kinase phosphatase 3 (MKP3) interacts with and is phosphorylated by protein kinase CK2alpha. J Biol Chem 2004, 279:44731-44739.

60. Cancer Genome Atlas Network: Comprehensive molecular portraits of human breast tumours. Nature 2012, 490:61-70.

61. Hilton HN, Graham JD, Kantimm S, Santucci N, Cloosterman D, Huschtscha LI, Mote PA, Clarke CL: Progesterone and estrogen receptors segregate into different cell subpopulations in the normal human breast. Mol Cell Endocrinol 2012, 361:191-201.

62. Fuqua SA, Wiltschke C, Zhang QX, Borg A, Castles CG, Friedrichs WE, Hopp T, Hilsenbeck S, Mohsin S, O'Connell P, et al.: A hypersensitive estrogen receptor-alpha mutation in premalignant breast lesions. Cancer Res 2000, 60:4026-4029.

63. Pooley KA, Healey CS, Smith PL, Pharoah PD, Thompson D, Tee L, West J, Jordan C, Easton DF, Ponder BA, et al.: Association of the progesterone receptor gene with breast cancer risk: a single-nucleotide polymorphism tagging approach. Cancer Epidemiol Biomarkers Prev 2006, 15:675-682.

64. De Vivo I, Huggins GS, Hankinson SE, Lescault PJ, Boezen M, Colditz GA, Hunter DJ: A functional polymorphism in the promoter of the progesterone receptor gene associated with endometrial cancer risk. Proc Natl Acad Sci U S A 2002, 99:12263-12268.

65. Terry KL, De Vivo I, Titus-Ernstoff L, Sluss PM, Cramer DW: Genetic variation in the progesterone receptor gene and ovarian cancer risk. Am J Epidemiol 2005, 161:442-451.

66. Pathiraja TN, Shetty PB, Jelinek J, He R, Hartmaier R, Margossian AL, Hilsenbeck SG, Issa JP, Oesterreich S: Progesterone receptor isoform-specific promoter methylation: association of PRA promoter methylation with worse outcome in breast cancer patients. Clin Cancer Res 2011, 17:4177-4186.

67. Romieu G, Maudelonde T, Ulmann A, Pujol H, Grenier J, Cavalie G, Khalaf S, Rochefort H: The antiprogestin RU486 in advanced breast cancer: preliminary clinical trial. Bull Cancer 1987, 74:455-461.

68. Klijn JG, de Jong FH, Bakker GH, Lamberts SW, Rodenburg CJ, Alexieva-Figusch J: Antiprogestins, a new form of endocrine therapy for human breast cancer. Cancer Res 1989, 49:2851-2856.

69. Perrault D, Eisenhauer EA, Pritchard KI, Panasci L, Norris B, Vandenberg T, Fisher B: Phase II study of the progesterone antagonist mifepristone in patients with untreated metastatic breast carcinoma: a National Cancer Institute of Canada Clinical Trials Group study. J Clin Oncol 1996, 14:2709-2712.

70. Jonat W, Bachelot T, Ruhstaller T, Kuss I, Reimann U, Robertson JF: Randomized phase II study of lonaprisan as second-line therapy for progesterone receptor-positive breast cancer. Ann Oncol 2013, 24:2543-2548.

71. Poole AJ, Li Y, Kim Y, Lin SC, Lee WH, Lee EY: Prevention of Brca1-mediated mammary tumorigenesis in mice by a progesterone antagonist. Science 2006, 314:1467-1470.

72. Encarnacion CA, Ciocca DR, McGuire WL, Clark GM, Fuqua SA, Osborne CK: Measurement of steroid hormone receptors in breast cancer patients on tamoxifen. Breast Cancer Res Treat 1993, 26:237-246.

73. Johnston SR, Saccani-Jotti G, Smith IE, Salter J, Newby J, Coppen M, Ebbs SR, Dowsett M: Changes in estrogen receptor, progesterone receptor, and pS2 expression in tamoxifen-resistant human breast cancer. Cancer Res 1995, 55:3331-3338.

74. Hayes E, Nicholson RI, Hiscox S: Acquired endocrine resistance in breast cancer: implications for tumour metastasis. Front Biosci (Landmark Ed) 2011, 16:838-848.

75. Cleator SJ, Ahamed E, Coombes RC, Palmieri C: A 2009 update on the treatment of patients with hormone receptor-positive breast cancer. Clin Breast Cancer 2009, 9:S6-S17.

76. Pierson-Mullany LK, Skildum A, Faivre E, Lange CA: Cross-talk between growth factor and progesterone receptor signaling pathways: implications for breast cancer cell growth. Breast Dis 2003, 18:21-31.

77. Mote PA, Balleine RL, McGowan EM, Clarke CL: Colocalization of progesterone receptors A and B by dual immunofluorescent histochemistry in human endometrium during the menstrual cycle. J Clin Endocrinol Metab 1999, 84:2963-2971.

78. Mote PA, Bartow S, Tran N, Clarke CL: Loss of co-ordinate expression of progesterone receptors A and B is an early event in breast carcinogenesis. Breast Cancer Res Treat 2002, 72:163-172.

79. Mote PA, Graham JD, Clarke CL: Progesterone receptor isoforms in normal and malignant breast. Ernst Schering Found Symp Proc 2007, 1:77-107.

80. Graham JD, Yeates C, Balleine RL, Harvey SS, Milliken JS, Bilous AM, Clarke CL: Characterization of progesterone receptor A and B expression in human breast cancer. Cancer Res 1995, 55:5063-5068.

81. Boonyaratanakornkit V, McGowan E, Sherman L, Mancini MA, Cheskis BJ, Edwards DP: The role of extranuclear signaling actions of progesterone receptor in mediating progesterone regulation of gene expression and the cell cycle. Mol Endocrinol 2007, 21:359-375.

82. Clemm DL, Sherman L, Boonyaratanakornkit V, Schrader WT, Weigel NL, Edwards DP: Differential hormone-dependent phosphorylation of progesterone receptor A and B forms revealed by a phosphoserine site-specific monoclonal antibody. Mol Endocrinol 2000, 14:52-65.

83. Qiu M, Lange CA: MAP kinases couple multiple functions of human progesterone receptors: degradation, transcriptional synergy, and nuclear association. J Steroid Biochem Mol Biol 2003, 85:147-157.

84. Qiu M, Olsen A, Faivre E, Horwitz KB, Lange CA: Mitogen-activated protein kinase regulates nuclear association of human progesterone receptors. Mol Endocrinol 2003, 17:628-642.

85. Faivre EJ, Lange CA: Progesterone receptors upregulate Wnt-1 to induce epidermal growth factor receptor transactivation and c-Src-dependent sustained activation of Erk1/2 mitogen-activated protein kinase in breast cancer cells. Mol Cell Biol 2007, 27:466-480.

86. Lanari C, Wargon V, Rojas P, Molinolo AA: Antiprogestins in breast cancer treatment: are we ready? Endocr Relat Cancer 2012, 19:R35-R50. |

CHAPTER 11

ARCHITECTURE OF EPIGENETIC REPROGRAMMING FOLLOWING TWIST1-MEDIATED EPITHELIAL-MESENCHYMAL TRANSITION

GABRIEL G. MALOUF, JOSEPH H. TAUBE, YUE LU, TAPASREE ROYSARKAR, SHOGHAG PANJARIAN, MARCOS R.H. ESTECIO, JAROSLAV JELINEK, JUMPEI YAMAZAKI, NOEL J-M RAYNAL, HAI LONG, TOMOMITSU TAHARA, AGATA TINNIRELLO, PRIYANKA RAMACHANDRAN, XIU-YING ZHANG, SHOUDAN LIANG, SENDURAI A. MANI, AND JEAN-PIERRE J. ISSA

11.1 BACKGROUND

Epithelial-mesenchymal transition (EMT) is known to promote cellular plasticity during the formation of the mesoderm from epiblasts and the neural crest cells from the neural tube in the developing embryo as well as during adult wound healing [1]. During EMT, epithelial cells lose their epithelial characteristics and acquire mesenchymal morphology, which facilitates cellular dissociation and migration. Similar to embryo development, neoplastic cells have been shown to reactivate EMT leading to

This chapter was originally published under the Creative Commons Attribution License. Malouf GG, Taube JH, Lu Y, Roysarkar T, Panjarian S, Estecio MRH, Jelinek J, Yamazaki J, Raynal NJ-M, Long H, Tahara T, Tinnirello A, Ramachandran P, Zhang X-Y, Liang S, Mani SA, and Issa J-PJ. Architecture of Epigenetic Reprogramming Following Twist1-Mediated Epithelial-Mesenchymal Transition. Genome Biology 14:R144 (2013). doi:10.1186/gb-2013-14-12-r144.

cancer metastasis [2]. Induction of EMT is also involved in the development of resistance to cytotoxic chemotherapy and targeted agents [3-5]. In addition, EMT imparts stem cell properties to differentiated cells [6]. Since cancer cells seem to acquire stem cell properties dynamically in response to the tumor microenvironment and become differentiated at distant sites, it has been suggested that major epigenetic remodeling would occur during EMT to facilitate metastasis. Although DNA methylation changes at specific loci have been established during EMT [7,8], changes in the global DNA methylation landscape are not well understood. Indeed, a recent report demonstrated that DNA methylation is largely unchanged during EMT mediated by transforming growth factor beta (TGF-β) [9], while another showed that EMT is associated with specific alterations of gene-related CpG-rich regions [10]. Moreover, another report showed a striking difference in DNA methylation in non-small cell lung cancers between mesenchymal-like tumors and epithelial-like tumors, which display a better prognosis and exhibit greater sensitivity to inhibitors of epidermal growth factor receptor [11].

In addition to DNA methylation, EMT mediates epigenetic reprogramming through widespread changes in post-translational modifications of histones [12]. However, it is unknown if switches in histone marks coordinate EMT and, in particular, whether genome regulation by Polycomb group (PcG) and Trithorax group (TrxG) proteins are critical regulators for this transition, as is the case for germ cell development and stem cell differentiation. Indeed, the TrxG complex activates gene transcription by inducing trimethylation of lysine 4 of histone H3 (H3K4me3) at specific sites, whereas the PcG complex represses gene transcription by trimethylation of lysine 27 on histone H3 (H3K27me3). Of note, a subset of promoters in embryonic stem cells are known to have methylation at both H3K4 and H3K27 (the bivalent state), which poise them for either activation or repression in different cell types upon differentiation [13]. However, the transcriptional dynamics and the role of those bivalent genes in differentiated cells and during EMT are still poorly understood.

The development of genome-wide sequencing expanded our understanding of the plasticity of DNA methylation during differentiation of embryonic stem cells, tumorigenesis and metastasis [14,15]. During the differentiation of embryonic stem cells into fibroblasts, the majority of DNA methylation changes occur outside of core promoters in partially methylated domains (PMDs), which represent large hypomethylated regions covering approximately 40% of our genome [14]. Using genome-wide DNA methylation analyses, these PMDs have been shown to be hypomethylated in adipose tissue [16], placenta [17,18], cultured breast cancer cells [19] and neuronal cells [20], as well as in several cancer types [15]. PMDs overlap with domains of H3K27me3 and/or H3K9me3, transcriptional-repression associated histone marks, in IMR90 fibroblasts [14]. In breast cancer, widespread DNA hypomethylation occurs primarily at PMDs in normal breast cells [21]. However, whether DNA methylome changes during EMT recapitulate tumor formation remains unknown.

EMT is often a transient process, with changes in gene expression, increased invasiveness, and acquisition of stem cell properties such as increased tumor initiation, metastasis and chemotherapeutic-resistance. It's transient nature suggests that significant features of an EMT could be regulated by epigenetic fluidity triggered by key transcription factors and signaling events in response to an alteration in the tumor microenvironment. We present genome-wide changes in DNA methylation and histone modifications in H3K4me3 and H3K27me3 following the induction of EMT by the ectopic expression of the transcription factor Twist1 using immortalized human mammary epithelial cells (HMLE) [2]. Additionally, we compared the Twist1-expressing HMLE cells, hereafter HMLE Twist, cultured in a monolayer to the same cells cultured as mammospheres (MS), which enriches for cells with stem cell properties [22]. We found that EMT is characterized by major epigenetic reprogramming required for phenotypic plasticity, with predominant alterations to polycomb targets. Moreover, we have shown that inhibition of the H3K27 methyltransferases EZH2 and EZH1—part of the polycomb repressive complex 2 (PRC2)—either by short hairpin RNA (shRNA) or pharmacologically, blocks EMT and stemness properties.

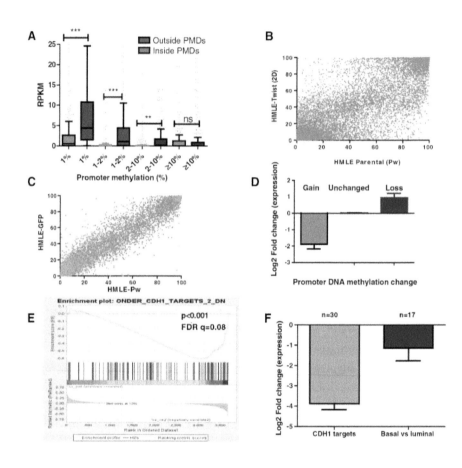

FIGURE 1: DNA Methylation changes occurring following epithelial-mesenchymal transition. (A) Box-plot for gene expression levels according to promoter DNA methylation level of genes located in partially methylated domains (PMDs) (lighter gray) and outside PMDs (darker gray) in human mammary epithelial cells (HMLE). The x-axis represents promoter % methylation and y-axis represents normalized expression level. *** P <0.0001, ** P <0.001, ns: non-significant. (B) Correlation of DNA methylation level of CpG sites in HMLE vector cells (Pw) (x-axis) and HMLE Twist cells (y-axis), showing dramatic changes in DNA methylation following EMT. (C) Correlation of DNA methylation level of CpG sites in HMLE vector cells transduced with two different 'control' vectors Pw (x-axis) and GFP (y-axis) showing no change in DNA methylation. (D) Box plots of average gene expression levels of genes with a gain, loss or no change in DNA methylation. Note that a gain of DNA methylation (increase to ≥2% DNA methylation from gene promoters which were fully unmethylated, ≤1%) was associated with 3.8-fold decrease of their expression

levels, while demethylation of gene promoters leading to fully unmethylated promoters (≤1%) was associated with an increase in their gene expression levels by about 2-fold. (E) GSEA showing that genes losing gene body methylation following EMT are enriched for genes which are down-regulated in a *CDH1*-knockdown model of EMT (P <0.0001). The bottom graph represents the rank-ordered, non-redundant list of genes. Genes on the far right correlated most strongly with decreased gene expression in the *CDH1*-knockdown model of EMT. FDR, false discovery rate. (F) Box plots showing decreased expression levels of genes losing gene body methylation following Twist-induced EMT in two different models: knockdown of *CDH1* and basal breast cancers compared to luminal breast cancers. y-axis represents the log2-fold change of gene expression.

11.2 RESULTS

11.2.1 ABERRANT PROMOTER DNA METHYLATION INDUCED BY EPITHELIAL-MESENCHYMAL TRANSITION IS CELL-TYPE SPECIFIC AND REGIONALLY COORDINATED

According to the EMT model of cancer progression, epithelial cells undergo a phenotypic change during the sequential progression of primary tumors towards metastasis, accompanied or not by DNA methylation changes [9,10]. Although aberrant promoter methylation on some specific promoters was previously reported [23], genomic distribution and genome-wide mapping of methylome changes during this process remains unclear. To identify DNA methylation changes in EMT, we used digital restriction enzyme analysis of methylation (DREAM), which yields highly quantitative genome-wide DNA methylation information [24]. Because small changes in DNA methylation could be important, we focused the analysis on sites with a threshold of 100-fold coverage per sample (average=1,178 tags per CpG site). By examining approximately 30,000 CpG sites spanning the promoters of around 5,000 genes (Table S1 in Additional file 1), we observed the expected relationships: lower DNA methylation at CpG islands (CGI) compared to non-CGI; lowest methylation around the transcription start site (TSS) (Figure S1A in Additional file 2); and a strong negative-correlation between promoter DNA methylation and gene expression (Spearman's $R < -0.50$, P <0.0001), independent of Twist1 expression. Interestingly, the quantitative nature of the data allowed us to establish that genes with completely unmethylated promoters (methylation $\leq 1\%$) were highly expressed in comparison to promoters with an appreciable level of methylation (>1%; Figure 1A). As there is an important overlap between PMD regions in different tissues [20,25], we analyzed gene expression according to the localization of genes in PMDs. As expected, genes located within PMDs had lower baseline expression in our model, regardless of methylation (Figure 1A), and average gene body methylation was lower for CpG sites located within PMDs compared to those located outside PMDs (21.5% versus 40% respectively, P <0.0001; Figure S1B in Additional file 2). Thus, we conclude that even low levels of DNA methylation at promoters are inhibitory for gene expression and genes within PMDs tend towards lower expression.

Expression of Twist1 caused a dramatic change in DNA methylation both at CGIs and at non-CGIs (Figure 1B) whereas no changes were seen between cells with independent control vectors, suggesting that the methylation changes observed are related to Twist1 expression and not to random clonal drift (Figure 1C). To study the impact of these changes on gene expression, we focused on completely unmethylated genes (<1%) and identified 90 genes out of 3,008 (3%) that switched from <1% to >2% with an average gain of 5.4% DNA methylation. As expected, this was associated with about a four-fold decrease in the expression of these genes (P <0.0001; Figure 1D). The gain of methylation was higher in genes located within PMDs (12%; 37 out of 309) versus outside PMDs (2%; 53 out of 2,699; $\chi2$ test, P <0.0001). Conversely, there were 39 genes that become unmethylated upon Twist1 expression, concomitant with around a two-fold increased expression of the respective genes (Figure 1D), such as *FOXC2*, a master regulator of EMT [26-28]. In contrast with promoter methylation, promoter hypomethylation was more frequent outside PMDs (4.6%; 31 out of 670) than within PMDs (1.8%; 8 out of 455; P <0.02). Gene ontology (GO) analysis for genes with methylation change associated with gene expression change showed enrichment for cell adhesion genes such as *DSCAM, NID1* and *NID2* (P=0.002), consistent with the functional change of motility and migration of mesenchymal cells. Moreover, we found an enrichment of genes (P=$5e^{-05}$) involved in calcium binding protein coding genes (that is, *FBN1, NPNT*), suggesting a functional role for orchestrated calcium-binding proteins in EMT that may represent a novel therapeutic target for controlling cell plasticity. Collectively, these data suggest that induction of EMT by Twist1 results in a moderate change in the DNA methylation of core promoters.

11.2.2 TWIST1 PROMOTES GLOBAL DEMETHYLATION OUTSIDE OF CORE PROMOTERS

To understand the global methylation and demethylation changes that occur in response to induction of EMT by Twist1, we focused on 4,903 CpG sites with a threshold detection of a minimum of 100 tags that had a baseline methylation $\geq 70\%$, as is typical of most of the genome [14]. Among these 4,903 CpG sites, one fifth (18.6%) lost DNA methylation following EMT (Table S2 in Additional file 1). We obtained comparable results using thresholds of 10 tags, and three tags per CpG site, covering 7,081 and 11,117 CpG sites respectively (data not shown). This widespread hypomethylation was mainly observed in PMDs ($P < 0.0001$; Table S2 in Additional file 1) and was independent of the genomic CpG location in repeats and lamina-associated domains (Figure S2A-C in Additional file 3). Moreover, we found decreased methylation of repetitive elements at short interspersed nuclear elements, long interspersed nuclear elements and satellite repeats (Figure S2D in Additional file 3). Concomitant with global PMD demethylation, we also observed focal hypermethylation specific to those promoters (Figure S1C,D in Additional file 2), consistent with data recently reported in colon cancer [25]. These data suggest that methylome change during EMT is reminiscent of methylome changes observed in cancer.

To understand the functional relevance of gene body methylation changes following the induction of EMT by Twist1, we performed Gene Set Enrichment Analysis (GSEA). GSEA is a computational method that assesses whether a defined set of genes (herein, gene bodies) shows statistically significant difference between two conditions (herein, between epithelial and mesenchymal states) [29]. While there was no enrichment for any pathway associated with gain of gene body methylation, GSEA reveals enrichment for gene body hypomethylation for EMT targets in the *CDH1*-knockdown model (P <0.0001 [30]; Figure 1E), and for MIR34B and MIR34C targets [31] (Table S3 in Additional file 1). Concomitantly, average expression level of those hypomethylated genes was lower after knockdown of *CDH1*, as well as in basal-like compared to luminal-like breast cancer subtypes [32,33] (P <0.004; Figure 1F). Collectively, these data suggest that following the induction of EMT by Twist expression, Twist reprograms the genome by demethylating gene bodies of epithelial cell-specific genes, leading to a decrease of their expression levels.

11.2.3 TWIST1 INCREASES THE NUMBER OF PROMOTERS WITH H3K4ME3 BY MORE THAN ONE FIFTH

Overall, the number of genes marked by H3K4me3 and also by both H3K4me3 and H3K27me3 (bivalent) was increased following Twist1-induced EMT (Figure 2A,B). Specifically, we observed that more than 20% (3,253 out of 15,853) of tallied genes acquired H3K4me3 but less than 3% (424 out of 15,853) of genes lost H3K4me3 (Figure 2C). As expected, acquisition of H3K4me3 was associated with increased mRNA expression whereas loss of H3K4me3 led to reduced expression of the corresponding genes (Figure 2D). GO analysis indicated that the set of genes that lose H3K4me3 is significantly enriched for genes associated with cell adhesion and differentiation (Figure 2E). Conversely, gain of H3K4me3, which is mediated by the TrxG complex, was found in EMT-promoting transcription factors, including zinc-ion binding proteins (i.e. *ZNF75A*), highlighting the dramatic effect of TrxG machinery in chromatin remodeling during EMT (Figure 2F). GSEA showed enrichment for estrogen receptor (ESR1) targets (P <0.0001, false discovery rate (FDR) q <0.05) within genes losing H3K4me3 (Table S4 in Additional file 1). As a result, ESR1 targets in HMLE vector cells lose the active mark H3K4me3 consistent with the three-fold decrease of ESR1 expression in HMLE Twist cells (data not shown). Importantly, genes losing H3K4me3 were also enriched for genes down-regulated in blood vessel cells from the wound site, suggesting epigenetic conservation of the EMT process between wound healing and cancer (Table S4 in Additional file 1). Thus, EMT is accompanied by a widespread gain in H3K4me3-mediated gene activation, and loss of H3K4me3 at ESR1 targets.

11.2.4 SWITCHES BETWEEN H3K4ME3 AND H3K27ME3 MODULATE TRANSCRIPTIONAL DYNAMICS

Using chromatin immunoprecipitation sequencing (ChIP-seq) for the H3K27me3 repressive histone modification, we found that the genomic distribution of H3K27me3 was significantly reduced in HMLE Twist cells (250 Megabases in vector cells compared to 153 and 138 Megabases in HMLE Twist cells cultured in monolayer and spheres, respectively; data not shown). This is consistent with the notion that cells that have undergone EMT are less differentiated and have acquired stem cell properties [6]. Given these changes in the landscape of H3K27me3, we investigated switches between H3K27me3 and H3K4me3 during EMT. Expression of Twist1 caused a loss of H3K27me3 in more than 50% of the genes marked by H3K27me3 in HMLE cells (Figure 3A,B). Of the 2,070 genes that lost H3K27me3, approximately 11% (225 out of 2,070) switched to H3K4me3 (Figure 3C) and we found that transcription of these genes was dramatically induced (around five-fold; P <0.0001). Conversely, the genes that lost H3K27me3 without gain of H3K4me3 had no average change in their respective gene expression (Figure 3C). Overall, 102 genes switched from H3K4me3 to H3K27me3 after Twist1-induced EMT, and were transcriptionally repressed by more than 32 fold (Figure 3C). These chromatin switches were associated with differential gene expression, particularly at typical EMT markers. For example, the repression of E-cadherin expression during EMT correlated with a switch from H3K4me3 to H3K27me3 (Figure 3B). Conversely, gain of N-cadherin expression correlated with a switch from H3K27me3 to H3K4me3. Strikingly, the same interplay between H3K4me3 and H3K27me3 occurs for master genes involved in the EMT process, such as *PDGFRα*, which is essential for Twist1 to promote tumor metastasis via invadopodia [34], and the splicing regulator *ESRP1*, which is repressed by Snail1 to promote EMT [35] (Table S5 in Additional file 1). Among genes with highly altered expression during EMT increasing or decreasing at least nine fold), 23.1% of them switched between

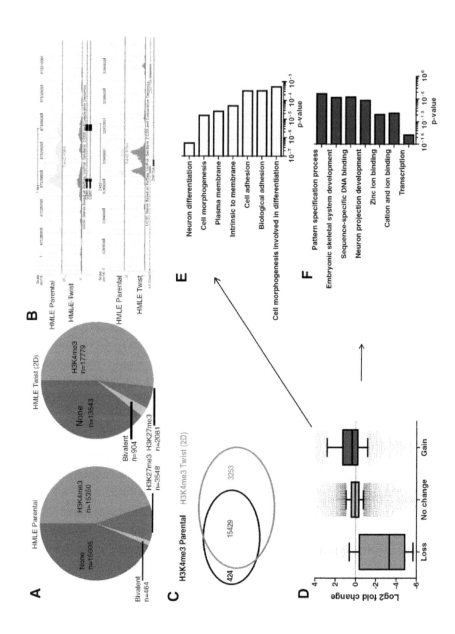

FIGURE 2: H3K4me3 dynamic modifications are coupled with transcriptional changes related to epithelial-mesenchymal transition genes. (A) Pie chart showing the distribution of H3K4me3 and H3K27me3 marks in human mammary epithelial cells (HMLE) vector cells and HMLE Twist cells. (B) Landscape of H3K4me3 for *CDH1* (loss of H3K4me3 in vector cells) and *ZNF75A* (gain of H3K4me3) in HMLE Twist cells. (C) Venn diagram of H3K4me3 at gene promoters in HMLE vector cells and HMLE Twist cells. (D) Box plots for gene expression changes in genes losing or gaining the H3K4me3 mark. (E) Gene ontology analysis using DAVID for genes losing the H3K4me3 mark. The x-axis represents the P-value levels and y-axis the gene ontology pathways. (F) Gene ontology analysis using DAVID for genes gaining the H3K4me3 mark. The x-axis represents the P-value levels and y-axis the gene ontology pathways.

H3K4me3 and H3K27me3 marks, as compared to only 2.8% for genes without highly altered expression (Figure S3 in Additional file 4). Altogether, these data suggest that an epigenetic program orchestrated by TrxG or PcG complexes regulate key EMT genes.

During EMT, parallel to the dramatic loss of H3K27me3 occupancy in nearly 50% of genes (n=2,070) marked in HMLE vector cells, mesenchymal cells gained H3K27me3 at 1717 genes. GSEA analysis showed that these genes were enriched for functional categories in a cell-type specific manner. Indeed, the set of genes which gained H3K27me3 is related to genes down-regulated in a previously described *CDH1* knockdown model of EMT [30] (Figure 4A) and to genes with low expression in basal-like as compared to luminal-like breast cancer cell lines [32] (P <0.0001; Figure 4B; Table S6 in Additional file 1). Of interest, the majority of genes down-regulated by *CDH1* knockdown and which gained H3K27me3 in Twist1-induced cells were pre-marked by H3K4me3 in HMLE cells, highlighting the importance of TrxG and PcG switches in defining cell identity during EMT (Figure 4C). Furthermore, GSEA revealed that genes belonging to pathways that gained H3K27me3 were associated with DNA repair and mRNA splicing (Table S6 in Additional file 1). Conversely, genes belonging to pathways that lost H3K27me3 were associated with mitotic pre-metaphase and undifferentiated cancer signature (Table S6 in Additional file 1).

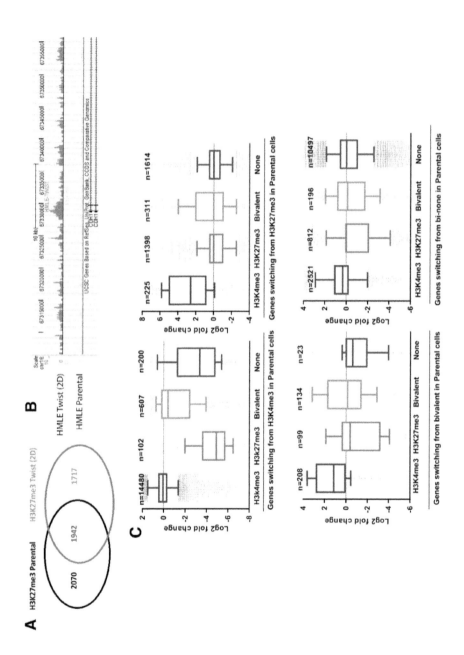

FIGURE 3: H3K27me3 switches orchestrate a mesenchymal cell-type specific gene expression signature. (A) Venn diagram of genes marked by H3K27me3 in human mammary epithelial cells (HMLE) vector cells and HMLE Twist cells. (B) Landscape of H3K27me3 mark in *CDH1* gene (gain of H3K27me3 mark in HMLE Twist cells). (C) Boxplot of gene expression (the bars represent 10% and 90% extremes) for genes switching in HMLE vector cells from H3K4me3, H3K27me3, bivalent or neither marks to other histone mark combinations in HMLE Twist cells.

Importantly, we sought to investigate if the changes we observed in Twist cells could be replicated in other EMT model systems such as Snail and TGF-β1-induced model systems. If we found similar findings across multiple EMT models, this would rule out adaptation and suggest that the effect we observed in Twist cells was due to EMT and not necessarily adaptation. In fact, we found that the majority of sites (14 out of 17) demonstrated the same directional change in H3K4me3 and/or H3K27me3 by ChIP-qPCR in HMLE Snail, TGF-β1 and Twist cells as we observed by ChIP-seq in HMLE Twist cells (Figure S4 in Additional file 5). The Pearson correlation coefficients for Snail versus Twist (r=0.8982, P <0.0001), for Snail versus TGF-β1 (r=0.4613, P=0.006) and for TGF-β1 versus Twist (r=0.1791, P=0.3108) point to close similarities between Snail- and Twist-induced EMTs in their effects on H3K4me3 and H3K27me3 whereas expression of TGF-β1 has a less similar effect (Figure S5 in Additional file 6). We also observed similar results for the methylation of DNA elements assessed using bisulfite sequencing in the promoters of seven genes randomly chosen out of genes switching between H3K27me3 and H3K4me3 (Figure S6 in Additional file 7, Figure S7 in Additional file 8 and Figure S8 in Additional file 9). Collectively, our data suggest that our a majority of changes due to Twist expression are not due to adaptation but rather shared with cells undergoing EMT through other means.

11.2.5 ENRICHMENT IN BIVALENT GENES UPON TWIST1 INDUCTION

Bivalent genes were characterized initially in stem cells by the co-occurrence of H3K27me3 (repression) and H3K4me3 (activation) at genes

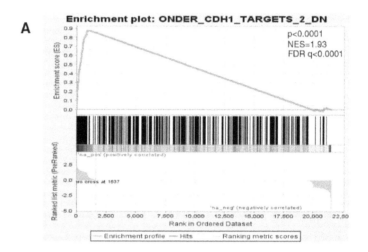

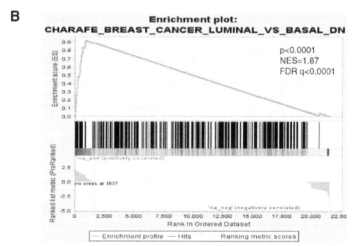

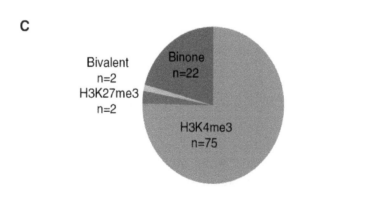

FIGURE 4: Gene set enrichment analysis of H3K27me3 marks. (A) Gene set enrichment analysis showing positive enrichment for H3K27me3 marks in genes which are down-regulated by the *CDH1*-knockdown model of EMT (P <0.0001). The bottom graph represents the rank-ordered, non-redundant list of genes. Genes on the far left (red) correlated the most with decreased gene expression following *CDH1*-knockdown. The vertical black lines show the position of each of the genes of the studied gene set in the ordered, non-redundant data set. The green curve is related to the enrichment score curve. (B) Gene set enrichment analysis showing positive enrichment for H3K27me3 marks in genes distinguishing luminal from basal like breast cancer (P <0.0001). (C) Pie chart showing that the majority of downregulated genes in the *CDH1*-knockdown model of EMT which gain H3K27me3 in Twist1-cells were also pre-marked by H3K4me3 in vector cells. FDR, false discovery rate; NES, normalized enrichment score.

which become either transcriptionally active (H3K4me3) or repressed (H3K27me3) upon differentiation [13]. We found that the number of biva-lent genes increased by more than 2.7-fold in HMLE Twist cells (n = 1,248; Figure 5A), as was the case for HOX genes (for example, HOX11; Figure 5B). The number of bivalent genes was further enriched by almost four fold in stem cell-enriched MS culture (n = 1,628) as compared to baseline of 464 genes in HMLE vector cells (Figure 5C). These data are consis-tent with the notion that the mesenchymal cells generated via EMT are less differentiated and that this feature is further enriched in MS culture. Overall, of all genes marked by H3K27me3, 34.1% were bivalent (1,249 out of 3,659) in HMLE Twist cells compared to 11.6% in HMLE cells (464 out of 4,012). The majority of bivalent genes in HMLE Twist cells were pre-marked by H3K4me3 or H3K27me3 in HMLE cells (Figure 5D). Thus, Twist-induced chromatin changes indicate a reverse differentiation state, with more 'poised' genes and, therefore, greater plasticity. Using GO analysis, we found that the newly bivalent genes in Twist1 cells were enriched for genes involved in neuron differentiation, cell morphogenesis, axonogenesis and cell fate commitment both in monolayer and sphere cul-tures (Figure 5E,F). This is consistent with the notion that genes involved in differentiation acquire a poised state in less differentiated cells. Of note, more than 50% (825 out of 1,628) of bivalent genes found in MSs were also bivalent in embryonic stem cells.

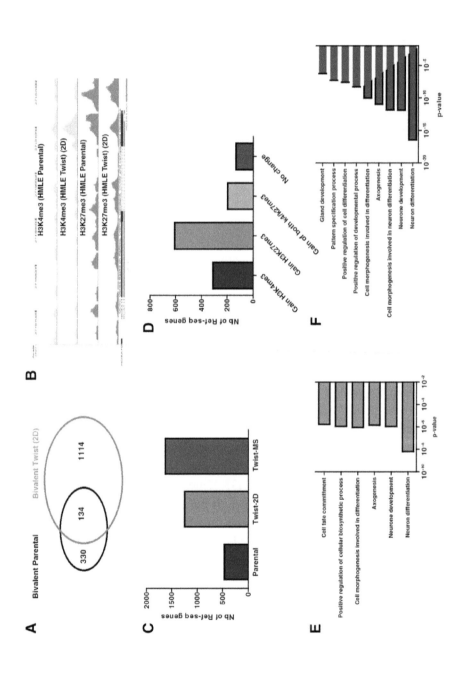

FIGURE 5: Bivalent genes are highly enriched in mesenchymal cells. (A) Venn diagram of bivalent genes in human mammary epithelial (HMLE) vector cells and HMLE Twist-cells. (B) Landscape of H3K4me3 and H3K27me3 mark for the homeobox gene *HOXA11*. Note that HOXA11 is marked by H3K27me3 in both epithelial and mesenchymal cells. In HMLE Twist cells, *HOXA11* gains H3K4me3 mark in addition to H3K27me3 mark. (C) Bar graph showing the number of bivalent genes in HMLE vector cells, and HMLE Twist cells in a monolayer (Twist-2D) and sphere culture (Twist-MS). (D) Changes in histone modifications among genes becoming bivalent in HMLE Twist cells (Twist-2D) as compared to HMLE vector cells. (E) Gene ontology of newly bivalent genes in HMLE Twist cells cultured in monolayer (Twist-2D). (F) Gene ontology of newly bivalent genes in HMLE Twist cells cultured in spheres (Twist-MS). 2D, monolayer; MS, mammosphere.

11.2.6 CHROMATIN CHANGES IN SPHEROID CULTURES

Cells with stem cells properties are known to initiate sphere formation in non-attachment cultures including the cells induced to undergo EMT. Whereas previous work has shown that sphere culture actually decreases the number of bivalent genes [36], we observed an increase in the number of bivalent genes from 464 to 1,628 (Figure 5C). Furthermore, we compared the expression of genes, DNA methylation and histone modifications of HMLE Twist cells cultured in monolayers (two dimensional) or in MS (three dimensional) and found that the DNA methylation in the two states was highly similar (Spearman's R >0.96, P <0.0001; Figure S9A in Additional file 10). By contrast, we found that 2.6% of the genes (849 out of 33,004) increased their expression more than four fold and 2.2% (737 out of 33,004) decreased their expression more than four fold when transitioned from monolayer to MS culture. GSEA analysis revealed positive enrichment for different pathways related to interferon responses (Table S7 in Additional file 1); by contrast, there was negative enrichment (exclusion) for pathways involved in proliferation [37], as well as for genes up-regulated in grade 3 versus grade 1 invasive breast cancer tumors [38] (Table S7 in Additional file 1). Consistent with earlier findings in an ovarian cancer model [36], there was a significant switch toward more genes marked by H3K27me3 in MS cells (3,607) compared to monolayer

(2,411; Figure S9B in Additional file 10), but the majority of these genes were already transcriptionally silenced in monolayer culture in response to the overexpression of Twist1. This was also the case for the 186 genes switching from H3K4me3 in monolayer to H3K27me3 in MS. Remarkably, there was a loss of H3K4me3 mark in 2,894 genes when cultured in MS compared to monolayer (Figure S9B in Additional file 10). We then asked whether histone switches between HMLE vector and HMLE Twist cells cultured in spheres were consistent with gene expression changes despite culture condition, and found that this was the case (Figure S9C in Additional file 10). These data suggest that changes in H3K4me3 and H3K27me3 distribution accompany a Twist1-driven EMT, either cultured in monolayer or spheres.

11.2.7 CHROMATIN INTERPLAY BETWEEN DNA METHYLATION AND HISTONE MODIFICATIONS

Because our data suggest that both DNA methylation and histone modifications are altered throughout the genome following Twist1-induced EMT, we examined the relationship between these different modifications. We found that low-level gain of DNA methylation in promoter regions was associated with a significant increase in H3K27me3 in both Twist1 monolayer and MS culture (Figure 6A). This was validated by bisulfite sequencing in seven selected gene promoters switching from H3K4me3 in HMLE vector cells to H3K27me3 in HMLE Twist cells (Table S8 in Additional file 1). Conversely, loss of DNA methylation was not associated with loss or gain of H3K27me3 (Figure 6B). Furthermore, both in HMLE and HMLE Twist cells, we found a strong positive correlation between low-level DNA methylation at the promoter and the presence of H3K27me3 (Spearman's R >0.34, P <0.0001). Interestingly, the association between gene silencing and enrichment of H3K27me3 around TSSs was much more pronounced in genes located in PMDs (Figure 6C) than outside PMDs (Figure 6D).

To assess whether there is an opposing correlation between H3K4me3 and DNA methylation, we focused on the gene promoters that gain DNA methylation and decrease gene expression by two fold or more. Overall,

22 out of 30 genes lost their H3K4me3 mark, highly confirming the opposing relationship between DNA methylation and H3K4me3 (data not shown). Conversely, out of the 19 gene promoters losing DNA methylation and gaining expression, six genes gained a de novo H3K4me3 mark, while the 13 other genes that already had a H3K4me3 mark in HMLE vector cells kept it in HMLE Twist cells.

11.2.8 EPIGENETIC PLASTICITY MEDIATED BY EZH2 IS REQUIRED FOR EPITHELIAL-MESENCHYMAL TRANSITION

We then asked if there is any chromatin regulator that may explain gene expression changes during EMT. Ingenuity Pathway Analysis was used to investigate chromatin regulators capable of altering major changes in gene expression during EMT. We identified EZH2 as one of the top upstream regulators to be inhibited following EMT (P=8.25e^{-11}; Figure 7A; Table S9 in Additional file 1). In our HMLE model, following EMT, we did not observe an alteration in expression of EZH2, which could account for a change in PRC2 activity. Nevertheless, our ChIP-seq revealed a decrease in H3K27me3 peak lengths, presumably mediated by EZH2 by more than half. Furthermore, out of 37 EZH2 target genes identified by IPA as repressed, approximately one third (n=11) were repressed through gain of H3K27me3; for the remaining 26 genes, this was independent of PRC2 activity. Thus, we reasoned that post-translational modification of EZH2 may account for a decrease in methyltransferase activity and consequent repression of its target genes. In fact, phosphorylation of EZH2 at serine residue 21 is known to significantly affect its function [39,40]. Thus, we used immunofluorescence to examine gain of phosphorylation in EZH2 at serine residue 21 in HMLE Twist cells relative to control cells. We observed a striking increase in phosphorylation of EZH2 in HMLE Twist cells (Figure 7B). This suggests that this post-translational modification may account for decreased trimethylation of H3K27 notwithstanding inhibition of EZH2 target genes through a PRC2-independent mechanism.

We then asked whether EZH2 is also required for the EMT. We examined the importance of EZH2 in Twist-induced EMT. While the expression of EZH2 was not significantly altered following the expression of Twist1

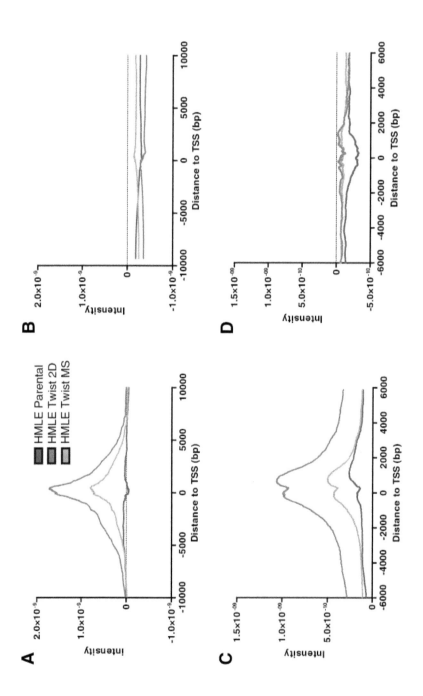

FIGURE 6: Correlation between DNA methylation and H3K27me3 mark. (A) During epithelial-mesenchymal transition (EMT), gain of DNA methylation (≥2%) at selected promoters in human mammary epithelial (HMLE) Twist cells cultured either in monolayer or spheres and which were completely unmethylated (≤1%) in HMLE vector cells is accompanied by gain of the H3K27me3 mark. The x-axis represents distance from the CpG sites gaining DNA methylation to the transcription start sites (TSS). The y-axis represents average intensity of H3K27me3 peaks. (B) During EMT, loss of DNA methylation (≤2%) of selected promoters in HMLE Twist cells cultured either in monolayer or spheres and which were methylated in HMLE vector cells is not associated with change of H3K27me3 distribution. The x-axis represents distance from the CpG sites losing DNA methylation to the TSSs. The y-axis represents average intensity of H3K27me3 peaks. (C) Distribution of H3K27me3 marks in partially methylated domain (PMDs) in HMLE Twist cells cultured in a monolayer and spheres as compared to HMLE vector cells. (D) Distribution of H3K27me3 marks outside PMDs in HMLE Twist cells either cultured in a monolayer or as spheres as compared to HMLE vector cells.

or other EMT-inducing factors in HMLE cells (Figure 8A), the suppression of EZH2 using shRNA was sufficient to reduce the level of H3K27me3 (Figure 8B,C). In addition, reduction of EZH2 significantly reduced the number of MSs compared to the control shRNA expressing cells (Figure 8D). This reduction in sphere formation was not due to changes in cell growth rates in monolayer culture (Figure 8E) or changes in the expression of EMT-promoting transcription factors (Figure S10 in Additional file 11). Interestingly, knockdown of the EZH2 homolog EZH1 (also a part of the PRC2 complex) mimics the sphere-formation defect similarly to EZH2 knockdown (Additional file 12: Figure S11). Furthermore, reduction of EZH2 protein using 3-deazaneplanocin A [41] was sufficient to reduce both H3K27me3 and sphere formation (Figure 8F,G). Notably, the sphere assay was performed in the absence of the drug, following an 8-day pretreatment and a 48-hour interim period to avoid cytotoxicity during sphere growth. Furthermore, the morphology of HMLE Twist cells following 3-deazaneplanocin A treatment was more epithelial than control-treated cells (Figure 8H). Together, these results indicate that EZH2 plays an essential role in the pathophysiology of EMT through PRC2-dependent and -independent mechanisms.

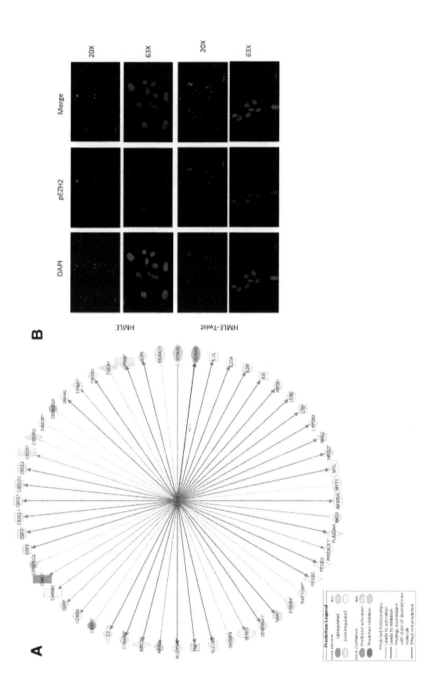

FIGURE 7: Phosphorylation of serine residue 21 of EZH2 is increased following epithelial-mesenchymal transition. (A) Ingenuity Pathway Analysis showing inhibition of EZH2 as a potential upstream regulatory event following Twist-induced epithelial-mesenchymal transition. Overall, 37 target genes were predicted to be inhibited by EZH2. (B) Immunofluorescence showing increase of phosphorylation of EZH2 at serine residue 21 in human mammary epithelial (HMLE) Twist cells as compared to HMLE vector cells.

11.3 DISCUSSION

The findings presented here provide the first comprehensive genome-wide demonstration of the remodeling of the epigenome following Twist1-induced EMT, in terms of DNA methylation as well as trithorax and polycomb-related histone modifications, concurrent with quantitative gene expression analysis by RNA-seq. In particular, we show evidence that genome-wide DNA methylation changes involve focal hypermethylation and global hypomethylation of PMDs, reminiscent of methylome changes observed between normal mammary cells and breast cancers. This highlights, for the first time, that EMT recapitulates DNA methylation changes observed during breast cancer carcinogenesis [21]. As with embryonic stem cell differentiation, the majority of changes occur outside core promoters, suggesting that cell plasticity, even in differentiated cells, involves a profound change in the DNA methylome and histone packaging. These data are in contrast with data reported recently showing unchanged DNA methylation during EMT mediated by TGF-β [9]. One explanation could be that the 36-hour exposure to TGF-β was too short to observe changes in DNA methylation. In addition, our analysis of H3K4me3 and H3K27me3 ChIP data from HMLE Snail, Twist and TGF-β cells indicates that TGF-β cells are more divergent than EMT induced by Snail and Twist, an observation that was initially made using microarray data in Taube et al. [33]. Nevertheless, our data provide powerful evidence to support the viewpoint recently brought forward by Pujadas and Feinberg that shifts in distinct regions of the epigenetic landscape, in our case PMDs, undergird cellular plasticity in both developmental and disease contexts [42].

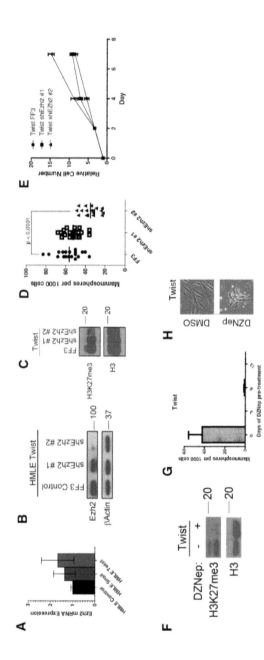

FIGURE 8: EZH2 is required for Twist-induced stemness. (A) Quantitative reverse transcription polymerase chain reaction expression of EZH2 in human mammary epithelial (HMLE) cells expressing the indicated epithelial-mesenchymal transition (EMT) inducers. (B) EZH2 protein levels assessed by western blot in HMLE Twist cells transduced by vectors expressing two different short hairpin RNAs (shRNAs) targeting EZH2 and a control vector (FF3). Note that ShRNA #2 was more efficient in reducing EZH2 protein levels as indicated by western blot. (C) Histone H3 and H3K27me3 protein levels assessed by western blot indicate that shRNA #2 reduces H3K27me3. (D) Mammospheres were counted in HMLE Twist FF3 cells (control) and shEZH2 cells grown in mammosphere-promoting conditions. Spheres greater in size than 50 micrometers after 10 days were counted and a Student's t-test was performed. (E) Count of HMLE Twist FF3 (control) and shEZH2 cells after multiple days in standard cell culture conditions. (F) Western blot analysis of H3K27me3 level in HMLE Twist cells before and after treatment with 3-deazaneplanocin A. (G) HMLE Twist cells were treated with 2.5 μM of 3-deazaneplanocin A for 8 days followed by 48 hours without drug. Surviving cells were then subjected to a sphere assay in the absence of drug for 8 days. (H) The morphology of HMLE Twist cells treated with 2.5 μM 3-deazaneplanocin A is depicted after 8 days of treatment. We observed a change toward an epithelial morphology.

Unexpectedly, our DREAM analysis found that a small gain of promoter DNA methylation is often coupled with a gain of H3K27me3, which strongly suggests that DNA methylation and PcG proteins act together. Indeed, DNA methylation is linked to polycomb protein-mediated repression; however, there is a debate as to whether DNA methylation and PcG proteins act together [43] or independently [44,45] to silence gene expression. Deep sequencing allows more accurate measurement in the low methylation range and the etiology and functional significance of small changes in DNA methylation is unknown. Although DNA methylation could be contributing to gene silencing, it is also possible that silencing by PcG proteins weakens protection against DNA methylation and thus indirectly promotes small increases. It is interesting that remodeling at PMDs was associated with both DNA hypermethylation of promoters and global demethylation; we speculate that a redistribution of TET (ten-eleven-translocation) and DNMT (DNA (Cytosine-5-)-Methyltransferase) proteins may facilitate plasticity during EMT [46].

Most strikingly, cells that had undergone EMT displayed a large increase in the number of bivalent genes. Certainly, these poised genes might help mediate the stem cell properties previously reported in this model. How Twist1 leads to this gain of bivalency deserves further investigation; however, this is the first description of remodeling involving differentiated cells having similarities with the differentiation of embryonic stem cells [13].

Beyond the pure description of the landscape of DNA methylation and histone changes, these data provide a holistic framework for studying EMT-mediated changes in chromatin and gene expression. Indeed, our data show that key EMT markers switch between H3K4me3 and H3K27me3. As an example, we found an opposing histone switch for E-cadherin and N-cadherin, and that the expression change was associated with histones switches but not with DNA methylation changes. In fact, E-cadherin lost H3K4me3 and gained H3K27me3, whereas N-cadherin did the opposite. Moreover, other EMT regulators (different from EMT markers) were also found to be subject to H3K4me3 and H3K27me3 switches, such as *ESRP1* and *PDGFα,* which are known to be involved in splicing and invadopodia formation respectively [34,47]. Therefore, we speculate that other genes that exhibit histone modification switches are key EMT genes and deserve more focus in the study of this process. Indeed, with the development of deep sequencing and the decreased cost of sequencing, we speculate that alterations in histone landscape may be used in the future as a tool for drug discovery.

Lastly, we have shown that both the epigenetic modifiers EZH2 and EZH1 are essential for the stemness property of cells that have undergone EMT in response to Twist1 expression. EZH2 is a known marker of aggressive breast cancer [48], specifically through an influence on cancer stem cells [49], but a link to EMT has not yet been described. Because the expression of either EZH2 or EZH1 was not significantly altered by EMT, we hypothesize that the EMT-induced changes in the H3K27me3 landscape are mediated primarily by changes in EZH2 or EZH1 localization and function. Of note, the suppression of EZH2 by shRNA and pharmacologically by 3-deazaneplanocin A was sufficient to reduce both H3K27me3 and sphere formation, opening avenues for the use of EZH2 inhibitors to reverse EMT-induced tumor resistance to hypoxia or chemotherapeutics. Further work is called for to detail the mechanisms leading to these changes. Currently, there are active efforts to develop EZH2 inhibitors for cancer therapy, and our data suggest that they may also be useful to suppress epigenetic plasticity and its physiological consequences, such as metastasis and drug resistance.

11.4 CONCLUSIONS

We show that induction of EMT results in dramatic alterations in the epigenetic landscape involving significant changes in both DNA methylation (mainly outside core promoters) and histone modifications (that is, an increase in bivalent genes, gene switching between H3K4me3 and H3K27me3) and that these changes contribute to the stem cell properties and increase cellular plasticity. Thus, inhibiting epigenetic remodeling may block plasticity which facilitates EMT and associated breast cancer metastasis.

11.5 METHODS

11.5.2 CHARACTERIZATION OF HUMAN MAMMARY EPITHELIAL CELLS

HMLE Twist cells were derived as shown in Yang et al. [2]. Briefly, we overexpressed Twist1 using retroviral vectors and the transduced cells were selected using puromycin. This method yields a very high transduction rate (>99%). To further confirm the homogeneity of this population of cells, we performed immunofluorescence for *VIM* and *FOXC2* markers, which are known to be induced following EMT (Figure S12 in Additional file 13). Similar results were obtained for HMLE Snail and HMLE TGF-β1 cells (Figure S13 in Additional file 14).

11.5.3 DIGITAL RESTRICTION ENZYME ANALYSIS OF METHYLATION METHODS

DREAM was performed as reported previously [50]. Briefly, genomic DNA was sequentially digested with a pair of enzymes recognizing the same restriction site (CCCGGG) containing a CpG dinucleotide, as previously reported. The first enzyme, SmaI, cut only at unmethylated CpG and left blunt ends. The second enzyme, XmaI, was not blocked by methylation and left a short 5′ overhang. The enzymes thus created methylation-specific signatures at the ends of digested DNA fragments. These were deciphered by next-generation sequencing using the Illumina Gene Analyzer II and Hiseq2000 platforms (Illumina, San Diego, CA, USA). Methylation levels for each sequenced restriction site were calculated based on the number of DNA molecules with the methylated or unmethylated signatures. Overall, we acquired around 36 million sequence tags per sample that were mapped to unique CpG sites in the human genome, using version hg18. Details of the DREAM method were previously reported by Challen et al. [50].

11.5.4 GENOME ANNOTATION OF DREAM DATA AND STATISTICAL ANALYSIS

Genomic regions were defined according to National Center for Biotechnology Information coordinates downloaded from the University of California Santa Cruz website [51] in April 2010. Promoters were defined as regions between −1,000 bp and +1000 bp from TSSs for each RefSeq transcript. Gene bodies were defined as the transcribed regions, +1,000 bp from TSS to the end of the transcription sites for each RefSeq transcript. To calculate promoter methylation, we averaged the methylation level of all CpG sites located between −1000 bp and 1000 bp of the TSS. To estimate the FDR of promoter methylation using our method, we reasoned that a comparison of HMLE cells transduced with Twist1 according to culture conditions (monolayer versus MS) could be used because their methylation was remarkably identical (Spearman's R >0.96, P <0.0001; Figure S3B in Additional file 4). The FDR was 0% for 5% gain of methylation for unmethylated genes (≤1%), and 0.001 (5 out of 4,655) for a difference of 2% of methylation for unmethylated genes (≤1%). Of note, a different transduction of HMLE with a different vector led to a minimal change of promoter methylation with 20 genes out of 3,933 genes gaining 2% methylation at unmethylated loci (≤1%). Using a minimum of 300 tags per SmaI site, only 7 out of 3,933 unmethylated genes gained more than 2% methylation, confirming the method's high level of precision.

Because the majority of the genome is heavily methylated in contrast to CpG sites related to promoters located in CGI, we used different criteria to analysis DNA methylation changes at the genome level. Arbitrarily, we considered that CpG sites with methylation level ≤10% were unmethylated and a threshold gain of 20% was defined as hypermethylation; conversely, CpG sites with methylation level ≥70% were considered methylated and a threshold loss of 20% was defined as hypomethylation.

For the localization of CpG sites in PMDs or outside PMDs, we downloaded data published for fetal lung fibroblasts (IMR90) [14]. The genes were considered to be located in PMDs if their promoters were located within PMDs. The graphs were prepared using GraphPad Prism 5.0 for Windows, GraphPad Software, San Diego California USA. For the graph of the distribution of CpG sites detected by DREAM, we average 25 to 50 neighbor points according to their distance to TSS, and the data were then smoothed using GraphPad Prism 5.0.

For gene body methylation, the average of CpG sites located +1000 bp from the start of transcription to the transcription end site was calculated. For GSEA, gene sets were downloaded from the Broad Institute's MSig-DB website [29]. Gene set permutations were used to determine statistical enrichment of the gene sets using the difference of methylation between Twist1-transduced cells (monolayer) and vector cells.

11.5.5 CHROMATIN IMMUNOPRECIPITATION-SEQUENCE GENERATION AND MAPPING

ChIP-seq experiments performed for H3K4m3 and H3K27me3 produced more than 10 million uniquely mapped tags per chromatin modification. ChIP was performed according to the Abcam protocol [52] with few modifications. Library preparation and sequencing were performed on an Il-lumina/Solexa Genome Analyzer II or Hiseq 2000 in accordance with the manufacturer's protocols. ChIP-seq reads were aligned to the human genome (hg18) using the Illumina Analyzer pipeline.

Unique reads mapped to a single genomic location were called peaks using the MACS software (version 1.3.7.1) for H3K4me3 marks (the window was 400 bp, and the P-value cutoff= $1e^{-5}$) [53].

For peak calling of H3K27me3, SICER (version 1.03) was used to detect peaks and enriched domains as the peaks were large and not as sharp as for H3K4me3 [54]. The window size was set as 200 bp as default. The gap size was determined as recommended by Zang et al. [54], or at most 2 kb, since the performance worsens as the gap size increases beyond more than 10 times the window size. Following Wang et al. [55], the E-value was set at FDR $\leq 5\%$, which was estimated as E-value (the expected number of significant domains under the random background) divided by the number of identified candidate domains. The FDR cutoff to further filter out the candidate domains by comparing to control was set as 5%.

Sequencing reads for histone H3 DNA were used as control for MACS and SICER. Annotated RefSeq genes with a peak located at their promoters (−1 kb to +0.5 kb of TSS) were identified as being marked by H3K4me3 or H3K27me3 modifications. For the pathway analysis, GO analysis was done using DAVID [56,57]. DAVID analyses were performed online using parameters of EASE value of $<1 \times 10^{-5}$, count of >10, fold enrichment of >2 and Bonfferroni of $<1 \times 10^{-2}$. For GSEA, gene sets were downloaded from the Broad Institute's MSigDB website [29]. Gene set permutations were used to determine statistical enrichment of the gene sets using the fold enrichment difference in histone modifications between H3K4me3 and H3K27me3 of mesenchymal cells (Twist1 cells) and vector cells.

To exclude the possibility of technical variations, we performed technical (independent IP) replicates for the ChIP of H3K4me3 and H3K27me3 in HMLE cells transduced with Twist1 and cultured in spheres followed by sequencing. Likewise, we performed a technical replicate for ChIP of H3K27me3 in HMLE vector cells. We obtained high correlations between the technical replicates (r >0.82; Table S10 in Additional file 1), suggesting that our findings were not due to chance. A list of primers used for ChIP-qPCR validation of selected genes is available in Table S11 in Additional file 1.

11.5.6 RNA-SEQUENCE LIBRARY GENERATION AND MAPPING

RNA extraction from vector cells and Twist1-transduced cells (monolayer and sphere) were done with Trizol reagent (Invitrogen, 15596–026). Library preparation was done using a SOLiD™ Total RNA-seq Kit according to the manufacturer's protocol (Life Technologies, Carlsbad, CA, USA). Reads sequenced produced by the SOLiD analysis pipeline were aligned with to the National Center for Biotechnology Information BUILD hg19 reference sequence. Short reads were mapped to the human reference genome (hg19) and exon junctions using the ABI Bioscope (version 1.21) pipeline with default parameters. Only the tags that mapped to the hg19 reference at full 35-nucleotide length were used. Reads that aligned to multiple positions were excluded. Tags mapped to RefSeq genes were counted to derive a measure of gene expression. To compare the gene expression values, we reasoned that cell type change associated with EMT could result in a change in the total amount of RNA. We therefore used the most conservative normalization by assuming most genes did not change their expression. This was done by constructing a histogram of expression ratio and by assuming that the maximum of the histogram corresponded to no change in gene expression. When compared to the normalization procedure where the total tags mapped to the genes were assumed to be constant, the differences were less than 10%.

11.5.7 DATA AVAILABILITY

All sequencing data and processed files are available on Gene Expression Omnibus accession number [GEO:GSE53026].

REFERENCES

1. Thiery JP, Acloque H, Huang RY, Nieto MA: Epithelial-mesenchymal transitions in development and disease. Cell 2009, 139:871-890.
2. Yang J, Mani SA, Donaher JL, Ramaswamy S, Itzykson RA, Come C, Savagner P, Gitelman I, Richardson A, Weinberg RA: Twist, a master regulator of morphogenesis, plays an essential role in tumor metastasis. Cell 2004, 117:927-939.
3. Ghoul A, Serova M, Astorgues-Xerri L, Bieche I, Bousquet G, Varna M, Vidaud M, Phillips E, Weill S, Benhadji KA, Lokiec F, Cvitkovic E, Faivre S, Raymond E: Epithelial-to-mesenchymal transition and resistance to ingenol 3-angelate, a novel protein kinase C modulator, in colon cancer cells. Cancer Res 2009, 69:4260-4269.
4. Wang Z, Li Y, Kong D, Banerjee S, Ahmad A, Azmi AS, Ali S, Abbruzzese JL, Gallick GE, Sarkar FH: Acquisition of epithelial-mesenchymal transition phenotype of gemcitabine-resistant pancreatic cancer cells is linked with activation of the notch signaling pathway. Cancer Res 2009, 69:2400-2407.
5. Creighton CJ, Li X, Landis M, Dixon JM, Neumeister VM, Sjolund A, Rimm DL, Wong H, Rodriguez A, Herschkowitz JI, Fan C, Zhang X, He X, Pavlick A, Gutierrez MC, Renshaw L, Larionov AA, Faratian D, Hilsenbeck SG, Perou CM, Lewis MT, Rosen JM, Chang JC: Residual breast cancers after conventional therapy display mesenchymal as well as tumor-initiating features. Proc Natl Acad Sci USA 2009, 106:13820-13825.
6. Mani SA, Guo W, Liao MJ, Eaton EN, Ayyanan A, Zhou AY, Brooks M, Reinhard F, Zhang CC, Shipitsin M, Campbell LL, Polyak K, Brisken C, Yang J, Weinberg RA: The epithelial-mesenchymal transition generates cells with properties of stem cells. Cell 2008, 133:704-715.
7. Hiraguri S, Godfrey T, Nakamura H, Graff J, Collins C, Shayesteh L, Doggett N, Johnson K, Wheelock M, Herman J, Baylin S, Pinkel D, Gray J: Mechanisms of

inactivation of E-cadherin in breast cancer cell lines. Cancer Res 1998, 58:1972-1977.

8. Vrba L, Garbe JC, Stampfer MR, Futscher BW: Epigenetic regulation of normal human mammary cell type-specific miRNAs. Genome Res 2011, 21:2026-2037.

9. McDonald OG, Wu H, Timp W, Doi A, Feinberg AP: Genome-scale epigenetic reprogramming during epithelial-to-mesenchymal transition. Nat Struct Mol Biol 2011, 18:867-874.

10. Ruike Y, Imanaka Y, Sato F, Shimizu K, Tsujimoto G: Genome-wide analysis of aberrant methylation in human breast cancer cells using methyl-DNA immunoprecipitation combined with high-throughput sequencing. BMC Genomics 2010, 11:137.

11. Walter K, Holcomb T, Januario T, Du P, Evangelista M, Kartha N, Iniguez L, Soriano R, Huw LY, Stern HM, Modrusan Z, Seshagiri S, Hampton GM, Amler LC, Bourgon R, Yauch RL, Shames DS: DNA Methylation Profiling Defines Clinically Relevant Biological Subsets of Non-small Cell Lung Cancer. Clin Cancer Res 2012, 18:2360-2373.

12. Wu CY, Tsai YP, Wu MZ, Teng SC, Wu KJ: Epigenetic reprogramming and posttranscriptional regulation during the epithelial-mesenchymal transition. Trends Genet 2012, 28:454-463.

13. Bernstein BE, Mikkelsen TS, Xie X, Kamal M, Huebert DJ, Cuff J, Fry B, Meissner A, Wernig M, Plath K, Jaenisch R, Wagschal A, Feil R, Schreiber SL, Lander ES: A bivalent chromatin structure marks key developmental genes in embryonic stem cells. Cell 2006, 125:315-326.

14. Lister R, Pelizzola M, Dowen RH, Hawkins RD, Hon G, Tonti-Filippini J, Nery JR, Lee L, Ye Z, Ngo QM, Edsall L, Antosiewicz-Bourget J, Stewart R, Ruotti V, Millar AH, Thomson JA, Ren B, Ecker JR: Human DNA methylomes at base resolution show widespread epigenomic differences. Nature 2009, 462:315-322.

15. Hansen KD, Timp W, Bravo HC, Sabunciyan S, Langmead B, McDonald OG, Wen B, Wu H, Liu Y, Diep D, Briem E, Zhang K, Irizarry RA, Feinberg AP: Increased methylation variation in epigenetic domains across cancer types. Nat Genet 2011, 43:768-775.

16. Lister R, Pelizzola M, Kida YS, Hawkins RD, Nery JR, Hon G, Antosiewicz-Bourget J, O'Malley R, Castanon R, Klugman S, Downes M, Yu R, Stewart R, Ren B, Thomson JA, Evans RM, Ecker JR: Hotspots of aberrant epigenomic reprogramming in human induced pluripotent stem cells. Nature 2011, 471:68-73.

17. Aran D, Toperoff G, Rosenberg M, Hellman A: Replication timing-related and gene body-specific methylation of active human genes. Hum Mol Genet 2011, 20:670-680.

18. Popp C, Dean W, Feng S, Cokus SJ, Andrews S, Pellegrini M, Jacobsen SE, Reik W: Genome-wide erasure of DNA methylation in mouse primordial germ cells is affected by AID deficiency. Nature 2010, 463:1101-1105.

19. Shann YJ, Cheng C, Chiao CH, Chen DT, Li PH, Hsu MT: Genome-wide mapping and characterization of hypomethylated sites in human tissues and breast cancer cell lines. Genome Res 2008, 18:791-801.

20. Schroeder DI, Lott P, Korf I, LaSalle JM: Large-scale methylation domains mark a functional subset of neuronally expressed genes. Genome Res 2011, 21:1583-1591.

21. Hon GC, Hawkins RD, Caballero OL, Lo C, Lister R, Pelizzola M, Valsesia A, Ye Z, Kuan S, Edsall LE, Camargo AA, Stevenson BJ, Ecker JR, Bafna V, Strausberg RL, Simpson AJ, Ren B: Global DNA hypomethylation coupled to repressive chromatin domain formation and gene silencing in breast cancer. Genome Res 2012, 22:246-258.

22. Dontu G, Abdallah WM, Foley JM, Jackson KW, Clarke MF, Kawamura MJ, Wicha MS: In vitro propagation and transcriptional profiling of human mammary stem/progenitor cells. Genes Dev 2003, 17:1253-1270.

23. Dumont N, Wilson MB, Crawford YG, Reynolds PA, Sigaroudinia M, Tlsty TD: Sustained induction of epithelial to mesenchymal transition activates DNA methylation of genes silenced in basal-like breast cancers. Proc Natl Acad Sci USA 2008, 105:14867-14872.

24. Jelinek J, Liang S, Lu Y, He R, Ramagli LS, Shpall EJ, Estecio MR, Issa JP: Conserved DNA methylation patterns in healthy blood cells and extensive changes in leukemia measured by a new quantitative technique. Epigenetics 2012, 7:1368-1378.

25. Berman BP, Weisenberger DJ, Aman JF, Hinoue T, Ramjan Z, Liu Y, Noushmehr H, Lange CP, van Dijk CM, Tollenaar RA, Van Den Berg D, Laird PW: Regions of focal DNA hypermethylation and long-range hypomethylation in colorectal cancer coincide with nuclear lamina-associated domains. Nat Genet 2011, 44:40-46.

26. Hader C, Marlier A, Cantley L: Mesenchymal-epithelial transition in epithelial response to injury: the role of Foxc2. Oncogene 2010, 29:1031-1040.

27. Mani SA, Yang J, Brooks M, Schwaninger G, Zhou A, Miura N, Kutok JL, Hartwell K, Richardson AL, Weinberg RA: Mesenchyme Forkhead 1 (FOXC2) plays a key role in metastasis and is associated with aggressive basal-like breast cancers. Proc Natl Acad Sci USA 2007, 104:10069-10074.

28. Hollier BG, Tinnirello AA, Werden SJ, Evans KW, Taube JH, Sarkar TR, Sphyris N, Shariati M, Kumar SV, Battula VL, Herschkowitz JI, Guerra R, Chang JT, Miura N, Rosen JM, Mani SA: FOXC2 expression links epithelial-mesenchymal transition and stem cell properties in breast cancer. Cancer Res 2013, 73:1981-1992.

29. Subramanian A, Tamayo P, Mootha VK, Mukherjee S, Ebert BL, Gillette MA, Paulovich A, Pomeroy SL, Golub TR, Lander ES, Mesirov JP: Gene set enrichment analysis: a knowledge-based approach for interpreting genome-wide expression profiles. Proc Natl Acad Sci USA 2005, 102:15545-15550.

30. Onder TT, Gupta PB, Mani SA, Yang J, Lander ES, Weinberg RA: Loss of E-cadherin promotes metastasis via multiple downstream transcriptional pathways. Cancer Res 2008, 68:3645-3654.

31. Toyota M, Suzuki H, Sasaki Y, Maruyama R, Imai K, Shinomura Y, Tokino T: Epigenetic silencing of microRNA-34b/c and B-cell translocation gene 4 is associated with CpG island methylation in colorectal cancer. Cancer Res 2008, 68:4123-4132.

32. Charafe-Jauffret E, Ginestier C, Monville F, Finetti P, Adelaide J, Cervera N, Fekairi S, Xerri L, Jacquemier J, Birnbaum D, Bertucci F: Gene expression profiling of breast cell lines identifies potential new basal markers. Oncogene 2006, 25:2273-2284.

33. Taube JH, Herschkowitz JI, Komurov K, Zhou AY, Gupta S, Yang J, Hartwell K, Onder TT, Gupta PB, Evans KW, Hollier BG, Ram PT, Lander ES, Rosen JM, Wein-

berg RA, Mani SA: Core epithelial-to-mesenchymal transition interactome gene-expression signature is associated with claudin-low and metaplastic breast cancer subtypes. Proc Natl Acad Sci USA 2010, 107:15449-15454.

34. Eckert MA, Lwin TM, Chang AT, Kim J, Danis E, Ohno-Machado L, Yang J: Twist1-induced invadopodia formation promotes tumor metastasis. Cancer Cell 2011, 19:372-386.

35. Reinke LM, Xu Y, Cheng C: Snail Represses the Splicing Regulator ESRP1 to Promote Epithelial-Mesenchymal Transition. J Biol Chem 2012, 287:36435-36442.

36. Bapat SA, Jin V, Berry N, Balch C, Sharma N, Kurrey N, Zhang S, Fang F, Lan X, Li M, Kennedy B, Bigsby RM, Huang TH, Nephew KP: Multivalent epigenetic marks confer microenvironment-responsive epigenetic plasticity to ovarian cancer cells. Epigenetics 2010, 5:716-729.

37. Rosty C, Sheffer M, Tsafrir D, Stransky N, Tsafrir I, Peter M, de Cremoux P, de La Rochefordiere A, Salmon R, Dorval T, Thiery JP, Couturier J, Radvanyi F, Domany E, Sastre-Garau X: Identification of a proliferation gene cluster associated with HPV E6/E7 expression level and viral DNA load in invasive cervical carcinoma. Oncogene 2005, 24:7094-7104.

38. Sotiriou C, Wirapati P, Loi S, Harris A, Fox S, Smeds J, Nordgren H, Farmer P, Praz V, Haibe-Kains B, Desmedt C, Larsimont D, Cardoso F, Peterse H, Nuyten D, Buyse M, Van de Vijver MJ, Bergh J, Piccart M, Delorenzi M: Gene expression profiling in breast cancer: understanding the molecular basis of histologic grade to improve prognosis. J Natl Cancer Inst 2006, 98:262-272.

39. Cha TL, Zhou BP, Xia W, Wu Y, Yang CC, Chen CT, Ping B, Otte AP, Hung MC: Akt-mediated phosphorylation of EZH2 suppresses methylation of lysine 27 in histone H3. Science 2005, 310:306-310.

40. Lee ST, Li Z, Wu Z, Aau M, Guan P, Karuturi RK, Liou YC, Yu Q: Context-specific regulation of NF-kappaB target gene expression by EZH2 in breast cancers. Mol Cell 2011, 43:798-810.

41. Tan J, Yang X, Zhuang L, Jiang X, Chen W, Lee PL, Karuturi RK, Tan PB, Liu ET, Yu Q: Pharmacologic disruption of Polycomb-repressive complex 2-mediated gene repression selectively induces apoptosis in cancer cells. Genes Dev 2007, 21:1050-1063.

42. Pujadas E, Feinberg AP: Regulated noise in the epigenetic landscape of development and disease. Cell 2012, 148:1123-1131.

43. Vire E, Brenner C, Deplus R, Blanchon L, Fraga M, Didelot C, Morey L, Van Eynde A, Bernard D, Vanderwinden JM, Bollen M, Esteller M, Di Croce L, de Launoit Y, Fuks F: The Polycomb group protein EZH2 directly controls DNA methylation. Nature 2006, 439:871-874.

44. Kondo Y, Shen L, Cheng AS, Ahmed S, Boumber Y, Charo C, Yamochi T, Urano T, Furukawa K, Kwabi-Addo B, Gold DL, Sekido Y, Huang TH, Issa JP: Gene silencing in cancer by histone H3 lysine 27 trimethylation independent of promoter DNA methylation. Nat Genet 2008, 40:741-750.

45. McGarvey KM, Greene E, Fahrner JA, Jenuwein T, Baylin SB: DNA methylation and complete transcriptional silencing of cancer genes persist after depletion of EZH2. Cancer Res 2007, 67:5097-5102.

46. Song SJ, Poliseno L, Song MS, Ala U, Webster K, Ng C, Beringer G, Brikbak NJ, Yuan X, Cantley LC, Richardson AL, Pandolfi PP: MicroRNA-antagonism regulates breast cancer stemness and metastasis via TET-family-dependent chromatin remodeling. Cell 2013, 154:311-324.

47. Brown RL, Reinke LM, Damerow MS, Perez D, Chodosh LA, Yang J, Cheng C: CD44 splice isoform switching in human and mouse epithelium is essential for epithelial-mesenchymal transition and breast cancer progression. J Clin Invest 2011, 121:1064-1074.

48. Kleer CG, Cao Q, Varambally S, Shen R, Ota I, Tomlins SA, Ghosh D, Sewalt RG, Otte AP, Hayes DF, Sabel MS, Livant D, Weiss SJ, Rubin MA, Chinnaiyan AM: EZH2 is a marker of aggressive breast cancer and promotes neoplastic transformation of breast epithelial cells. Proc Natl Acad Sci USA 2003, 100:11606-11611.

49. Chang CJ, Yang JY, Xia W, Chen CT, Xie X, Chao CH, Woodward WA, Hsu JM, Hortobagyi GN, Hung MC: EZH2 promotes expansion of breast tumor initiating cells through activation of RAF1-beta-catenin signaling. Cancer Cell 2011, 19:86-100.

50. Challen G, Sun D, Jeong M, Luo M, Jelinek J, Berg J, Bock C, Vasanthakumar A, Gu H, Xi Y, Liang S, Lu Y, Darlington GJ, Meissner A, Issa JP, Godley LA, Li W, Goodell MA: Dnmt3a is essential for hematopoietic stem cell differentiation. Nat Genet 2011, 4:23-31.

51. UCSC genome browser [http://genome.ucsc.edu/]

52. Abcam protocol for chromatin immunoprecipitation [http://www.abcam.com/ps/pdf/protocols/x_chip_protocol.pdf]

53. Zhang Y, Liu T, Meyer CA, Eeckhoute J, Johnson DS, Bernstein BE, Nusbaum C, Myers RM, Brown M, Li W, Liu XS: Model-based analysis of ChIP-Seq (MACS). Genome Biol 2008, 9:R137.

54. Zang C, Schones DE, Zeng C, Cui K, Zhao K, Peng W: A clustering approach for identification of enriched domains from histone modification ChIP-Seq data. Bioinformatics 2009, 25:1952-1958.

55. Wang Z, Zang C, Cui K, Schones DE, Barski A, Peng W, Zhao K: Genome-wide mapping of HATs and HDACs reveals distinct functions in active and inactive genes. Cell 2009, 138:1019-1031.

56. da Huang W, Sherman BT, Lempicki RA: Systematic and integrative analysis of large gene lists using DAVID bioinformatics resources. Nat Protoc 2009, 4:44-57.

57. da Huang W, Sherman BT, Lempicki RA: Bioinformatics enrichment tools: paths toward the comprehensive functional analysis of large gene lists. Nucleic Acids Res 2009, 37:1-13.

There are several supplemental files that are not available in this version of the article. To view this additional information, please use the citation information cited on the first page of this chapter.

CHAPTER 12

STABILITY AND PROGNOSTIC VALUE OF SLUG, SOX9, AND SOX10 EXPRESSION IN BREAST CANCERS TREATED WITH NEOADJUVANT CHEMOTHERAPY

COSIMA RIEMENSCHNITTER, IVETT TELEKI, VERENA TISCHLER, WENJUN GUO, AND ZSUZSANNA VARGA

12.1 INTRODUCTION

Prognosis for breast cancer has improved continuously during the last decades but it is still one of the most frequent causes of tumor related death in women in the western world. Possible reasons of breast cancer mortality are tumor dormancy after treatment followed by local, regional or distant recurrence. Several factors as hormone receptors, HER2 status, proliferation fraction predicting outcome in preoperative setting were extensively examined in previous studies (Chen et al. 2013; Denkert et al. 2010; Lips et al. 2013; Payne et al. 2008; Teleki et al. 2013; van Nes et al. 2012; Varga et al. 2005; Yoshioka et al. 2013). The role of transcription factors (TF) in breast cancer prognosis has been the subject of some previous studies (Ablett et al. 2012; Cimino-Mathews et al. 2013; Giordano et al. 2012;

This chapter was originally published under the Creative Commons Attribution License. Riemenschnitter C, Teleki I, Tischler V, Guo W, and Varga Z. Stability and Prognostic Value of Slug, Sox9 and Sox10 Expression in Breast Cancers Treated with Neoadjuvant Chemotherapy. SpringerPlus 2,695 (2013). doi:10.1186/2193-1801-2-695.

Guo et al. 2012; Mego et al. 2012). Slug (SNAI2), a transcriptional repressor, is member of the Snail family of zinc finger proteins and capable to act as a master regulator, altering expression of a number of genes including E-cadherin, a transmembrane protein which plays an important role in cell adhesion (Guo et al. 2012). The role of Slug in cancer developmental processes has been highlighted in several publications (Markiewicz et al. 2012; Mego et al. 2012; van Nes et al. 2012). It has been discovered that Slug is involved in an early developmental phenomenon known as 'Epithelial to Mesenchymal Transition' or EMT, which results in the acquisition of an invasive, mesenchymal phenotype by epithelial cells. It has been postulated to play an important role in cancer growth and metastases spreading (Guo et al. 2012). Slug and Sox9 were shown to induce epithelial mesenchymal transition (EMT) and their expression consistently defines mammary stem cell state (Guo et al. 2012). Sox9 and Sox10 are two of the 20 different human Sox genes that also encode family of transcription factors (Chakravarty et al. 2011; Muller et al. 2010; Smalley et al. 2013; Soady & Smalley 2012). In cooperation with Slug, Sox9 can be used to determine the mammary stem cell state (Soady & Smalley 2012). Sox9, a nuclear TF, is often localized in cytoplasm of invasive and metastatic breast cancer (Chakravarty et al. 2011). It has been seen that patients with elevated Sox9 levels in cytoplasm suffer from faster tumor cell proliferation and significantly shorter overall survival (Chakravarty et al. 2011).

Sox10 was described last year to be present in invasive breast cancer especially in triple negative phenotype (Cimino-Mathews et al. 2013). Sox9, Sox10 and Slug were shown to be associated with poor overall survival in breast cancer (Cimino-Mathews et al. 2013; Guo et al. 2012).

In surgical pathology Sox10—a TF that allows the survival and differentiation of neural crest cells into mature cells—is honored to support the diagnosis of melanoma and nerve sheath tumors (Mohamed et al. 2013; Shin et al. 2012). Recently Sox10 labeling has been documented in myoepithelial differentiated cells in salivary gland neoplasms and also in metaplastic triple negative breast cancers (Cimino-Mathews et al. 2013; Ivanov et al. 2013).

The aim of this study was to discover the prognostic role of Slug, Sox9 and Sox10 in neoadjuvantly treated breast cancer and the correlation to pathological response and overall survival. Furthermore, we tested stability of these markers during chemotherapy.

12.2 MATERIALS AND METHODS

12.2.1 PATIENT COHORT

96 breast cancer patients, diagnosed by tissue core needle or fine needle aspiration biopsy (FNAB) and treated with preoperative neoadjuvant chemotherapy were selected consecutively between 1998 and 2009 out of the archives of the Institute of Surgical pathology, University Hospital Zürich, Switzerland. The treatment regimens encompassed different modalities including Herceptin combined with Taxol or 2 to 6 cycles of Fluorouracil-Epirubicin and/or Cyclophosphamid/Epi-docetaxel.

For 64 out of 96 patients, formalin-fixed, paraffin-embedded (FFPE) tumor blocks from preoperative core biopsies and corresponding postoperative operation specimens were available. Core biopsies prior to neoadjuvant chemotherapy without postoperative surgical specimens were available in 17 patients. For 15 patients, postoperative surgical specimens were available without previous core biopsies. Clinico-pathological data and follow up information (2-10 years) on all 96 patients could be retrieved from the pathological and clinical files.

Seventy-five tumors were histologically/cytologically diagnosed as an invasive ductal carcinoma (75/96: 78%), 18 cases as invasive lobular carcinoma (18/96: 19%), 2 more were categorized as metaplastic squamous cell carcinoma (2/96: 2%) and 1 case as a small cell carcinoma (1/96: 1%). The age of the patients was ranged between 30 and 74 years; mean age 52 years. Histological grading could be done in 81 cases using the modified Bloom and Richardson score: 38 of 96 carcinomas were poorly differentiated (grade 3), (40%), 42 cases were moderately differentiated (grade 2) (44%) and one case was well differentiated (grade 1) (1%). In 15 cases correct histological grading on the core biopsy was not possible due to the too small amount of tumor tissue.

TABLE 1: Clinico-pathological parameter of the breast cancer samples in the tissue micro arrays

N=96	Prior to chemotherapy		After chemotherapy	
Tumor size	cT1	1 (1%)	ypT0	6 (6%)
	-a		ypT1	20 (21%)
	-b		-a	2
	-c	1	-b	11
	cT2	25 (26%)	-c	7
	cT3	22 (23%)	ypT2	31 (32%)
	cT4	41 (43%)	ypT3	24 (25%)
	-b	19	ypT4	7 (7%)
	-d	22	-b	6
	NA	7 (7%)	-d	1
			No surgery	8 (9%)
Nodal status	cN0	10 (11%)	pN0	25 (26%)
	cN1	62 (64%)	pN1	27 (28%)
	cN2		pN2	12 (12%)
	cN3	3 (3%)	pN3	14 (15%)
			no surgery	8 (9%)
	NA	21 (22%)	NA	10 (10%)
ER status	Positive	68 (71%)	Positive	59 (62%)
	Negative	25 (26%)	Negative	16 (17%)
	NA	3 (3%)	NA	21 (21%)
			(ypT0 or no surgery)	
PR status	Positive	59 (62%)	Positive	46 (48%)
	Negative	34 (35%)	Negative	29 (31%)
	NA	3 (3%)	NA	21 (21%)
			(ypT0 or no surgery)	
HER2 status	Positive	28 (29%)	Positive	18 (19%)
	Negative	65 (68%)	Negative	57 (60%)
	NA	3 (3%)	NA	21 (21%)
			ypT0, no surgery	

Abbreviations:NA not available, ER estrogen receptors, PR progesteron receptors.

Surgery was performed as follows: Mastectomy in 60 patients, segmentectomy, in 28 patients. Breast surgery was combined with axillary dissection in all but 5 of these patients. Residual tumor tissue after completing preoperative chemotherapy was determined as recommended in the residual cancer burden guidelines as the percentage of residual tumor cells distributed in the tumor bed area (Symmans et al. 2007).

Details of clinic-pathological parameter are shown in Table 1. Table with patient's collective in more details previously published (Teleki et al. 2013).

The study and the construction of the TMA was approved by the Ethical Committee of the Canton Zürich (KEK- ZH NR: 2009-0065) and also by the Internal Review Board of the Institute of Surgical Pathology.

12.2.2 DETECTION OF HORMONE RECEPTORS (ER/PR) AND HER2 STATUS

Estrogen receptors (ER, clone 6F11) and progesterone receptor (PR, clone 1A6) expression was determined using the iVIEW DAB detection kit in Ventana Benchmark (all from Ventana, Basel, Switzerland) immunostainer following heat induced epitope retrieval in CC1 solution. According the current guidelines, at least 1% nuclear positive tumor cells were considered to be positive (Hammond et al. 2010).

HER2 status was defined according to the initial and the modified ASCO criteria using immunohistochemistry (IHC) and/or fluorescence in situ hybridization (FISH) (between 1998-2004 IHC complemented with FISH, between 2004-2009 FISH only methdology).

For immunohistochemistry, the CB11 clone of Anti- Her2 monoclonal antibody (Ventana) was used for automated immunostaining as mentioned above. Scoring was used in agreement with the time current FDA and ASCO/CAP guidelines (Lebeau et al. 2001; Wolff et al. 2007). Cases with tumor cells of > 10% strong and complete membrane staining were considered 3+, cases with moderate and complete membrane staining tumor cells were defined to be 2+. For FISH, the HER2 gene amplification was tested using the dual color FISH kit of PathVision (Vysis, Abbott AG, Baar, Switzerland) following the manufacturer's protocol. FISH reactions were evaluated using an Olympus computer guided fluorescence micro-scope (BX61, Olympus AG, Volketswil, Switzerland). Scoring was done following the time current FDA and ASCO/CUP guidelines: amplified sta-tus was diagnosed when ratio (between HER2 gene and chromosome 17) was >2.0 (until 2007) resp. >2.2 (from 2008).

12.2.3 TISSUE MICROARRAY CONSTRUCTION

All cases were re-evaluated on hematoxylin-eosin (HE) stained sections of the FFPE tumors for suitability for the tissue microarray (TMA). Tumor tissues from 81 patients prior to chemotherapy and tumor samples from 79 patients after chemotherapy were arrayed into two TMA blocks us-ing methodology described earlier (Kononen et al. 1998; Theurillat et al. 2007). Matched tissue samples before and after neoadjuvant chemothera-py were available for 64 patients. From every patient duplicated cores of tissue samples were arrayed into the cores.

12.2.4 IMMUNOHISTOCHEMISTRY DETECTION SLUG, SOX9 AND SOX10

Slug, Sox9 and Sox10 were detected using immunohistochemistry on the fully automated Ventana Benchmark autostainer following the manufacturers' instructions. Following antibodies was used: Slug (Cell Signaling Technology, C19G7, 1:100), Sox9 (Millipore, AB5535, 1:400), Sox10 (1:50, Santa Cruz Biotechnology, Santa Cruz CA). IHC stains for Slug and Sox9 were homog-enous across entire tumor areas. Expression for all three markers was evalu-ated in invasive tumor cells and in tumor stroma and scored as 0, 1+, 2+ 3+. Expression profile prior to and after chemotherapy was correlated to overall survival (Kaplan Meier) and with established clinico-pathological parameter.

12.2.5 STATISTICAL ANALYSES

SPSS 15.0 software was used for statistical comparisons (SPSS, Inc., Chi-cago, IL, USA). Categorical data were analyzed using Chi-Square test.

Spearman rank correlation was used to correlate Slug, Sox9, Sox10 expression and clinic-pathological parameters (stage and grade, hormone receptor status, Her2 status). Kaplan-Meier method was used to calculate overall survival with evaluation of statistical significance by log rank test. Multivariate analysis was performed using Cox regression method with 95% confidence intervals by including Sox9 and hormone receptor status both of before and after chemotherapy. These parameters did not correlate directly each other. Results were statistically significant at p values of < 0.05. Bonferroni correction was not applied.

12.3 RESULTS

12.3.1 SOX9, SOX10 AND SLUG EXPRESSION BEFORE AND AFTER NEOADJUVANT TREATMENT AND STABILITY OF EXPRESSION PROFILE DURING CHEMOTHERAPY

Sox9: Before treatment 87% (73 of 84 cases) of the tumors cells were Sox9 positive. 23% (19 of 84 cases) showed expression of Sox9 in the stroma. After chemotherapy there were 88% (72 of 82 cases) tumors with Sox9 expression in the tumor cells and 21% (19 of 84 cases cases) with expression in the tumor stroma. Sox10 was positive in 94% of the cases (79 of 94 cases) in the tumor cells before chemotherapy and 91% (75 of 82 cases) after treatment. There was no positivity in stroma with Sox10.

Sox9 and Sox10 showed no significant change in expression profile after treatment and remained stable stable during chemotherapy.

Slug was expressed in 82% of the tumor cells (69 of 84) and in 97% of the cases (81 of 84) in the tumor stroma prior to chemotherapy. Slug showed relevant changes after chemotherapy: 51% of the cases (42 of 82) were Slug positive in the tumor cells. 74 of 82 cases were positive for Slug in the stroma (90%). Strong (score 3) positivity in stroma decreased form 57% (48 of 84) to 3% (3 of 82).

Details of scores of immunohistochemistry are shown in Table 2. and in in graphical illustration in Figure 1. Examples of immunohistochemical stains are illustrated in Figure 2.

TABLE 2: Distribution of marker expression in tumor and stroma, distinguishing between 4 different expression groups such as 0=no expression, 1+ low expression, 2+ intermediate expression, 3+ strong expression

n=84	SOX9	SOX9	SLUG	SLUG	SOX10
(pre)	Tumor	Stroma	Tumor	Stroma	Tumor
pre 0	11	65	15	3	5
	(13%)	(77%)	(18%)	(3%)	(6%)
pre 1+	15	8	35	8	15
	(18%)	(10%)	(41%)	(10%)	(18%)
pre 2+	35	7	30	25	32
	(41%)	(8%)	(36%)	(30%)	(38%)
pre 3+	23	4	4	48	32
	(28%)	(5%)	(5%)	(57%)	(38%)
n=82					
(post)					
post 0	10	65	40	8	7
	(12%)	(79%)	(49%)	(10%)	(55)
post 1+	22	14	24	27	22
	(26%)	(17%)	(29%)	(33%)	(26%)
post 2+	35	3	18	44	27
	(42%)	(3%)	(22%)	(54%)	(33%)
post 3+	17	1	0	3	28
	(20%)	(1%)	(0%)	(3%)	(34%)

12.3.2 STATISTICAL ANALYSIS

12.3.2.1 CORRELATION OF SLUG, SOX9 AND SOX10 EXPRESSION TO OVERALL SURVIVAL (OS)

Sox9, Sox10 and Slug prior to and after chemotherapy were correlated with overall survival (Kaplan-Meier). Only the stromal Sox9 expression prior to and after chemotherapy showed correlation with OS Cases with 0, 1+, 2+ expression of stromal Sox9 pre-chemotherapy had a nearly significant better overall survival than those of 3+ ($p=0.065$).

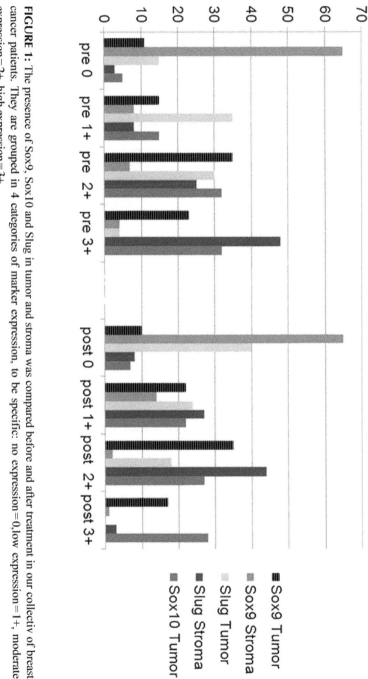

FIGURE 1: The presence of Sox9, Sox10 and Slug in tumor and stroma was compared before and after treatment in our collectiv of breast cancer patients. They are grouped in 4 categories of marker expression, to be specific: no expression=0,low expression=1+, moderate expression=2+, high expression=3+.

SOX9 score 3+ in tumor   SOX9 score 3+ in stroma

SLUG score 2+ in tumor   SLUG score 3+ in stroma

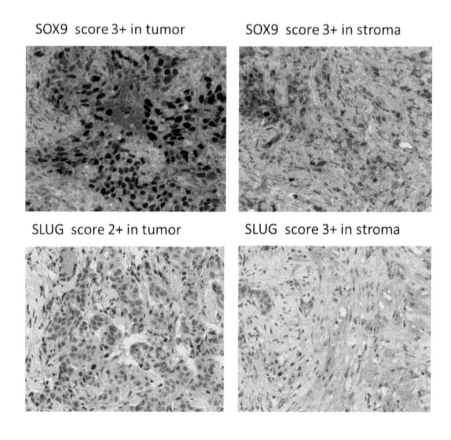

FIGURE 2: Immunohistochemical expression of Sox9 and Slug in tissue microarrays showed in tumor compared to the expression in stroma.

15.3.2.2 CORRELATION OF SLUG, SOX9 AND SOX10 EXPRESSION TO ER/PR AND HER2 STATUS

ER and PR status prior to and after chemotherapy showed correlation with overall survival (pER-pre=0.025; pER-post=0.044; pPR-pre=0.045; pPR-post=0.067). ER and PR status did not show correlation with Sox9 (chi square). ER and PR expression showed positive correlation with each other (p=0.002). Hormone receptor status and stromal Sox9 prior to and after chemotherapy were involved separately in the multivariate analysis (Cox regression). HER2 status did not correlate with overall survival in this cohort, so it was ignored from multivariate analysis.

According to multivariate analysis, stromal Sox9 expression after chemotherapy proved to be an independent and better (but negative) prognostic marker than hormone receptor status. Prior to chemotherapy the prognostic power of stromal Sox9 is similar than the hormone receptor status (Table 3). Cases with 0, 1+, 2+ expression of stromal Sox9 post-chemotherapy had significantly better overall survival than those of 3+ (p=0.004) Figure 3.

TABLE 3: Multivariate Cox regression analysis of hormone receptor status and stromal Sox9 expression before and after chemotherapy

Parameters	P-value	HR	95% CI	
			Lower	Upper
ER_{pre}	0.052	3.139	0.988	9.969
$Sox9_{pre}$ stroma	0.063	0.234	0.050	1.081
PR_{pre}	0.015	0.227	0.069	0.746
$Sox9_{pre}$ stroma	0.029	0.169	0.034	0.834
ER_{post}	0.077	2.796	0.893	8.753
$Sox9_{post}$ stroma	0.006	0.037	0.004	0.384
PR_{post}	0.223	2.062	0.644	6.609
$Sox9_{post}$ stroma	0.034	0.082	0.008	0.833

pre: before chemotherapy; post: after chemotherapy. HR: hazard ratio; CI: confidence intervals.

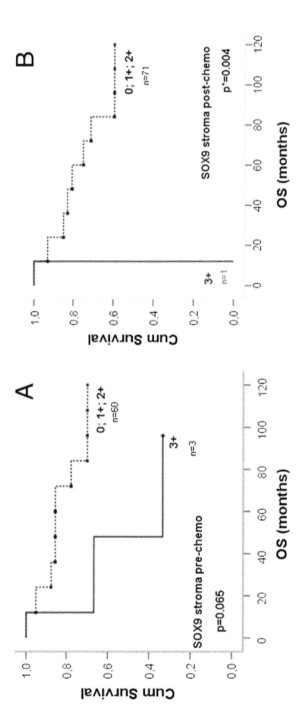

FIGURE 3: Kaplan Meier curves in stromal SOX9 expression. A: Cumulative survival in months of Sox9 expression prior to chemotherapy was compared to high expression (3+) und moderate/ no expression (0, 1+, 2+), log-rank P= 0.065. B: Post-chemotherapy survival in months, P= 0.004.

15.3.2.3 CHANGE IN SLUG, SOX9 AND SOX10 EXPRESSION PROFILE PRIOR TO AND AFTER CHEMOTHERAPY

Slug stromal expression (score 3) showed highly significant changes after chemotherapy (dropped from 57% to 3%), $p=0.0001$ (Fishers' exact test).

Lack of Slug expression in the tumor cells became more frequent after chemotherapy and proved also significant (increased from 18% to 49%), $p=0.0029$ (Fishers' exact test).

There was no relevant or significant change in the expression profile of Sox9 and Sox10 after chemotherapy.

15.4 DISCUSSION

In our study we analyzed the prognostic role and stability of transcription factors as Sox9, Sox10 and Slug in a cohort of preoperative chemotherapeutically treated breast cancers. We could show in this cohort that strong stromal Slug expression was instable during chemotherapy and postoperative stromal Sox9 expression was significantly associated with shortened overall survival.

We are not aware of any studies in the literature, showing association between these transcription factors and prognosis in a preoperative treated breast cancer cohort. In the last two years different studies addressed to the Sox genes (Chakravarty et al. 2011; Muller et al. 2010; Soady & Smalley 2012). It was demonstrated that these genes play an important role in cancer development as well. Differentiated mammary luminal cells expressing both Sox9 and Slug, an EMT-associated transcription factor, were able to activate an endogenous autoregulatory network in adult stem cells and reconstitute an entire mammary gland (Guo et al. 2012). Meanwhile, showing stem cell characteristics such as self-renewal and multi- potency, Sox10 is one of the most studied Sox family genes, mainly because of their role in the survival of neural crest cells and compilation into neural-crest-derived melanocytes and glia (Cimino-Mathews et al. 2013; Ivanov et al. 2013). Invasive breast carcinomas, especially those with triple negative hormone expression and metastases seem to be supportive for myoepithelial differentiation (Cimino-Mathews et al. 2013; Ivanov et al. 2013). Expression of Sox9, Sox10 and Slug was seen in 82-96% of the tumor cells prior to chemotherapy in our study, further supporting the fact that the transcription factors are highly active in the breast cancer tissue.

It was described in a recent work of A. Cimino-Mathews et al., that Sox10 is also expressed in breast carcinomas, mainly in triple negative cases, changing the previous view on Sox10 being exclusively expressed by melanocytic lesions (Cimino-Mathews et al. 2013).

Our cohort is a mainly hormone receptor dominated tumor collection. Interestingly, Sox10 was detected in a high percentage of tumor samples (up to 91-94%). The high expression of Sox10 might represent a rather unspecific staining property of unknown biological significance as staining intensity did not have any prognostic impact in this patients' cohort. Moreover, Sox10 remained stable during chemotherapy showing neither correlation to overall survival nor to response to chemotherapy. This is most likely due to the fact that hormone receptor positive carcinomas of the breast, as was the case in our study, can benefit from different therapeutic options than triple negative or basal like cancers. Further prognostic role of Sox10 in breast cancer, needs to be addressed in future studies.

Co-expression of Slug and Sox9 in breast cancer was found to correlate with rapid tumor growth and metastasis spreading in a study earlier (Guo et al. 2012). In other studies, the high expression of Sox9 was found to be associated with higher histological grade and/or triple negative phenotype and thus considered as prognostic marker in adjuvant oncological settings (Chakravarty et al. 2011). In our cohort of neoadjuvantly treated breast cancers, strong Sox9 expression (score 3) in tumor stroma significantly correlated with shortened overall survival after completion of preoperative chemotherapy, further supporting the hypothesis described by Guo et al (Guo et al. 2012). The lack of correlation between weaker Sox9 scores and survival can be most likely explained by the fact that our cohort included mainly hormone receptor positive cases. Strong stromal Sox9 expression may be potentially used as a prognostic marker additionally to conventional prognostic parameter when assessing residual tumor burden and response after chemotherapy.

As to expression of Slug, we could show a drastic drop in the stromal Slug (score 3) expression after chemotherapy. Furthermore, we detected significantly more tumors lacking Slug expression in the tumor cell postoperatively. As Slug expression could not be correlated to any known prognostic factors in our cohort, the biological relevance of this phenomenon remains unclear at the moment.

The correlation of Sox9, Sox10 and Slug to other classical clinicopathological parameters as hormone receptors, HER2 and proliferation index could confirm that Sox9 is an independent negative prognostic factor. This finding may be used as additional information on overall survival and therapy response in neoadjuvant treatment setting.

In summary, our study shows, that strong stromal SOX9 expression in breast cancer after neoadjuvant chemotherapy bears negative prognostic information and is associated with shortened overall survival. The biological relevance of change in phenotype change resp. in expression profile of Slug, both in stroma and tumor cells, needs to be addressed in further studies.

REFERENCES

1. Ablett MP, Singh JK, Clarke RB (2012) Stem cells in breast tumours: are they ready for the clinic? Eur J Cancer 48:2104-2116 doi:10.1016/j.ejca.2012.03.01 9S0959-8049(12)00303-6 [pii]
2. Chakravarty G, Moroz K, Makridakis NM, Lloyd SA, Galvez SE, Canavello PR, Lacey MR, Agrawal K, Mondal D (2011) Prognostic significance of cytoplasmic SOX9 in invasive ductal carcinoma and metastatic breast cancer. Exp Biol Med (Maywood) 236:145-155 doi:10.1258/ebm.2010.010086236/2/145 [pii]
3. Chen JH, Pan WF, Kao J, Lu J, Chen LK, Kuo CC, Chang CK, Chen WP, McLaren CE, Bahri S, Mehta RS, Su MY (2013) Effect of taxane-based neoadjuvant chemotherapy on fibroglandular tissue volume and percent breast density in the contralateral normal breast evaluated by 3T MR. NMR Biomed 26:1705-1713 doi:10.1002/nbm.3006
4. Cimino-Mathews A, Subhawong AP, Elwood H, Warzecha HN, Sharma R, Park BH, Taube JM, Illei PB, Argani P (2013) Neural crest transcription factor Sox10 is preferentially expressed in triple-negative and metaplastic breast carcinomas. Hum Pathol 44:959-965 doi:10.1016/j.humpath.2012.09.005S0046-8177(12)00333-4 [pii]

5. Denkert C, Loibl S, Noske A, Roller M, Muller BM, Komor M, Budczies J, Darb-
    Esfahani S, Kronenwett R, Hanusch C, von Torne C, Weichert W, Engels K, Solbach
    C, Schrader I, Dietel M, von Minckwitz G (2010) Tumor-associated lymphocytes as
    an independent predictor of response to neoadjuvant chemotherapy in breast cancer.
    J Clin Oncol 28:105-113 doi:10.1200/JCO.2009.23.7370JCO.2009.23.7370 [pii]
6. Giordano A, Gao H, Anfossi S, Cohen E, Mego M, Lee BN, Tin S, De Laurentiis
    M, Parker CA, Alvarez RH, Valero V, Ueno NT, De Placido S, Mani SA, Esteva FJ,
    Cristofanilli M, Reuben JM (2012) Epithelial-mesenchymal transition and stem cell
    markers in patients with HER2-positive metastatic breast cancer. Mol Cancer Ther
    11:2526-2534 doi:10.1158/1535-7163.MCT-12-0460
7. Guo W, Keckesova Z, Donaher JL, Shibue T, Tischler V, Reinhardt F, Itzkovitz S,
    Noske A, Zurrer-Hardi U, Bell G, Tam WL, Mani SA, van Oudenaarden A, Wein-
    berg RA (2012) Slug and Sox9 cooperatively determine the mammary stem cell
    state. Cell 148:1015-1028 doi:10.1016/j.cell.2012.02.008S0092-8674(12)00165-1
    [pii]
8. Hammond ME, Hayes DF, Dowsett M, Allred DC, Hagerty KL, Badve S, Fitzgib-
    bons PL, Francis G, Goldstein NS, Hayes M, Hicks DG, Lester S, Love R, Mangu
    PB, McShane L, Miller K, Osborne CK, Paik S, Perlmutter J, Rhodes A, Sasano H,
    Schwartz JN, Sweep FC, Taube S, Torlakovic EE, Valenstein P, Viale G, Visscher D,
    Wheeler T, Williams RB (2010) American Society of Clinical Oncology/College Of
    American Pathologists guideline recommendations for immunohistochemical test-
    ing of estrogen and progesterone receptors in breast cancer. J Clin Oncol 28:2784-
    2795 doi:10.1200/JCO.2009.25.6529JCO.2009.25.6529 [pii]
9. Ivanov SV, Panaccione A, Nonaka D, Prasad ML, Boyd KL, Brown B, Guo Y,
    Sewell A, Yarbrough WG (2013) Diagnostic SOX10 gene signatures in salivary ade-
    noid cystic and breast basal-like carcinomas. Br J Cancer 109:444-451 doi:10.1038/
    bjc.2013.326bjc2013326 [pii]
10. Kononen J, Bubendorf L, Kallioniemi A, Barlund M, Schraml P, Leighton S,
    Torhorst J, Mihatsch MJ, Sauter G, Kallioniemi OP (1998) Tissue microarrays for
    high-throughput molecular profiling of tumor specimens. Nat Med 4:844-847
11. Lebeau A, Deimling D, Kaltz C, Sendelhofert A, Iff A, Luthardt B, Untch M, Lohrs
    U (2001) Her-2/neu analysis in archival tissue samples of human breast cancer:
    comparison of immunohistochemistry and fluorescence in situ hybridization. J Clin
    Oncol 19:354-363
12. Lips EH, Mulder L, de Ronde JJ, Mandjes IA, Koolen BB, Wessels LF, Rodenhuis
    S, Wesseling J (2013) Breast cancer subtyping by immunohistochemistry and histo-
    logical grade outperforms breast cancer intrinsic subtypes in predicting neoadjuvant
    chemotherapy response. Breast Cancer Res Treat 140:63-71 doi:10.1007/s10549-
    013-2620-0
13. Markiewicz A, Ahrends T, Welnicka-Jaskiewicz M, Seroczynska B, Skokowski
    J, Jaskiewicz J, Szade J, Biernat W, Zaczek AJ (2012) Expression of epithelial
    to mesenchymal transition-related markers in lymph node metastases as a surro-
    gate for primary tumor metastatic potential in breast cancer. J Transl Med 10:226
    doi:10.1186/1479-5876-10-2261479-5876-10-226 [pii]
14. Mego M, Mani SA, Lee BN, Li C, Evans KW, Cohen EN, Gao H, Jackson SA, Gior-
    dano A, Hortobagyi GN, Cristofanilli M, Lucci A, Reuben JM (2012) Expression of

epithelial-mesenchymal transition-inducing transcription factors in primary breast cancer: the effect of neoadjuvant therapy. Int J Cancer 130:808-816 doi:10.1002/ijc.26037

15. Mohamed A, Gonzalez RS, Lawson D, Wang J, Cohen C (2013) SOX10 expression in malignant melanoma, carcinoma, and normal tissues. Appl Immunohistochem Mol Morphol 21:506-510 doi:10.1097/PAI.0b013e318279bc0a

16. Muller P, Crofts JD, Newman BS, Bridgewater LC, Lin CY, Gustafsson JA, Strom A (2010) SOX9 mediates the retinoic acid-induced HES-1 gene expression in human breast cancer cells. Breast Cancer Res Treat 120:317-326 doi:10.1007/s10549-009-0381-6

17. Payne SJ, Bowen RL, Jones JL, Wells CA (2008) Predictive markers in breast cancer–the present. Histopathology 52:82-90 doi:10.1111/j.1365-2559.2007.02897.xHIS2897 [pii]

18. Shin J, Vincent JG, Cuda JD, Xu H, Kang S, Kim J, Taube JM (2012) Sox10 is expressed in primary melanocytic neoplasms of various histologies but not in fibrohistiocytic proliferations and histiocytoses. J Am Acad Dermatol 67:717-726 doi:10.1016/j.jaad.2011.12.035S0190-9622(12)00013-8 [pii]

19. Smalley M, Piggott L, Clarkson R (2013) Breast cancer stem cells: obstacles to therapy. Cancer Lett 338:57-62 doi:10.1016/j.canlet.2012.04.023S0304-3835(12)00271-6 [pii]

20. Soady K, Smalley MJ (2012) Slugging their way to immortality: driving mammary epithelial cells into a stem cell-like state. Breast Cancer Res 14:319 doi:bcr3188 [pii]10.1186/bcr3188

21. Symmans WF, Peintinger F, Hatzis C, Rajan R, Kuerer H, Valero V, Assad L, Poniecka A, Hennessy B, Green M, Buzdar AU, Singletary SE, Hortobagyi GN, Pusztai L (2007) Measurement of residual breast cancer burden to predict survival after neoadjuvant chemotherapy. J Clin Oncol 25:4414-4422 doi:10.1200/JCO.2007.10.6823

22. Teleki I, Krenacs T, Szasz MA, Kulka J, Wichmann B, Leo C, Papassotiropoulos B, Riemenschnitter C, Moch H, Varga Z (2013) The potential prognostic value of connexin 26 and 46 expression in neoadjuvant-treated breast cancer. BMC Cancer 13:50 doi:10.1186/1471-2407-13-50

23. Theurillat JP, Zurrer-Hardi U, Varga Z, Storz M, Probst-Hensch NM, Seifert B, Fehr MK, Fink D, Ferrone S, Pestalozzi B, Jungbluth AA, Chen YT, Jager D, Knuth A, Moch H (2007) NY-BR-1 protein expression in breast carcinoma: a mammary gland differentiation antigen as target for cancer immunotherapy. Cancer Immunol Immunother 56:1723-1731 doi:10.1007/s00262-007-0316-1

24. van Nes JG, de Kruijf EM, Putter H, Faratian D, Munro A, Campbell F, Smit VT, Liefers GJ, Kuppen PJ, van de Velde CJ, Bartlett JM (2012) Co-expression of SNAIL and TWIST determines prognosis in estrogen receptor-positive early breast cancer patients. Breast Cancer Res Treat 133:49-59 doi:10.1007/s10549-011-1684-y

25. Varga Z, Caduff R, Pestalozzi B (2005) Stability of the HER2 gene after primary chemotherapy in advanced breast cancer. Virchows Arch 446:136-141 doi:10.1007/s00428-004-1164-4

26. Wolff AC, Hammond ME, Schwartz JN, Hagerty KL, Allred DC, Cote RJ, Dowsett M, Fitzgibbons PL, Hanna WM, Langer A, McShane LM, Paik S, Pegram MD, Perez EA, Press MF, Rhodes A, Sturgeon C, Taube SE, Tubbs R, Vance GH, van de

Vijver M, Wheeler TM, Hayes DF (2007) American Society of Clinical Oncology/ College of American Pathologists guideline recommendations for human epidermal growth factor receptor 2 testing in breast cancer. Arch Pathol Lab Med 131:18-43 doi:10.1043/1543-2165(2007)131[18:ASOCCO]2.0.CO;2

27. Yoshioka T, Hosoda M, Yamamoto M, Taguchi K, Hatanaka KC, Takakuwa E, Hatanaka Y, Matsuno Y, Yamashita H (2013) Prognostic significance of pathologic complete response and Ki67 expression after neoadjuvant chemotherapy in breast cancer. Breast Cancer. doi:10.1007/s12282-013-0474-2

CHAPTER 13

COMBINATION THERAPY OF ANTI-CANCER BIOACTIVE PEPTIDE WITH CISPLATIN DECREASES CHEMOTHERAPY DOSING AND TOXICITY TO IMPROVE THE QUALITY OF LIFE IN XENOGRAFT NUDE MICE BEARING HUMAN GASTRIC CANCER

XIULAN SU, CHAO DONG, JIALING ZHANG, LIYA SU, XUEMEI WANG, HONGWEI CUI, AND ZHONG CHEN

13.1 BACKGROUND

Cancer is a genetic disease that is developed due to accumulated multiple genetic defects along human life span. The heterogenous natures of genetic and malignant phenotypes within each type of cancer create a great challenge for cancer diagnosis and treatment [1]. At present the time, a high percentage of cancer patients are incurable, especially for solid tumors in late stage [2]. So, the quality of life (QOL) is particularly important to patients with advanced cancer, because most of them are symptomatic when diagnosed [3,4]. In addition, repeated utilization of chemotherapy agent

This chapter was originally published under the Creative Commons Attribution License. Su X, Dong C, Zhang J, Su L, Wang X, Cui H, and Chen Z. Combination Therapy of Anti-Cancer Bioactive Peptide with Cisplatin Decreases Chemotherapy Dosing and Toxicity to Improve the Quality of Life in Xenograft Nude Mice Bearing Human Gastric Cancer. Cell & Bioscience *4,7 (2014). doi:10.1186/2045-3701-4-7*

nonspecifically kills proliferating cells, which leads to significantly toxic side effects and decreased patient QOL [5]. Such nonspecific therapeutic strategy induces chemo-resistance, which creates a situation of either further increasing chemotherapy dosage with more severe toxicity or making patients intolerable for treatment. Neither is a best therapeutic strategy for cancer patients. Therefore, when treating cancer patients at the intermediate or advanced stage, the evaluation of patient outcomes should not just focus on the elimination of tumor burden at the expense of cancer patient QOL.

Cisplatin is one of the first line chemotherapy agents for treating advanced gastric cancer [6]. Previously, many clinical trials have been conducted to find the best combinatory regiment of Cisplatin with other chemotherapy agents, such as Docetaxel and Fluorouracil [7]. However, cancer patient QOL was ignored in many of the clinical trials because the traditional way to evaluate the efficacy of cancer therapy usually only relies on the index of cure rate and survival rate. As recent bio-psyco-social models have been developed to evaluate total cancer patent condition as a whole biological system, more attention has been drawn to the QOL of cancer patients during treatment [8]. Currently, QOL is one of several important indicators used to evaluate the efficacy of treatment, which is not only based on clinical objective indexes as evaluation standard, but also emphasizes subjective conditions of cancer patients [9]. For example, prospectively assessed QOL (even after completion of protocol treatment) was proposed and performed as one of the secondary end points of the phase III trial of combined therapy of Docetaxel plus Cisplatin and Fluorouracil for advanced gastric or gastroesophageal adenocarcinoma [10].

Anti-cancer treatments using biologically active materials, including bioactive peptides, have recently been identified with potent anti-cancer activity and lack of side effects [11-13]. Bioactive peptides exist naturally in living beings such as animals, plants, and microorganisms. In addition, these peptides can be produced through artificial modification of biological materials, such as proteolysis of tissues and serum from animals or plants; or synthetic methods, such as chemical synthesis or biological engineering. The naturally existing bioactive peptides play a crucial role in regulating biological activities, including molecular recognition, signaling transduction, cell proliferation, and differentiation. Previously, we identified anti-cancer bioactive peptide (ACBP), which is such a naturally existing peptide [14]. ACBP is a mixture of several polypeptides with a molecular weight of about 8 kD, and isolated and purified from goat spleens or livers after immunization with human gastric cancer protein extracts. ACBP was analyzed using high performance capillary electrophoresis (HPCE) and matrix-assisted laser desorption-ionization time-of-flight mass spectrometry (MALDI-TOF-MS) [15,16]. ACBP exhibited multiple biological activities, including anti-tumor activity in vitro and in vivo[14]. In addition, acute and chronic toxicological tests in mice and rats showed no measurable toxicities or side effects interfering with normal physiological functions and enzyme metabolism activities [17,18].

Based on the low toxicity of ACBP, we hypothesized that ACBP could potentiate chemotherapy agent, enhance the efficacy of treatment, and lower the chemotherapy dosage to decrease drug induced side effects and toxicity. However, the combinatory effects of ACBP and chemotherapy have never been investigated in animal tumor models. In this study, we used a combined regiment of ACBP-L (from liver) with a lower dosing Cisplatin, which could reach the same anti-tumor efficacy as the higher dosing Cisplatin alone in the xenograft nude mouse model bearing human gastric cancer MGC-803. Decreased chemotherapy dosage leads to improved QOL of tumor-bearing nude mice. The molecular mechanisms of the combined therapy could be regulated through the modulation of apoptotic molecules, such as BAX, Bcl-2, Casapse 3, and Caspase 8. Our study suggests that the combination of ACBP-L and chemotherapy could be a new anti-cancer strategy, which is capable of concurrently suppressing tumor growth and improving host QOL.

13.2 RESULTS

13.2.1 ACBP-L EXHIBITED ANTI-TUMOR ACTIVITY AGAINST MGC-803 CANCER CELLS IN A DOSE AND TIME-DEPENDENT MANNER IN VITRO

To determine the effect of ACBP-L on cell proliferation, MGC-803 cancer cells were treated with increasing concentrations of ACBP-L for 24

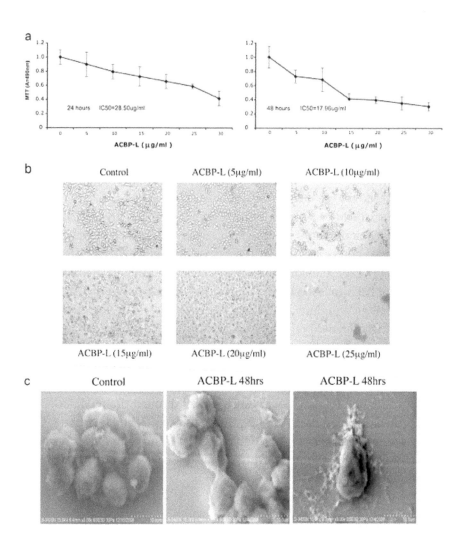

FIGURE 1: ACBP-L exhibits anti-tumor activity against MGC-803 cancer cells in a dose and time-dependent manner in vitro. (a) ACBP-L suppressed MGC-803 gastric cancer cell proliferation at a dose-dependent manner by MTT assay measured at the absorbance of 490 nm (A =490 nm) during 24 hr (left) and 48 hr (right) time points. The samples were measured in triplicates and the data were presented as mean ± standard deviation (SD). The median concentration of inhibition (IC50) at 24 hr is 28.50 μg/ml, and at 48 hr is 17.96 μg/ml. (b) ACBP-L inhibited cell growth and induced cell apoptosis in a dose dependent manner observed under light microscopy (400X). (c) Scanning electron microscopy showed cell membrane damage after 48 hr ACBP-L treatment (shown in two different fields). Scale indicates 10 μM.

hrs and 48 hrs. Increased ACBP-L (5.0-30.0 μg/mL) inhibited cell prolif-eration in a dose dependent manner measured by MTT assay (Figure 1a). The survival of MGC-803 cancer cells was decreased by approximately 10.3%, 20.7%, 27.6%, 34.5%, 41.4%, and 58.6% after a 24 hr exposure to 5, 10, 15, 20, 25, or 30 μg/ml ACBP-L, respectively. The median concen-tration of inhibition (IC_{50}) at 24 hrs was 28.50 μg/ml. In addition, ACBP-L anti-proliferative effect on MGC-803 cells was persistent and increased with prolonged treating time. The inhibitory rates at 48 hrs were: 27.3%, 31.8%, 59.1%, 60.6%, 65.2%, and 69.7%, respectively, and IC_{50} = 17.96 μg/ml. We compared the inhibitory rates at the different time point and found that, a 48-hour treatment of MGC-803 cancer cell with the 25 μg/ml concentration of ACBP-L resulted in a 65.2% decrease in cell viability, compared with a 41.4% decrease with 24-hour exposure to the same con-centration of ACBP-L (Figure 1a).

The cell morphology after different doses of ACBP-L treatment was observed (Figure 1b). The cell morphology resembled cell growth inhi-bition and an induction of cell apoptosis. At the lower dose (10 μg/ml), fewer cells were observed in the culture, and the remaining cells exhibited the morphology of bleb, loss of cell membrane asymmetry, and detach-ment. At the higher concentration (15 μg/ml), cell shrinkage and nuclear condensation were more apparent. In addition, under the treatment of the highest concentrations of the two doses, cells completely lost membrane and exhibited condensed nucleus, or only cell debris was left (Figure 1b). Scanning electron microscope revealed a consistent morphology that cell membrane was damaged and lost asymmetry (Figure 1c).

13.2.2 ACBP-L POTENTIATED THE LOW DOSE CISPLATIN TREATMENT TO SUPPRESS GASTRIC TUMOR GROWTH IN A XENOGRAFT TUMOR MODEL

A xenograft nude mouse model was established with subcutaneous inocu-lation of human gastric MGC-803 cancer cells. The tumor growth rate was measured and calculated at each time point, and the statistical significance of the tumor growth rate was examined (Figure 2a). In the control group, tumor volume increased significantly shown by each measurement, indi-

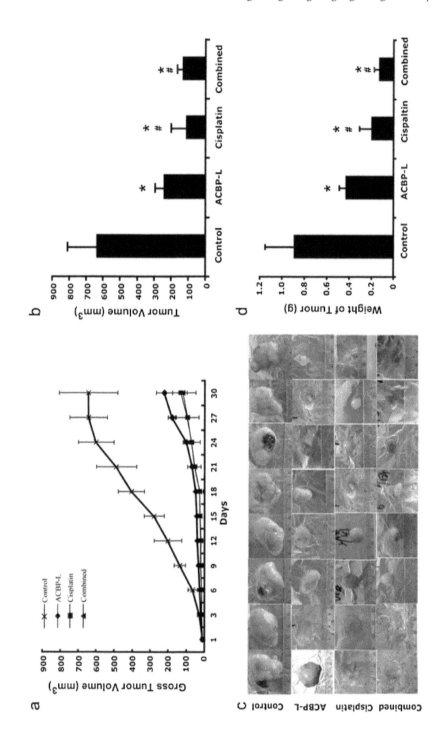

FIGURE 2: ACBP-L and low dose Cisplatin combined treatment suppresses gastric tumor growth in the xenograft tumor model. ACBP-L suppresses tumor growth in vivo. The in vivo tumor growth experiment was established by subcutaneous injection of 2×10^7 MGC-803 gastric cancer cells. After tumors were palpable and when the sizes of tumors were about ~10 mm³, the tumor bearing mice were randomized with eight mice (n =8) each into four groups: control with daily intraperitoneal injection of saline, daily injection of ACBP-L (7 µg/mouse) alone, Cisplatin alone (5 mg/kg every five days for 4 times, given the drug at day 6, 11, 16, 21), and Cisplatin (5 mg/kg, twice giving at day 6 and 16) plus daily ACBP-L. The tumor volumes were measured every three days (a). The final tumor volumes were calculated before the end of experiments (b). The data were calculated and presented as mean±standard deviation (SD). Statistical significance was determined by Student t-test (statistical difference was indicated as p <0.05, when compared with the control group (*), or compared with ACBP-L treated alone group (#). (c and d) Tumor weight was measured after harvest, and the data were calculated and presented as mean ± standard deviation (SD). Statistical significance (p <0.05) was determined by Student t-test when compared with the control (*) and ACBP-L treated alone group (#).

cating an aggressive malignant phenotype with a fast tumor growth rate. The anti-tumor activity of ACBP-L was tested by daily injection of 7 µg/mouse, and compared with traditional regiment of Cisplatin alone at 5 mg/kg four times every five days. The regiments are the essential dosages with anti-tumor activity identified from pilot experiments (data not shown). To test if ACBP-L could potentiate Cisplatin anti-tumor effects, we decreased dose of Cisplatin to 5 mg/kg for every 10 days, with a total of two injections, and combined with ACBP-L (7 µg/mouse for daily injection). When compared the dynamic growth rate of the control and the three treated groups, the tumor growth rates were statistically different after treatment at day 7. At the end of treatment, ACBP-L significantly inhibited tumor growth by 61.3%, Cisplatin at the higher dose exhibited strongest anti-tumor activity with an inhibitory rate of 81.6% (Figure 2b, Student t test, p <0.05). When ACBP-L was combined with low dose of Cisplatin, the treatment suppressed tumor growth at the similar rate as the high dose of Cisplatin alone (Figure 2a). At the end of the treatment, the combinatory treatment decreased tumor growth by 78.3% (Figure 2b, Student t test, p <0.05). After harvesting tumors at the end of the experiment, the tumor weights of four experimental groups were examined and compared: control, 0.90±0.25 g; ACBP-L, 0.44±0.05 g; Cisplatin, 0.21±0.10 g; combination, 0.14±0.04 g (Figure 2c, d). The inhibition rate was ACBP-L, 51.1%; Cisplatin, 76.7%; combined therapy 84.4%. The statistical significance of tumor weights resembled those of tumor volumes (Figure 2b, d). There are statistical differences observed when comparing each treated groups with the control, however, high dose Cisplatin alone or combinatory treatment with low dose of Cisplatin exhibited the strongest anti-tumor effect (Student t test, $p < 0.05$).

13.2.3 ACBP-L ALONE OR IN COMBINATION WITH THE LOW DOSE CIAPLATIN IMPROVE QUALITY OF LIFE IN THE XENOGRAFT TUMOR MODEL

When we examined the ACBP-L and Cisplatin anti-tumor activity in vivo, we also observed the quality of life (QOL) of tumor bearing animals. The QOL of ACBP-L and combinatory treated groups were significantly improved over that of the high dose Cisplatin group, indicated by body weight and food intake (Figure 3). The mice in the ACBP-L or combinatory treated groups were more active, had good appetite, and their appearance and body weight were close to that of a normal mouse. The mice in the group with high dosage of Cisplatin exhibited strong gastrointestinal toxicity (diarrhea), systemic toxicity (piloerection and lethargy), and a consistent decline in body weight (Figure 3a). At the end of experiment, there was no body weight loss in ACBP-L treatment when compared to control. The body weight of the high dose Cisplatin group was the lowest, and the body weight of the combinatory treated group was slightly lower than the control and ACBP-L treated groups (Student t test, $p < 0.05$, Figure 3b).

QOL was also examined by daily food intake of the experimental animals (Figure 3c). There were three large decreases of food intake in the Cisplatin alone group after the mice were given the 1st, 2nd, 3rd doses of Cisplatin at day 6, 11, and 16. In the combinatory treated group, there were two decreased food intake corresponding to the Cisplatin dosing at day 6 and 16. At the end of the experiment, daily food intake by mice in the Cisplatin group was significantly lower than that of the other three groups (Figure 3d, Student t test $p < 0.05$).

13.2.4 HIGH DOSE OF CISPLATIN ALONE DECREASED SPLEEN WEIGHT IN THE XENOGRAFT TUMOR MODEL

The spleen index was measured as an indication of systemic toxicity induced by Cisplatin (Figure 4a). There was a significant decrease of spleen index in the high dose of Cisplatin group when compared with the other three experimental groups, while the combination treatment group did not show significant decrease of spleen index, indicating less toxicity. However, no significant decrease of the liver weight was observed in any of the experimental groups (data not shown).

13.2.5 ACBP-L AND CISPLATIN ALONE, OR IN COMBINATION, SUPPRESSED BIOLOGICAL AND METABOLIC ACTIVITIES IN LIVE TUMOR CELLS BY PET-CT IMAGE

Viability and metabolic activity of xenograft tumors from four experimental groups were evaluated by ^{18}F-FDG PET/CT imaging (Figure 4b). PET with ^{18}F-FDG is a noninvasive approach for determination of the glycolytic status, and enhanced glycolysis is one of the most important characteristics of energy metabolism in cancer cells. The higher the ratio of radioactivity uptake in tumor versus normal area (Target/Non-target) indicates the stronger glycolysis in tumor cells, suggesting more energetic and aggressive tumor cell status. In this study, we observed a significant difference in radioactivity uptake when comparing the Target/Non-target ratio of tumor/spine (T/NT:1.39) in the control group with mice from three treated groups, ACBP-L (T/NT:1.26), Cisplatin (T/NT:1.18), and combined therapy (T/NT:1.19). Our data suggested that all three treatment regiments decreased the viability and metabolic activity of live tumor cells, where comparable inhibitory effects were observed in higher dosing Cisplatin treatment alone as the combined therapy of ACBP-L with lower dosing of Cisplatin (Figure 4b).

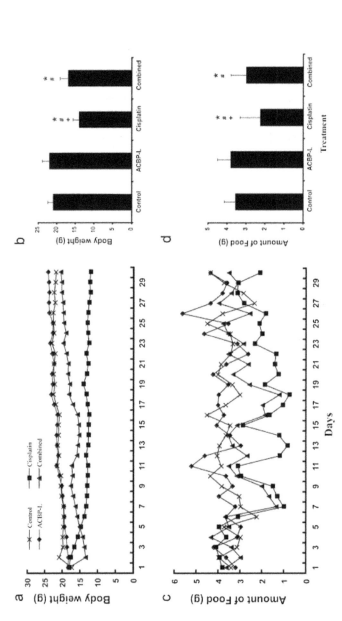

FIGURE 3: ACBP-L and low dose Cisplatin combined treatment improves quality of life in the xenograft tumor model. (a) Body weight of xenografted nude mice (n = 8) was measured every two days for thirty days, and means of the body weight of each group were presented. (b) Body weight was presented by mean ± (SD). Statistical significance were determined by Student t-test when compared with the control (*), ACBP-L (#), and combined treatment (+), p <0.05. (c) Food intake by tumor bearing mice was measured every two days, and the means of each group were presented. (d) Data of food intake were presented as mean ± SD. Statistical significance was determined by Student t-test when compared with the control (*), ACBP-L (#), and combined treatment (+), p <0.05.

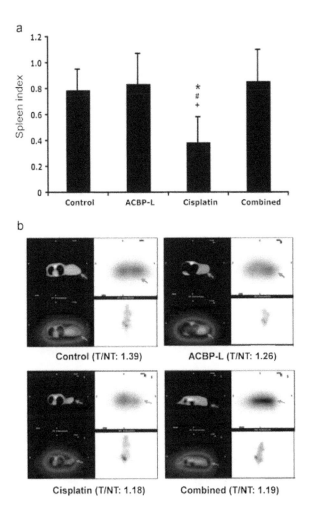

FIGURE 4: ACBP-L or Cisplatin alone, or combined treatment suppresses metabolic activities in tumor baring animals and in live tumor cells in vivo. (a) Spleen weight of each mouse (n = 8) was measured after euthanasia. Mouse spleen index was calculated by mouse spleen weight (mg) divided by the mouse body weight (g). Data are presented as mean ± SD from eight mice in each group. Statistical significance was determined by Student t-test when compared with the control (*), ACBP-L (#), and combined treatment (+), p < 0.05. (b) ^{18}F-FDG PET/CT fuse imaging (cross section) from a representative nude mouse of each group is presented (CT image, left; PET image, right). Radioactivity uptake represents the biological and metabolic activities of tumors (Target, T, black arrows) were compared with spine counts (Non-Target, NT) within the identical section, as the T/NT ratio. The T/NT ratio is significantly decreased in treated groups when compared with the control.

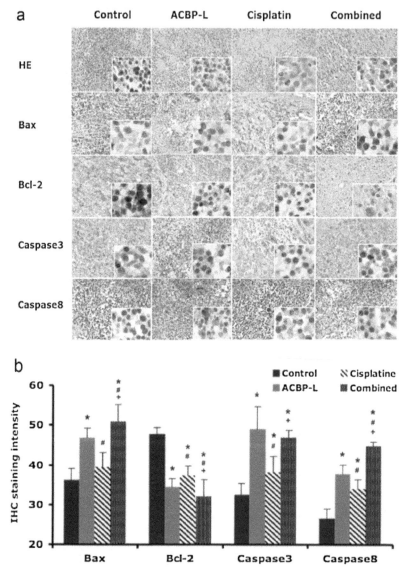

FIGURE 5: ACBP-L or Cisplatin alone, or combined treatment induces molecules promoting cell apoptosis in vivo. (a) Paraffin embedded tumor specimens (n =8/each group) were stained with HE and immunohistochemistry of Bax, Bcl-2, Caspase 3, and Caspase 8, (low magnification 40X, high magnification 400X). (b) The quantitation of IHC scores were analyzed, the statistical significance was determined by Student t-test (p <0.050), and data are presented as compared with control (*), ACBP-L (#), and Cisplatin treated group (+).

13.2.6 ACBP-L AND CISPLATIN ALONE, OR IN COMBINATION, INDUCED MOLECULES PROMOTING CELL APOPTOSIS

We hypothesized that the anti-tumor activity of ACBP-L could be mediated through promotion of tumor cell apoptosis. Pathological analysis after H&E staining of harvested tumors showed more cells with apoptotic features in all treated groups (Figure 5). Immunohistochemical staining of proteins involved in apoptosis were performed and quantified, that significant differences in Bax, Bcl-2, Caspase 3, and Caspase 8 expression were observed between the experimental groups with controls (Figure 5). Stronger Bax staining was observed in ACBP-L alone and in the combined treatment groups when compared with the control. Decreased Bcl-2 expression was observed in all treated groups. Induction of Caspase 3 and Casapse 8 were observed in all treated groups, and a stronger response was observed in the ACBP-L alone or the combinatory group (Student t test, $p < 0.05$).

To further evaluate the molecular regulatory mechanism induced by ACBP-L and Cisplatin alone, or by the combinatory treatment, we analyzed the important apoptotic genes, such as Bax and Caspase 3 in tissue specimens by semi-quantitative RT-PCR (Figure 6). Consistent with the protein expression, Bax expression was significantly increased in all groups after treatment, and the ACBP-L alone and combinatory treated group exhibited a higher induction (Figure 6b, left panel, Student t test, $p < 0.05$). Caspase 3 expression was increased in all treated groups, and ACBP-L induced the highest level of Caspase 3 expression (Figure 6b, right panel, Student t-test, $p < 0.05$). All data strongly suggested that the molecular mechanism of apoptosis was involved by treatments.

13.3 DISCUSSION

In this study, the anticancer effect of ACBP-L in human gastric cancer is demonstrated by suppression of cell line MGC-803 proliferation in vitro by MTT assay and morphological observations under light or electron microscope (Figure 1). The IC_{50} of ACBP-L was in the range of 18-28 μg/ml,

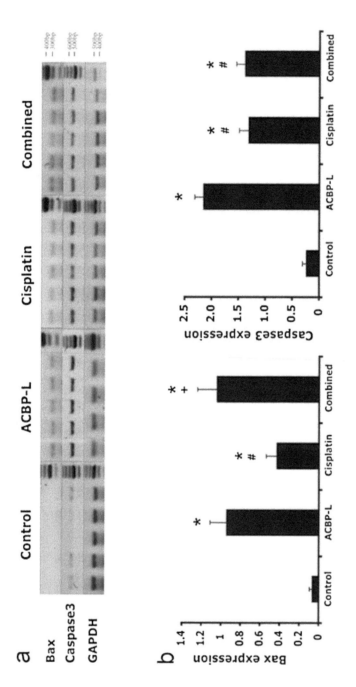

FIGURE 6: Combined ACBP-L and Cisplatin treatment strongly induces anti-tumor activity in tumors Specimens. (a) Semi-quantitative RT-PCR was performed to detect gene expression of Bax and Caspase 3 in tumor specimens (n = 5). RT-PCR products were run on 2% agarose gel, and gel images were shown with the indicated molecular weight. (b) The ratio of Bax and Caspase 3 gene expression were calculated and presented as mean ± standard deviation (SD), and the significant difference was examined by Student t-test (p < 0.050), and presented as compared with control (*), ACBP-L (#), and Cisplatin (+).

with a dose and time dependent manner. In vivo, ACBP-L exhibited potent anti-tumor effects when used alone, and was able to potentiate Cisplatin chemotherapeutic effect at the lower dose (Figures 2 and 4b). Such combinatory regiment significantly improved the QOL with lower systemic toxicity (Figures 3 and 4a). The tumor specimens harvested from treated groups exhibited increased apoptosis, detected by IHC and RT-PCR, suggesting that the anticancer effect of ACBP-L was due to induced apoptosis through Bax and Caspase mediated mechanisms (Figures 5 and 6).

The novel finding of this study is that ACBP-L alone exhibited potent anti-tumor activity, and additionally, to potentiate Cisplatin chemotherapeutic effects with lower systemic toxicity in vivo (Figures 2 and 3). Cisplatin is one of the most broadly used chemotherapeutic agents for treating cancers from the gastric region, the head and neck, and other sites of the aerodigestic tract. However, the high dosing and long-term administration of this medication creates severe systemic toxicity, including gastrointestinal problems, such as intense nausea and vomiting, hair loss, myelosuppression, renal toxicity, and hearing loss. To balance the efficacy of eliminating cancer cells while maintaining the QOL of cancer patients is a great challenge for the oncologist and drug development. Our current study provides a new strategy of utilization of natural existing bioactive peptides, such as ACBP-L used in this study, to potentiate chemotherapeutic effects and minimize toxic effects. In natural resources, broad spectrums of bioactive peptides exhibit regulatory activities that are involved in different biological processes. These natural peptides exist, but in the relatively low amounts or at the low levels of activities. Using immunization protocol with cancer tissue extracts is one way to enrich or activate such naturally exiting peptides. During the process of the isolation and purification of ACBP, we observed significantly higher peaks of these peptides from the induced liver or spleen, when compared with those isolated from normal organs. The enrichment of the recovered peptides is more than three folds after the purification (Su X, unpublished observations). In addition, we compared the anti-tumor activity between the peptides isolated from normal or induced spleens, using the same amount of materials. We observed both normal or induced peptides exhibited anti-tumor proliferation activity in vitro using MTT assay. However, a stronger inhibition of tumor cell proliferation, near two-fold increase, was observed in the peptides from induced spleens when compared with the peptides isolated from normal spleens (Su X, unpublished observations). We also compared the peptide biochemical profiles of ACBP isolated from goat spleens [14] or livers, which exhibited similar elution time and characteristics of elution peaks through MPLC (data not shown). Furthermore, ACBP not only exhibited anti-tumor activity of gastric cancer lines MGC803 as shown in this manuscript and SGC-7901 (unpublished data), but also exhibited a broad activity of anti-tumor effects in different cancer types in vitro and in vivo, regardless immunized cancer types, suggesting it seems not related to specific immune recognition. We have previously published several studies to show the broad anti-tumor activity of ACBP in human myelogenous leukemia line K562 [19], human nasopharyngeal carcinoma line CNE [20], human cholangiocarcinoma cell line QBC939 [21], murine hepatocarcinoma line H22 [22], and human breast cancer line nm231 [23], human colon adenocarcinoma cell line HT29 [24]. However, ACBP showed minimal inhibitory effects on normal cells that we tested, such as human skin fibroblast HS-68 cells and rabbit bone marrow mesenchymal stem cells (Su X, unpublished data). The exact molecular mechanisms of the broad anti-tumor activities and preference against tumor but not normal cells are under investigation. Consistent with the observations in vitro, ACBP exhibited minimal side effects in vivo (Figure 2). This supports to test ACBP anti-tumor activity in future clinical trials for cancer patients. In addition, one of the advantages for isolating ACBP-L from liver is the higher yield, which is more than 10 times harvest per animal with similar potency than those harvested from spleen (data not shown). The high yield enables us to improve production efficiency and feasibility study of future clinical trials.

In this manuscript, we examined whether ACBP-L could be used as a new adjuvant agent to enhance chemotherapeutic efficacy and reduce toxicity in the treatment of cancer patients. Consistent with the potent anti-cancer effects as previously seen using ACBP from the spleen, in this study, the effects of ACBP-L on the QOL has been demonstrated intuitively by the maintenance of normal body weight when using ACBP-L alone (Figure 3a, b), in contrast to a continuing decreased body weight observed due to the toxic effects of high dosing of Cisplatin alone. Further, the combinatory treated group with lower dosing of Cisplatin exhibited similar anti-tumor effects with a continuing recovery of the body weight that eventually reached normal level (Figure 3a, b). In addition, the QOL was also indicated by average daily food intake, which showed that the amount of food intake in mice treated with high dosing Cisplatin significantly dropped after each treatment. Although, in the combinatory treated group, significant decreased food intakes occurred similarly to the Cisplatin alone group, but the mice recovered quickly and gradually returned to a normal level of food intake. We also observed the suppression of the spleen index in the high dosing of Cisplatin group, but not in the combinatory treated group (Figure 4a). The data suggest that high dosing of Cisplatin could suppress the host immune and defense systems, but the combinatory treated group did not suffer from side effects. Our data strongly support the notion that the combination of chemotherapeutic agents with natural anti-cancer bioactive peptides could be a new strategy for more efficiently treating cancer patients while maintain QOL.

We observed that the viability and metabolism in live tumors exhibited significantly decreased in all three treated groups measured by ^{18}F-FDG PET/CT in the live animals (Figure 4b). Integrated PET/CT with ^{18}F-FDG is a hybrid of radiation and imaging modalities, which has been recently established in the staging, restaging and therapy response assessment of oncology patients [25]. The machine is capable to determine the glycolytic status in tissues, through an image that shows the tissue distribution of the positron emitter from ^{18}F-FDG, a structural analog of glucose. Enhanced glycolysis is one of the most important characteristics of live cancer cells. ^{18}F-FDG PET/CT has been proven to be successful as a diagnostic instrument for many solid tumors, including gastric cancer, especially for follow-up after cancer treatment, because it indicates the energy metabolism and viability of the tumor cells [25,26]. In this study, ^{18}F-FDG PET/CT method clearly showed consistent results with other biological and molecular measurements, with an advantage of monitoring the anti-tumor effect of different treatments without interrupting the ongoing animal experiment.

Further, ACBP-L induced cell death of tumors also well documented at the cellular level, with morphology of bleb, loss of cell membrane asymmetry and detachment, cell shrinkage, and nuclear condensation (Figure 1b, c). The altered cell morphology is consistent with the molecular mechanism of triggering apoptotic pathways involving BCL and Caspase family members. At the protein level, a significant increase of Bax in ACBP-L alone and in the combinatory groups, and a significantly decreased Bcl-2 in the ACBP-L and Cisplatin groups were observed (Figure 5), suggesting that the mechanisms involved could be slightly different. Consistent with the protein expression, Bax expression at the mRNA level was also strongly induced in the ACBP-L alone and in the combinatory treated groups (Figure 6). The Bcl family, whose members may be antiapoptotic (such as Bcl-2) or proapoptotic (such as Bax), regulates cell death by controlling the mitochondrial membrane permeability during apoptosis [27]. Bax is a 21-kD program partner associated with Bcl-2, and exhibits an extensive amino acid homology with Bcl-2 and forms homo- or heterodimers with Bcl-2 in cells. When BAX predominates, programmed cell death is accelerated, and the death repressor activity of Bcl-2 is countered, such that the ratio of Bcl-2 to BAX determines survival or death following an apoptotic stimulus [28].

While anti-apoptotic effect of Bcl-2 is through the inhibition of the release of Cytochromum C from mitochondria for Caspase activation [29-31], all three treated groups also exhibited an induction of Caspase protein and mRNA expression (Figures 5 and 6). A relatively stronger induction of Caspase 3 was observed in the ACBP-L treated group, while a stronger induction of Caspase 8 was observed in the combinatory treated group (Figure 5). In apoptosis cell, Caspase initiates the opening of the Permeability Transition (PT) aperture of the mitochondria and regulates apoptosis via regulating transmembrane electrochemical gradient [32]. The mechanisms of ACBP-L induced apoptosis are consistent with the apoptotic morphology of altered and destroyed cell surface membrane and structure observed through light and scanning electron microscopes (Figure 1b, c) [33-35].

13.4 CONCLUSIONS

We showed that ACBP-L potently inhibited gastric cancer cell proliferation and induced apoptosis in vitro. ACBP-L alone exhibited potent anticancer effects and potentiated Cisplatin chemotherapeutic effects in vivo. ACBP-L alone, or combined with lower dosing of Cisplatin, significantly improved host QOL without compromising therapeutic effects. ACBP-L and combinatory therapy induced cell apoptosis through modulation of BCL and Caspase pathways. Our study suggests that ACBP-L alone, or in combination with chemotherapy agents, could enhance anti-cancer effects and improve patient QOL, which could be a new therapeutic approach for further development against gastric or other neoplasms.

13.5 MATERIAL AND METHODS

13.5.1 PRODUCTION AND PURIFICATION OF ACBP-L FROM IMMUNIZED GOAT LIVER

The research involved with human subjects were performed following "Ethical Principles for Medical Research Involving Human Subjects". Anonymously primary human gastric tissue samples were obtained from The Affiliated Hospital, Inner Mongolia Medical University, with the approval of the Ethics Committee of Inner Mongolia Medical University, (2012-SWLL-001). All animal experiments were carried out under protocols approved by the Animal Care and Use Committee of the Inner Mongolia Medical University, and were in compliance with the international guidelines (Guide for the Care and Use of Laboratory Animal Resource, National Research Council, USA). ACBP-L was produced and extracted using the following steps. Goats were immunized five times with an interval of one week through a series of injections with human gastric cancer extracts. The livers were harvested from immunized animals. The tissues were subjected to several rounds of ultrasonication (Ultrasonic Disrupter, Model F525 from FLUKO Company, China). After centrifugation at 14,000 rpm for 10 min, the supernatants were collected and ACBP-L was isolated through mesolow preparative liquid chromatography (MPLC, Model YFLC-AI-580 from YAMAZEN corporations, Japan). MPLC is a commonly used protein and peptide purification chromatography, which can separate relatively large amount of materials. The molecular weight of eluted ACBP-L is ~8000 Dalton based on measurement with SDS-PAGE gels [14]. ACBP-L was included in an invention patent of Dr. Su Xiulan's ACBP-L laboratory and is protected by the Chinese national patent bureau (patent number: ZL96122236.0, and international patent: A61K35/28).

13.5.2 CELL CULTURE

Gastric adenocarcinoma cell line MGC-803 was kindly provided by Professor Ke Yang (Beijing University, Health Center, Beijing, China). Cells were maintained in RPMI1640 culture medium (Invitrogen, USA), which was supplemented with 10% heat-inactivated fetal bovine serum (FBS, TBD Science, China), 100U/ml penicillin, and100 U/ml streptomycin, and cultured in a humidified atmosphere of 5% CO_2 at 37°C.

13.5.3 MTT ASSAY

Cell proliferation was measured by MTT assay [3-(4,5-dimethylthiazol-2-yl)-2,5-diphenyltetrazolium bromide, TBD Science Co., China]. MTT was dissolved in sterile PBS at room temperature, sterilized by passing through a 0.22 μm filter, and stored in the dark at 4°C. MGC-803 human gastric cells (5×10^3/well) were placed in 200 μl of culture medium and incubated overnight. After 24 h, cultures were treated with various doses of ACBP-L in triplicates. MTT reagent (20 μl) was added at different time points and then incubated at 37°C for 4 h. Following vibrating on a shaker for 10 min, the plates were measured for absorbance at 490 nm wavelength using a microtiter plate reader. Drug concentrations that inhibited proliferation by 50% (IC_{50} values) were calculated from dose-response plots by linear regression modeling of the logarithmic form of the equation.

13.5.4 SCANNING ELECTRON MICROSCOPE

MGC-803 cells were cultured in RPMI-1640 medium containing 20 μg/ml of ACBP-L for 48 hrs. All cells were fixed in 2.5% glutaraldehyde in 0.1 M cacodylate buffer (pH 7.2) at 4°C for 1 hr. The samples were rinsed in 0.1 M cacodylate buffer several times, and then dehydrated in graded concentrations of alcohol. The ultrathin sections were dried with Vacuum plating apparatus and treated by spray-gold with ion sputtering equipment (JEOL Company, USA). Then the specimens were examined with a S-3400 N scanning electron microscope (JEOL Company, USA) operated at 15 KV.

13.5.5 XENOGRAFT TUMOR MODEL AND ADMINISTRATION OF ACBP-L AND CISPLATIN

All animal experiments were carried out under the protocol approved by the Animal Care and Use Committee of the Inner Mongolia Medical University, as previously described. Five-week-old athymic nude female mice (BALB/c nu/nu, Institute of Laboratory Animal Sciences, Chinese Academy of Medical Sciences, Beijing, China) were housed in a sterile animal facility and inoculated with MGC-803 cells (2×10^7) in 0.2 ml of PBS subcutaneously. All mice developed single palpable tumors, with the average volume of the tumors equaling ~10 mm^3 at day 3, following inoculation. The mice were then randomized into four groups, including control, ACBP-L alone, Cisplatin alone, and ACBP-L plus Cisplatin. Each group contained 8 mice (n = 8). The drugs were administered via intra-peritoneal injection as follows: control: 0.2 ml of 0.9% NaCl daily for 30 days; ACBP-L alone: 7 µg/mouse ACBP-L in 0.2 ml daily for 30 days (ACBP-L batch#2008-10); Cisplatin alone: 5 mg/kg every 4 days for total 4 doses, at day 6, 11, 16, 21, according to the manufacturer's protocol (batch# 806027CF, Qilu Medicine Ltd, Jinan, China). The combinatory treatment group of ACBP-L and Cisplatin: 7 µg/mouse ACBP-L in 0.2 ml daily for 30 days plus 5 mg/kg Cisplatin twice every 10 days at day 6 and 16. The mouse body weight, the amount of water and food intake, as well as vital signs and living status, were checked daily. Tumor volume was measured and calculated as follows: the longer diameter, designated as "a", and the shorter one as "b" of the tumors were measured by Vernier calipers, and the tumor volume (TV) was calculated using the equation: TV = $ab^2/2$. The growth curve of tumors was plotted using the mean and standard deviation.

13.5.6 MEASUREMENT OF BODY WEIGHT, FOOD INTAKE AND COLLECTION OF TUMOR SPECIMENS

Body weight was measured once every other day, and food intake was measured by weighing the food left in the cage. To calculate the average daily food intake per mouse in a cage, the amount of food present was subtracted from that of the previous day, and then divided by the number of mice in the cage. After euthanasia, tumors, livers, and spleens were collected, weighed, and dissected. The spleen index was calculated by the spleen weight measured in mg versus the mouse body weight measured in grams. Portions of tissues were frozen at -80°C, and portions of tissues were fixed in formalin and embedded in paraffin.

13.5.7 POSITRON-EMISSION TOMOGRAPHY-COMPUTED TOMOGRAPHY (PET/CT)

^{18}F-deoxyglucose (^{18}F-FDG) Positron radioactive tracer was synthesized by medical cyclotron (MINITrace, General Electric Co., Milwaukee, USA) and FX-FN chemosynthesis system (TraceLab, General Electric Co.). Mice were anesthetized by Ether 60 min after tail intravenous injection of 0.2 millicurie (mci) ^{18}F-FDG and scanned systemically by 2D positron-emission tomography-computed tomography (PET-CT). Figures from PET-CT were analyzed in Xeleris by two experienced radiologists. The abnormal thick, thin, or defect of radioactivity uptake in inoculation site was excluded. Radioactivity uptake from tumor (Target) was divided by uptake from spine (Non-Target) and presented as Target/Non-Target ratio (T/NT).

13.5.8 IMMUNOHISTOCHEMISTRY (IHC)

Paraffin embedded tissues were sectioned at a thickness of 4um, and stained with hematoxylin and eosin, and immunohistochemistry (IHC). IHC protocol was modified from S-P Method (Maixin Biological Technology Development Co., Fuzhou, China). Briefly, after deparaffin, antigen retrieval was carried out in a microwave oven for 10 min. The antibodies used were mouse anti-human Bax antibody (90107254D1); mouse anti-human Bcl-2 antibody (90104014 F1), from Maixin Biological Technology Development Co.; rabbit anti-Caspase3 (KGA717, Keygene Biological Technology Development Co., Nanjing, China); and rabbit anti-Caspase 8, (Santa Cruz Biotechnology Co., Santa Cruz, CA, USA). S-P hypersensitive kits (mouse and rabbit, 812059710) and DAB reagents (806180031) were from Maixin Biological Technology Co. Sections were stained with 3, 3-diaminobenzidine (DAB), counterstained with haematoxylin, and dehydrated in xylene, and mounted. The immunohistochemical staining was observed under light microscope (Olympus, Tokyo, Japan), and quantified by Olympus CX41 image analysis system.

13.5.9 TOTAL RNA EXTRACTION AND RT-PCR

Total RNA was extracted from tumor specimens using TRIZOL reagent (Invitrogen, USA) according to the manufacturer's protocol. The quality and concentrations of RNA were measured by a Du-800 UV spectrophotometer (Beckman, USA). Reverse transcription was performed using an RNA PCR Kit (AMV) Version 3.0 (TaKaRa Co., Japan), and the reaction mixture contained: 1 µg RNA sample, 1 µl 10 × buffer, 1 µl dNTP (2 mmol/l), 0.5 µl oligo-dT primer (0.25 µM), AMV RTase 10 U, RNase inhibitor 5 U and DEPC ddH2O up to 10 µl. The mixture was kept at room temperature for 10 min, then was incubated at 42°C for 30 min and at 99°C for 5 min. PCR was performed in 25 µl reaction mixture which consisted of 2 µl reverse transcription products, 5 µl 5X buffer, 0.5 µl each gene specific primers (20 pmol/L), 0.5 µl 10 mM dNTP, 0.75 U Taq DNA polymerase (TaKaRa Co, Japan), with ddH2O added up to 25 µl. PCR cycle parameters were conducted with a pre-amplification denaturation at 94°C for 2 min, followed by 35 cycles of denaturation at 94°C for 30 sec, annealing at 58°C to 65°C for 30 sec, and extension at 72°C for 1 min, with a final extension at 72°C for 5 min. GAPDH was used as an internal control. Amplified PCR products were visualized in a 2% agarose gel electrophoresis containing ethidium bromide. Ratios amplified gene/GAPDH was calculated by software Imagetool 2.0 (University of Texas Health Science

Center, San Antonio, Texas, USA). PCR results were confirmed by three repeat amplifications. The sequences of PCR primers were designed using GenBank database and the BLAST program. All primers were synthesized commercially by the TaKaRa Company (Dalian, China), and the experimental condition is presented in Additional file 1: Table S1.

13.5.10 DATA PROCESSING AND STATISTIC ANALYSIS

The calculation of tumor inhibition rate was based on the equation described as follows: Tumor inhibition rate = (tumor weight of control group of saline − tumor weight of medicine group)/tumor weight of control group of saline × 100%.

Data are presented as the mean ± standard deviation (SD). Statistical analysis was performed using t test for two groups, $p < 0.05$ was considered statistically significant. All statistical analyses were performed using an SPSS program (version 13.0).

REFERENCES

1.   Liotta L, Petricoin E: Molecular profiling of human cancer. Nat Rev Genet 2000, 1:48-56.
2.   Lee N, Harris J, Garden AS, Straube W, Glisson B, Xia P, Bosch W, Morrison WH, Quivey J, Thorstad W, Jones C, Ang KK: Intensity-modulated radiation therapy with or without chemotherapy for nasopharyngeal carcinoma: radiation therapy oncology group phase II trial 0225. J Clin Oncol 2009, 27:3684-3690.
3.   Cella D, Li JZ, Cappelleri JC, Bushmakin A, Charbonneau C, Kim ST, Chen I, Motzer RJ: Quality of life in patients with metastatic renal cell carcinoma treated with sunitinib or interferon alfa: results from a phase III randomized trial. J Clin Oncol 2008, 26:3763-3769.
4.   Joly F, Vardy J, Pintilie M, Tannock IF: Quality of life and/or symptom control in randomized clinical trials for patients with advanced cancer. Ann Oncol 2007, 18:1935-1942.
5.   Kayl AE, Meyers CA: Side-effects of chemotherapy and quality of life in ovarian and breast cancer patients. Curr Opin Obestet Gynecol 2006, 18:24-28.
6.   Takashima A, Yamada Y, Nakajima TE, Kato K, Hamaguchi T, Shimada Y: Standard first-line chemotherapy for metastatic gastric cancer in Japan has met the global standard: evidence from recent phase III trials. Gastrointest Cancer Res 2009, 3:239-244.

7. Fujii M, Kochi M, Takayama T: Recent advances in chemotherapy for advanced gastric cancer in Japan. Surg Today 2010, 40:295-300.
8. Conroy T, Marchal F, Blazeby JM: Quality of life in patients with oesophageal and gastric cancer: an overview. Oncology 2006, 70:391-402.
9. Kassam Z, Mackay H, Buckley CA, Fung S, Pintile M, Kim J, Ringash J: Evaluating the impact on quality of life of chemoradiation in gastric cancer. Curr Oncol 2010, 17:77-84.
10. Ajani JA, Moiseyenko VM, Tjulandin S, Majlis A, Constenla M, Boni C, Rodrigues A, Fodor M, Chao Y, Voznyi E, Awad L, van Cutsem E: Quality of life with docetaxel plus cisplatin and fluorouracil compared with cisplatin and fluorouracil from a phase III trial for advanced gastric or gastroesphageal adenocarcinoma: the V-325 study group. J Clin Oncol 2007, 25:3210-3216.
11. Li ZJ, Cho CH: Development of peptides as potential drugs for cancer therapy. Curr Pharm Des 2010, 16:1180-1189.
12. Sookraj KA, Bowne WB, Adler V, Sarafraz-Yazdi E, Michl J, Pincus MR: The anticaner peptide, PNC-27, induces tumor cell lysis as the intact peptide. Cancer Chemother Pharmacol 2010, 66:325-331.
13. Zou Y, Chen Y, Jiang Y, Gao J, Gu J: Targeting matrix metalloproteinases and endothelial cells with a fusion peptide against tumor. Cancer Res 2007, 67:7295-7300.
14. Su L, Xu G, Shen J, Tuo Y, Zhang X, Jia S, Chen Z, Su X: Anticancer bioactive peptide suppresses human gastric cancer growth through modulation of apoptosis and cell cycle. Oncol Rep 2010, 23:3-9.
15. Yang ZY, Wang WL, Su XL: HPCE analysis of polypeptides isolated from goat spleens. Chin J Pharm Anal 2005, 10:1248-1249.
16. Yang ZY, Zhang ZP, Su XL: MALDI-TOF-MS analysis of polypeptides isolated from goat spleens. Chin Pharm J 2005, 10:797.
17. Yan MR, Su XL, Liu QP: Effect of anti-gastric cancer biological peptide on lactic dehydrogenase isoenzyme. China J Cancer Prev Treat 2002, 9:382-383.
18. Quan XH, Su X: Effect of anti-gastric cancer biological peptide on mice and induction of TNF. Chin J Cancer Biother 2005, 12:301-302.
19. Hou JF, Yan MR, Yan XH, Rong YN, Jiao TM, Su XL: Effect of Anti-cancer bioactive peptide on Leukemia mice. J Inner Mongolia Med Sch 2004, 26(1):3-6.
20. Zhao YY, Peng SD, Su XL: Effects of anti-cancer bioactive peptide on cell cycle in human nasopharyngeal carcinoma strain CNE. Chin J Otorhinolaryngol Head Neck Surg 2006, 41(8):607-611.
21. Geertu DL, Chen K, Su XL: Gene chips are used for studying how anti-carcinoma bioactive peptides influence the expression of cholangiocarcinoma QBC939 apoptotis gene. Chin J Cliniclans (electronic Version) 2009, 3(10):1636-1644.
22. Wang ZY, Yang CW, Oy XH, Su XL: Anti-cancer bioactive peptide-S induced apoptosis of hepatoma cells and cell cycle regulation. Chin J Lab Diagn 2010, 14(1):1-6.
23. Jia SQ, Wang WL, Su XL: Inhibitory effect of anti-cancer bioactive peptide on proliferation of human breast cancer cell line nm231. Chin Med Biotechnol 2007, 2(4):270-275.
24. Yang ZY, Su XL: Study about the effect of polypeptides from goat spleen to HT29 cell. J Med Pharm Chin Minorities 2004, 10(3):29-30.

25. Herbertson RA, Scarsbrook AF, Lee ST, Tebbutt N, Scott AM: Established, emerging and future roles of PET/CT in the management of colorectal cancer. Clin Radiol 2009, 64:225-237.

26. Yoshioka T, Yamaguchi K, Kubota K, Saginoya T, Yamazaki T, Ido T, Yamaura G, Takahashi H, Fukuda H, Kanamaru R: Evaluation of 18F-FDG PET in patients with advanced, metastatic, or recurrent gastric cancer. J Nucl Med 2003, 44:690-699.

27. Shimuzu S, Narita M, Tsujimoto Y: Bcl-2 family proteins regulate the release of apoptogenic cytochrome c by the mitochondrial channel VDAC. Nature 1999, 399:483-487.

28. Oltvai ZN, Milliman CL, Korsmeyer SJ: Bcl-2 heterodimers in vivo with a conserved homolog, Bax, that accelerates programmed cell death. Cell 1993, 74:609-619.

29. Keller U, Doucet A, Overall CM: Protease research in the era of systems biology. Biol Chem 2007, 388:1159-1162.

30. Wong WW, Puthalakath H: Bcl-2 family proteins: the sentinels of the mitochondrial apoptosis pathway. IUBMB Lif 2008, 60:390-397.

31. Ow YP, Green DR, Hao Z, Mak TW: Cytochrome c: functions beyond respiration. Nat Rev Mol Cell Biol 2008, 9:532-542.

32. Bozhkov PV, Filonova LH, Suarez MF, Helmersson A, Smertenko AP, Zhivotovsky B, von Arnold S: VEIDase is a principal caspase-like activity involved in plant programmed cell death and essential for embryonic pattern formation. Cell Death Differ 2004, 11:75-182.

33. Oberst A, Bender C, Green DR: Living with death: the evolution of the mitochondrial pathway of apoptosis in animals. Cell Death Differ 2008, 15:1139-1146.

34. Kim H, Hsieh JJ, Cheng EH: Deadly splicing: Bax becomes Almighty. Mol Cell 2009, 33:145-146.

35. Nickells RW, Semaan SJ, Schlamp CL: Involvement of the Bcl2 gene family in the signaling and control of retinal ganglion cell death. Prog Brain Res 2008, 173:423-435.

There are several supplemental files that are not available in this version of the article. To view this additional information, please use the citation information cited on the first page of this chapter.

CHAPTER 14

COMBINATION OF SULINDAC AND DICHLOROACETATE KILLS CANCER CELLS VIA OXIDATIVE DAMAGE

KASIRAJAN AYYANATHAN, SHAILAJA KESARAJU, KEN DAWSON-SCULLY, AND HERBERT WEISSBACH

14.1 INTRODUCTION

Sulindac is an FDA-approved non-steroidal anti-inflammatory drug (NSAID), which has also been shown to have anti-cancer activity [1]–[6]. Recent studies from our laboratory have demonstrated that RKO, A549 and SCC25 cancer cell lines exhibited sensitivity towards a combination of sulindac and an oxidizing agent, such as TBHP or H_2O_2 [7]. The data indicated that the sulindac effect was not related to its NSAID activity but that sulindac made cancer cells more sensitive to oxidative stress resulting in mitochondrial dysfunction and loss of viability. In contrast, normal cells did not show enhanced killing under similar conditions [7]. In the past 10 years there have been scattered reports of enhanced cancer kill-

This chapter was originally published under the Creative Commons Attribution License. Ayyanathan K, Kesaraju S, Dawson-Scully, K and Weissbach H. Combination of Sulindac and Dichloroacetate Kills Cancer Cells via Oxidative Damage. PLoS ONE 7,7 (2012), e39949. doi:10.1371/journal. pone.0039949.

ing using sulindac in combination with a variety of compounds including arsenic trioxide, bortezomib, difluoromethylornithine (DFMO) and suberoylanilide hydroxamic acid (SAHA) [8]–[14]. Although these compounds have different sites of action, a common mechanism for the sulindac/drug combination enhanced killing might involve oxidative damage, as was clearly demonstrated in our previous studies using sulindac and an oxidizing agent [7], [15]. In fact, ROS have been implicated in the studies using sulindac in combination with arsenic trioxide, bortezomib and SAHA [10], [12], [14].

Our previous results suggested that the enhanced killing of cancer cells by the combination of sulindac and an oxidizing agent might be due to a defect in respiration in cancer cells, as first described by Warburg more than 50 years ago [16], who noted that cancer cells favor glycolysis, not respiration, to obtain energy, unlike normal cells. Some cancer cells obtained as much as 50% of their energy from glycolysis, whereas glycolysis in normal cells account for less than 5% of the energy requirement [16]. To obtain further evidence for the possible roles of altered respiration and ROS in the killing of cancer cells by sulindac and oxidative stress, we initiated studies with sodium dichloroacetic acid (DCA). DCA is an ideal candidate as it is known to inhibit a kinase that down regulates the activity of pyruvate dehydrogenase, resulting in a shift of pyruvate metabolism away from lactic acid formation, towards respiration [17], [18]. DCA has been used clinically to treat patients with lactic acidosis [19], and based on its biochemical properties DCA has also been tested as an anticancer agent. Bonnet et al. 2007 have shown that DCA reverses the Warburg effect in cancer cells by redirecting cancer cell metabolism from glycolysis to oxidative phosphorylation. In these previous studies it was shown that DCA increases levels of ROS from complex I. This in turn triggers "remodeling" of mitochondrial metabolism (reduces $\Delta\Psi m$, opens mitochondrial transition pore) in cancer cells pushing them towards apoptosis. Furthermore, several recent studies have verified that DCA can increase ROS levels in cancer cells and depolarize the mitochondria membrane in lung, endometrial, and glioblastoma cell lines resulting in apoptosis both in vitro and in vivo [18], [20]–[22]. Of interest was the observation that under the conditions used DCA did not appear to significantly affect mitochondrial metabolism or viability in normal cells [18], [23].

Based on our previous observations on the cancer killing effect of su-lindac and an oxidizing agent that affected mitochondrial metabolism [7], we postulated that the combination of sulindac and DCA could synergisti-cally enhance cancer killing and have important therapeutic value. In the present study we have examined the effect of using sulindac in combina-tion with DCA on the viability of A549 and SCC25 cancer cell lines. We have also studied the role of mitochondrial function and apoptosis in the cancer killing observed with this drug combination.

14.2 MATERIALS AND METHODS

14.2.1 MATERIALS

Sulindac, N-acetylcysteine and Tiron were purchased from Sigma (St. Louis, MO). DCA sodium salt was obtained from Acros Organics (Geel, Belgium). H2DCFDA and JC-1 were purchased from Molecular Probes (Eugene, OR). MTS assay reagent and Deadend Tunel Kit were obtained from Promega (Madison, WI). Cytosol/mitochondria fractionation kit and the CBA077 InnoCyte™ Flow Cytometric Cytochrome c Release kit were from Calbiochem, Gibbstown, NJ. All cell culture media, fetal bovine se-rum, and other supplements such as penicillin/streptomycin, glutamine, etc. were purchased from American Type Culture Collection (ATCC; Rockville, MD).

14.2.2 CELL CULTURE

A non small cell lung carcinoma cell line (NSCLC), A549, the normal hu-man lung cell line, MRC-5, and a tongue-derived squamous cell carcinoma line, SCC25 were purchased from ATCC (Rockville, MD) and maintained in F12-K medium supplemented with 10% fetal bovine serum, 2 mM glu-tamine, 100 IU/ml penicillin, and 100 μg/ml streptomycin in a humidified, 5% CO_2 incubator at 37°C. Normal human epidermal keratinocytes were

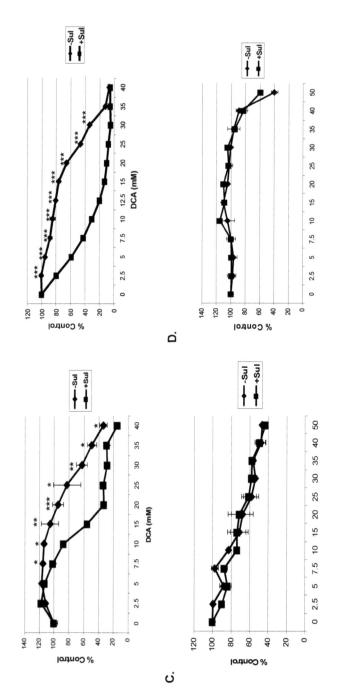

FIGURE 1: Sulindac in combination with DCA selectively kills cancer cells. The A549 and SCC25 cancer cells, normal lung cells, and human epidermal keratinocytes were treated with the indicated concentrations of DCA in the presence or absence of sulindac (Sul) for 48 hours. The cell viability was monitored by MTS assay as mentioned in the Methods. The cell viability is expressed as % of control (cells not treated with sulindac). Error bars are standard error of the mean (SEM) expressed as % of the mean value of quadruplicates from a representative experiment. Cell viabilities are illustrated for A549 cancer cells (A), SCC25 cancer cells (B), normal lung cells (C) and normal human epidermal keratinocytes (D). ♦, -Sul; ■, + Sul. * p<0.05, ** p<0.005, ***p<0.0005.

obtained from Promocell GmbH (Heidelberg, Germany) and maintained in the recommended culture medium. Early passage, non-immortalized normal cells were used for the experiments.

14.2.3 CELL VIABILITY ASSAY

The A549 cancer and lung normal cells were plated at 3×10^3 cells per well while SCC25 cancer cells and normal keratinocytes were plated at 7.5×10^3 cells per well in a 96-well plate. The cells were grown for 18–20 hours, the medium discarded in aseptic conditions and replaced with fresh culture medium containing the indicated drug combinations. Where indicated 500 μM sulindac was used with the A549 cancer and lung normal cells and 100 μM sulindac was used with SCC25 cancer and normal keratinocyte cells. The plates were incubated for 48 hours at 37°C in a 5% CO_2 incubator. The culture medium was discarded and the cells were thoroughly rinsed in $1 \times$ PBS. Cell viability was determined by using the CellTiter 96 Aqueous One Cell Proliferation Assay (Promega) according to the manufacturer's instructions. The assay utilizes a tetrazolium compound that is converted into a water-soluble formazan by the action of cellular dehydrogenases present in the metabolically active cells [24]. The formazan was quantified by measuring the absorbance at 490 nm using a colorimetric microtiter plate reader (SpectraMax Plus; Molecular Devices). Background absorbance was subtracted from each sample.

14.2.4 INTRACELLULAR MEASUREMENT OF ROS

The A549 and SCC25 cancer cell lines were plated as above. Following the 48 hr drug treatment, the cells were incubated with 50 μM of dichloro-dihydrofluorescein diacetate (H_2DCFDA, Molecular Probes) in indicator free medium for 30 min at 37°C. Cells were rinsed with PBS and ROS levels were visualized by fluorescence microscopy. The images were captured using the Qcapture software and processed in Adobe photoshop. Image analysis was done using the slidebook software. Data obtained from a representative experiment were used for the quantification of DCF-positive cells as measured by the green fluorescence due to oxidized DCF.

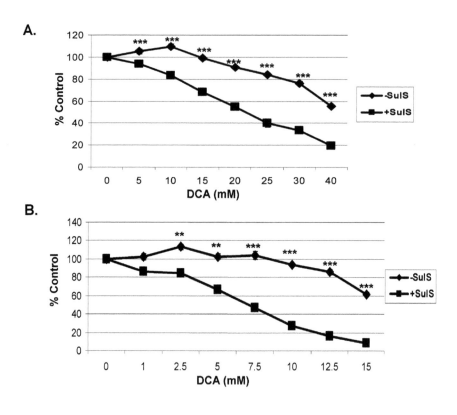

FIGURE 2: Sulindac sulfone in combination with DCA also kills cancer cells. The A549 and SCC25 cancer cells were treated with the indicated concentrations of DCA in the presence or absence of sulindac sulfone (SulS) for 48 hours. SulS was used at a final concentration of 250 μM for A549 cancer cells and 75 μM for SCC25 cancer cells. The cell viability was monitored by MTS assay as described in Methods. The cell viability is expressed as % of control (cells not treated with sulindac sulfone). Error bars are standard error of the mean (SEM) expressed as % of the mean value of triplicates from a representative experiment. Cell viabilities are illustrated for A549 cancer cells (A) and SCC25 cancer cells (B). ♦, -SulS; ■, + SulS. * p<0.05, ** p<0.005, ***p<0.0005.

A. A549 Cells:

DCA (10 mM):

Sulindac (500 μM):

B. SCC25 Cells:

DCA (10 mM):

Sulindac (100 μM):

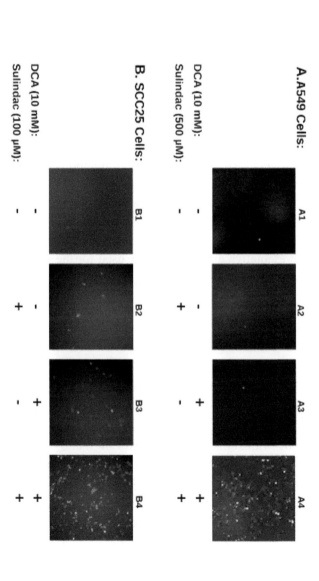

FIGURE 3: The combination of sulindac and DCA increases intracellular ROS levels in A549 and SCC25 cancer cells. Top panels (A) illustrate the results for A549 cancer cells while the bottom panels (B) depict the results for SCC25 cancer cells. The cells were treated with the indicated concentrations of drugs and processed for fluorescent microscopy as described in the Methods. The extent of intracellular ROS levels are illustrated as intensity of green fluorescence observed in cells treated with no drugs (sub-panels A1 and B1), sulindac alone (sub-panels A2 and B2), DCA alone (sub-panels A3 and B3), and sulindac and DCA combination treatment (sub-panels A4 and B4). Several independent fields were photomicrographed and representative fields for each condition are shown.

14.2.5 JC-1 STAINING TO MONITOR MITOCHONDRIAL MEMBRANE POTENTIAL

Mitochondrial membrane potential was determined using the JC-1 dye (Molecular Probes). The A549 and SCC25 cancer cell lines were plated as above. Following the 48 hr drug treatment, the cells were incubated with 5 ng/ml of JC-1 dye in indicator free medium for 30 min at 37°C. Cells were rinsed with PBS and visualized by fluorescence microscopy. Normal mitochondria actively take up JC-1 dye in a potential-dependent manner and form J-aggregates, which gives a red fluorescence. Disruption and subsequent loss of mitochondrial membrane potential leads to increased green fluorescence in the cytosol due to monomeric JC-1, which is determined by following the appearance of green fluorescence using an FITC filter (Zeiss inverted microscope-Axiovert 40 CFL). Image capturing, processing, and analysis were performed as above. Data obtained from a representative experiment were used for the quantification of JC-1-green positive cells.

14.2.6 EFFECT OF ROS SCAVENGERS ON CELL VIABILITY IN THE PRESENCE OF SULINDAC AND DCA

The A549 and SCC25 cancer cell lines were plated as described above. To scavenge ROS, either 2 mM N-acetylcysteine (NAC) or 2 mM Tiron (4,5-dihydroxy-1,3-benzenedisulfonic acid disodium salt) was added along with sulindac and DCA for 48 h at 37°C. Cell viability was monitored by the MTS assay and statistical analysis performed as mentioned above.

14.2.7 TUNEL STAINING TO MONITOR CELLS UNDERGOING APOPTOSIS

TUNEL assay was performed in 96 well plates using the DeadEnd colorimetric tunel assay kit (Promega) following the manufacturer's protocol. The A549 and SCC25 cancer cell lines were plated as above and treated

A. A549 Cells:

DCA (30 mM):
Sulindac (500 µM):

A1	A2	A3	A4
–	+	–	+
–	–	+	+

B. SCC25 Cells:

DCA (10 mM):
Sulindac (100 µM):

B1	B2	B3	B4
–	+	–	+
–	–	+	+

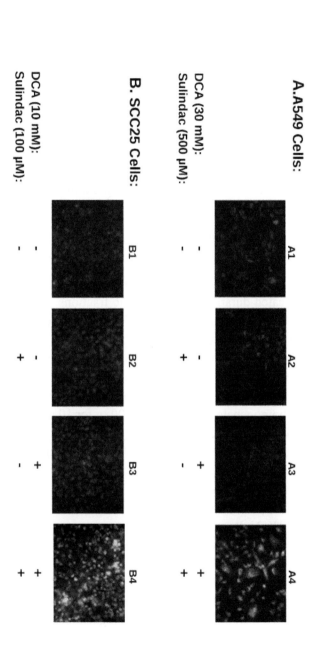

FIGURE 4: The combination of sulindac and DCA causes disruption of the mitochondrial membrane potential in cancer cells. Top panels (A) illustrate the results for A549 cancer cells while the bottom panels (B) depict the results for SCC25 cancer cells. Mitochondrial membrane potential loss was detected by a change in JC-1 distribution resulting in an increase in green fluorescence (see Methods). The experimental conditions for JC-1 staining and fluorescent microscopy are explained in detail under Methods and the drug treatment regimens are depicted below the panels. Untreated cells (sub-panels A1 and B1), cells treated with sulindac (sub-panels A2 and B2), cells treated with DCA (sub-panels A3 and B3), and cells treated with sulindac and DCA (sub-panels A4 and B4). Several independent fields were photomicrographed and representative fields for each condition are shown.

for 48 hr with no drug, sulindac, DCA, or drug combination. Subsequent to drug treatment, the cells were fixed with formalin and permeabilized with 0.2% Triton X-100 in PBS. Cells were incubated with recombinant terminal deoxynucleotidyl transferase (TdT) and biotinylated nucleotides. Endogenous peroxidases were blocked with 0.3% H_2O_2 prior to the incubation with horseradish peroxidase-streptavidin (HRP-streptavidin) that binds to the biotinylated nucleotides incorporated into the nicked ends present in cells undergoing apoptosis. HRP-streptavidin labeled cells were detected by hydrogen peroxide and diaminobenzidine (DAB). Cells that show dark brown nuclear staining are indicative of apoptosis.

14.2.8 WESTERN BLOT ANALYSIS

Cells were grown to 70% confluency, treated with specified drugs for the indicated durations, and cytosolic fractions were isolated using the cytosol/mitochondria fractionation kit (Calbiochem, Gibbstown, NJ) following the manufacturer's protocol. Briefly, cells were harvested at different time points and were then centrifuged at 600×g for 5 min at 4°C. The pelleted cells were suspended into the supplied buffer and incubated for 10 min on ice. The cells were then homogenized using a glass douncer and the homogenate centrifuged at 700×g for 10 min at 4°C to sediment nuclei and cell debris. The supernatant was spun at 10, 000×g for 30 min at 4°C to obtain the mitochondrial pellet and the supernatant was considered as the cytosolic fraction. Protein concentration was determined using a standard Bradford assay.

Sixty micrograms of total protein was loaded and separated on a 4–12% NuPage Bis-Tris gels (Invitrogen, Eugene, OR) and transferred onto a PVDF membrane that was probed by the primary antibodies. The primary antibodies, JNK, pJNK, cytochrome c, and PARP (Cell Signaling Technology, Danvers, MA), were used at 1:1000 dilution. β-actin, (Santa Cruz Biotechnologies, Santa Cruz, California), was used at 1:4000 dilution. Horseradish peroxidase conjugated secondary antibodies were used and bands were visualized using an enhanced chemiluminescence method (GE Healthcare, Piscataway, NJ).

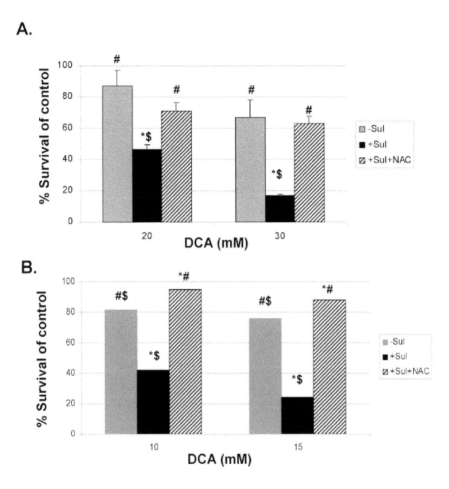

FIGURE 5: The ROS scavenger NAC reverses the killing of cancer cells by the combination of sulindac and DCA. The A549 and SCC25 cancer cells were treated with the indicated concentrations of DCA in the absence (grey bar) or presence of sulindac (black bar) or presence of sulindac and N-acetylcysteine (striped bar) for 48 hours. The cell viability was monitored by MTS assay as mentioned in the Methods. The cell viability is expressed as % of control (cells not treated with sulindac). Error bars are standard error of the mean (SEM) expressed as % of the mean value of quadruplicates from a representative experiment. Inhibition of cancer cell growth occurred in a dose dependent manner during combination treatment of DCA and sulindac (black bars) in both A549 cancer (A) and SCC25 cancer cells (B). However, this enhanced killing was prevented when N-acetylcysteine was present with the drug combination treatment (striped bars in A and B).

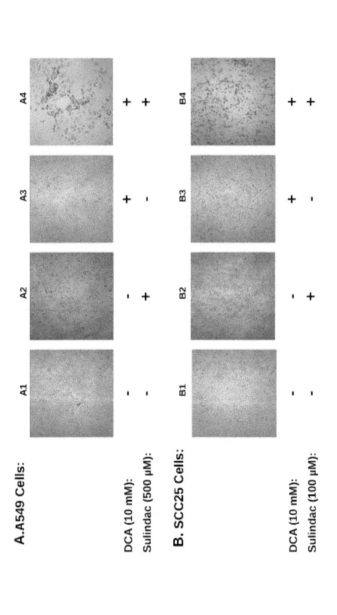

A. A549 Cells:

	A1	A2	A3	A4
DCA (10 mM):	-	-	+	+
Sulindac (500 µM):	-	+	-	+

B. SCC25 Cells:

	B1	B2	B3	B4
DCA (10 mM):	-	-	+	+
Sulindac (100 µM):	-	+	-	+

FIGURE 6: Sulindac in combination with DCA induce apoptosis in cancer cells. Top panels (A) illustrate the results for A549 cancer cells while the bottom panels (B) depict the results for SCC25 cancer cells. The extent of cells undergoing apoptosis was monitored by TUNEL staining of cells treated with no drugs (sub-panels A1 and B1), sulindac alone (sub-panels A2 and B2), DCA alone (sub-panels A3 and B3), and sulindac and DCA (sub-panels A4 and B4). The cells were treated with the indicated drugs as mentioned in the panels, subjected to TUNEL staining, and processed for fluorescent microscopy as described in the Methods. Several independent fields were photomicrographed and representative fields for each condition are shown. Brown-stained cells are indicative of cells undergoing apoptosis.

14.2.9 LIGATION-MEDIATED PCR BASED DNA LADDERING ASSAY TO MONITOR EXTENT OF CELLS UNDERGOING APOPTOSIS

To confirm the extent of apoptosis, ligation-mediated PCR based nucleosomal DNA laddering assay was performed as described [25]. The A549 and SCC25 cancer cell lines were plated at 5×10^4 and 1×10^5 cells per well in 35 mm dishes. The A549 cancer cells were treated for 48 hours with a) no drug, b) 500 µM sulindac, c) 20 mM DCA, and d) 500 µM sulindac plus 20 mM DCA. Similarly, SCC25 cancer cells were treated with the abovementioned four different drug combinations except that sulindac and DCA were used at 100 µm and 10 mM concentrations, respectively. After treatment, total cellular DNA were extracted, ligated to the adaptor constructed from 27-mer 5'-GACGTCGACGTCGTACACGTGTCGACT-3' and 12-mer 5'- AGTCGACACGTGTAC-3'. Subsequent to ligation, the DNA was heated to release the 12-mer, filled with Taq polymerase, subjected to semi-quantitative PCR, and analyzed on a 1.2% agarose gel along with size markers.

14.2.10 IN SITU LOCALIZATION OF CYTOCHROME C BY IMMUNOFLUORESCENCE

Intracellular location of cytochrome c was monitored by immunofluorescence by using the CBA077 InnoCyte™ Flow Cytometric Cytochrome c Release Kit according to the manufacturer's instructions. Briefly, the SCC25 cells were plated at 3.5×10^5 cells per 35 mm glass bottom dish and treated with the indicated drugs for 15 h. Cells were rinsed in 5 ml of 1× PBS and permeabilized on ice for 10 min in 300 µl of supplied buffer. The cells were fixed at RT for 20 min in 500 µl of 4% paraformaldehyde. Subsequent to washing and blocking, the cells were incubated with 250 µl of anti-cytochrome c antibody (1:500 dilution) for 1 hr at RT. After washing, the cells were incubated with 300 µl of FITC-IgG (1:300 dilution) for 1 hr at RT. Finally, the cells were stained with 300 µl of DAPI (1 mg/ml) for 10 min at RT. Cells were visualized using an Olympus inverted fluorescent microscope. Images were captured and processed as mentioned above. Several fields were analyzed and representative micrographs showing the localization patterns of cytochrome c under each treatment condition were obtained. Quantitative values are presented in the text.

14.2.11 STATISTICAL ANALYSIS AND DETERMINATION OF COMBINATION INDICES

Data are presented as mean ± SEM for the cell viability assays. For statistical analysis, Minitab statistical software was used to perform Student's t-test and values with $p < 0.05$ were considered statistically significant. To ascertain the synergistic effect of sulindac and DCA on A549 and SCC25 cancer cell lines, a quantitative analysis of dose-effect relationship was performed to determine the combination indices [26]. Both sulindac and DCA were tested alone on A549 and SCC25 cells at the concentrations indicated. For A549 cells, a ratio of 1:50 was maintained for the sulindac:DCA drug combinations ranging from 0.2 mM:10 mM up to 1 mM:50 mM, respectively. For SCC25 cells, a ratio of 1:100 was maintained for the sulindac:DCA drug combinations ranging from 0.05 mM:5 mM up to 0.3 mM:30 mM, respectively. Our experimental results and the determined combination index values are included in the text.

14.3 RESULTS

14.3.1 SULINDAC AND DCA CAUSE ENHANCED KILLING OF A549 AND SCC25 CANCER CELLS, BUT NOT NORMAL CELLS

For these studies we tested the combination of sulindac and DCA on A549 and SCC25 cancer cells. The cells were incubated with each compound alone or in combination for 48 hours before assaying for viability (see Methods). A sulindac dose response curve under these conditions indicated that A549 and SCC25 cancer cells can tolerate a maximum concentration of 500 µM and 100 µM of sulindac, respectively, without exhibiting any significant killing (data not shown), and these concentrations were used in all the studies. DCA, when added, was used at concentrations from 0–40 mM, as indicated. We used these concentrations based on previous reports, which indicated that above 5 mM is required to cause mitochondrial dysfunction in in vitro experiments [27]. As shown in Figure 1A, DCA alone (no sulindac) is somewhat toxic to A549 cancer cells, especially above concentrations of 20 mM, but in the presence of sulindac there is enhanced killing of these cells at DCA concentrations above 5 mM. In the case of the SCC25 cancer cells some loss of cell viability with DCA alone was seen even at DCA concentrations below 10 mM (Figure 1B). However, in the presence of sulindac there was again a marked increase in cell death that was clearly evident between DCA concentrations of 2–10 mM. Previously we showed that the combination of sulindac and an oxidizing agent was selective for cancer cells and did not enhance the killing of normal cells [7]. Sulindac and DCA also did not enhance the killing of normal lung and skin cells under the experimental conditions used, as shown in Figures 1C and D. It should be noted that the MRC-5 (lung normal) cells are especially sensitive to DCA, as reported previously [28], for reasons that are not known.

To verify that there was a synergistic effect when the drug combination was used, we determined the combination indices by performing a quantitative analysis of dose-effect relationship [26] on two different cancer cell lines (Figure S1). The combination indices were 0.84 for the A549 and 0.73 for the SCC25 cancer cells, respectively. A value less than 1.00 indicates a synergistic cancer killing effect (Figure S2).

14.3.2 THE SULINDAC EFFECT IS NOT DUE TO ITS NSAID ACTIVITY

In previous studies using sulindac and an oxidizing agent it was shown that the enhanced and selective killing of cancer cells by sulindac and an oxidizing agent was not related to the known NSAID ability of sulindac. To determine the role of COX inhibition a sulindac metabolite, sulindac sulfone, can be used, since it does not inhibit COX 1 or 2 [7], [29]. As shown in Figure 2, using both A549 (A) and SCC25 (B) cancer cells, the combination of sulindac sulfone and DCA showed a similar killing effect as seen above with sulindac. These results indicate that the sulindac enhanced cancer killing effect in the presence of DCA is not related to its known anti-inflammatory activity.

14.3.3 THE COMBINATION OF SULINDAC AND DCA GENERATE ROS

The synergistic effect on viability observed with sulindac and dichloroacetate with both A549 and SCC25 cancer cells is strikingly similar to previous studies using the combination of sulindac and TBHP [7]. To determine whether ROS production was involved in the selective killing observed in the present studies, the production of ROS, using the indicator dye H_2D-CFDA (see Methods), was determined in the cancer cell lines exposed to sulindac and DCA. The results are summarized in Figure 3. Figure 3A shows the results with A549 cancer cells. It is evident from the results depicted in Figure 3A that untreated A549 cancer cells (panel A1), or cells treated with sulindac alone (panel A2), or DCA alone (panel A3), show only a few positively stained cells. However, when the cells were exposed to both sulindac and DCA (panel A4), a large increase in positively stained cells for ROS (green fluorescence) is seen, showing that the presence of both sulindac and DCA results in the generation of significant levels of ROS. As shown in Figure 3B similar results are seen with the SCC25 cancer cells. Sulindac or DCA alone result in a small increase in ROS producing cells (panels B2 and B3), but a large increase in ROS production is again observed when both drugs are added (panel B4). Quantification using SCC25 cells shows that the number of DCF-positive cells (see Methods) is 9–10× more when the cells are treated with sulindac and DCA as compared to each of the drugs alone (see Figure S3A). It appears from these results and earlier studies that ROS production may be a common feature in the enhanced killing of cancer cells when sulindac is used in combination with compounds that affect mitochondrial function.

A. JNKI reversal:

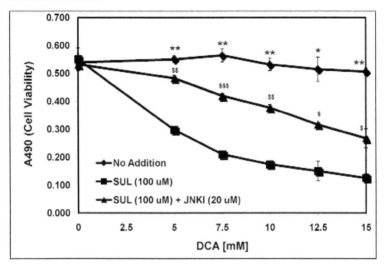

B. Western blotting:

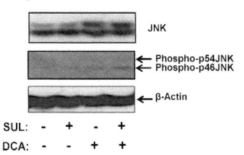

FIGURE 7: Combination of sulindac and DCA induced apoptosis involves JNK activation. A. SCC25 cells were treated for 48 h with sulindac (100 μM) and the indicated concentrations of DCA and where indicated 20 μM of the JNK inhibitor, SP600125. ◆, No drug; ■, sulindac; ▲, sulindac and SP600125. The cell viability was monitored by MTS assay as mentioned in the Methods. Error bars are standard error of the mean (SEM) expressed as % of the mean value of quadruplicates from a representative experiment. See text for details. Statistical analysis between ◆, No addition and ■, Sul; *$p<0.05$, **$p<0.005$, ***$p<0.0005$. Statistical analysis between ■, Sul and ▲, sulindac and JNKI; $p<0.05$, $p<0.005$, $p<0.0005$. B. Representative western blots depicting the effect of sulindac and DCA on the levels of total JNK, phopsho-p46JNK and phopsho-p54JNK. β-actin levels were used as an internal control. Cytosolic fractions from SCC25 cells were isolated at the 12 h time point and the levels of JNK and phospho-JNK were determined by western blotting. In the SCC25 cell line, total JNK showed two distinct bands at 46 and 54 kDa.

14.3.4 SULINDAC IN COMBINATION WITH DCA RESULTS IN A LOSS OF MITOCHONDRIAL MEMBRANE POTENTIAL

If ROS production is involved in the sulindac/DCA enhanced killing effect one would expect that the production of ROS by the drug combination would affect mitochondrial function. In order to determine this, mitochondrial membrane potential was measured using JC-1 staining as described in Methods. A loss of membrane potential is indicated by an increase in green fluorescence as described in Methods. A typical result is summarized in Figure 4. Both A549 and SCC25 cancer cells were exposed to sulindac and DCA either alone or in combination for 48 hrs and stained with JC-1 in order to monitor the mitochondrial membrane potential. Figure 4A shows the results with the A549 cancer cell line. In the absence of any drug, the mitochondria appear intact and maintain their membrane potential as indicated by little green fluorescence (panel A1). In the presence of sulindac alone (panel A2) or DCA alone (panel A3) there is a small increase in green fluorescence, indicating some loss of mitochondrial membrane potential. However, when both sulindac and DCA are present there is a striking loss of mitochondrial membrane potential as evidenced by a large increase in the green fluorescence (panel A4). We observed the same pattern when several independent fields were analyzed by fluorescent microscopy. Figure 4B shows similar results with the SCC25 cancer cells. Once again a significant loss of mitochondrial membrane potential was only seen when the cells were exposed to both sulindac and DCA (panel B4). Quantification of the effect is shown in Figure S3B. It can be seen that the percent of JC1-green positives cells when the drug combination was used is 3–4× that seen with either drug alone.

14.3.5 ROS ARE INVOLVED IN THE KILLING OF CANCER CELLS BY THE COMBINATION OF SULINDAC AND DCA

To provide more direct evidence that the ROS produced are involved in the enhanced killing of the cancer cells by sulindac and DCA, we have used two known ROS scavengers, N-acetylcysteine (NAC) and Tiron (see

Methods). The results using NAC are shown in Figure 5. Figure 5, panel A, shows that at both 20 and 30 mM DCA, the enhanced killing of A549 cancer cells observed in the presence of sulindac, is largely prevented if NAC (2 mM) is present during the 48 hour incubations. Very similar results are seen with the SCC25 cancer cells as shown in Figure 5, panel B. Comparable results were obtained when Tiron was used in place of NAC (Figure S4).

14.3.6 SULINDAC AND DCA KILLING OF CANCER CELLS INVOLVES APOPTOTIC DEATH

The results above (Figures 3, 4, 5) show that the enhanced killing of the cancer cell lines involves mitochondrial dysfunction, which suggest that the observed cell death is via apoptosis. Previous studies have indicated that sulindac and its derivatives are proapoptotic drugs [5], [6]. There are also reports that DCA can cause cell death by apoptosis [20], [23]. To determine whether killing of the cancer cells by the combination of these two drugs, mediated by ROS, involves apoptotic death we performed TUNEL staining to measure apoptosis (see Methods). Multiple replicates were tested for sulindac and DCA alone, or in combination, for the TUNEL staining experiments. A typical result is illustrated in Figure 6, where the top panels (Figure 6A, panels A1–A4) represent the results with the A549 cancer cells and the bottom panels (Figure 6B, panels B1–B4) depict the results with the SCC25 cancer cells. When the cells were treated with no drug, sulindac alone, or DCA alone (Figure 6, panels A1–A3 and B1–B3), only a few TUNEL-positive cells are observed. However, when the cells were exposed to both sulindac and DCA, there is a significant increase in TUNEL-positive apoptotic cells (Figure 6, panels A4 and B4), indicating a large induction of apoptosis. To verify the TUNEL results, a more sensitive ligation-mediated PCR-based DNA laddering assay was also used to monitor apoptosis [25]. The results also showed the presence of an enriched strong nucleosomal ladder only when both sulindac and DCA were used in combination (Figure S5; lanes 4 and 8), which strongly supports the TUNEL assay data.

14.3.7 SULINDAC AND DCA KILLING INVOLVES PROAPOPTOTIC JNK SIGNALING

Of the known mitogen activated protein kinases (MAP kinases), the stress-induced kinase, c-Jun N-terminal kinase (JNK/SAPK) has been directly implicated in apoptotic cell death [11]. Therefore we investigated the role of JNK signaling in sulindac-DCA mediated apoptosis by using SP600125, a JNK-specific inhibitor (JNKI) and these results are presented in Figure 7A. As shown above, SCC25 cells treated with sulindac showed enhanced death in the presence of increasing DCA concentrations. However, when these cells were incubated with sulindac along with SP600125, sulindac-DCA mediated cell death was largely prevented. These results indicate the participation of JNK mediated proapoptotic signaling in the sulindac-DCA mediated cell death.

By western blot analysis, we also determined that the combination of sulindac and DCA significantly increased the levels of phospho-JNK in cytosolic fractions 12 h after the cells were exposed to sulindac and DCA (Figure 7B). An increase in the levels of total JNK (protein bands at 46 and 54 kDa) was seen when the cells were treated with DCA alone as well as when the cells were treated with the combination of sulindac and DCA. It should be noted that both phospho-p46JNK and phospho-p54JNK isoforms were induced by the combination of sulindac and DCA treatment, although the increase in phospho-p46JNK was more significant (Figure 7B).

There is a body of evidence suggesting that JNK initiates release of apoptosis inducing factors from mitochondria, such as cytochrome c, that lead to cleavage of caspases and PARP (poly(ADP-ribose) polymerase) [30], [31]. Studies have also shown that during apoptosis, the cytochrome c released from mitochondria into the cytoplasm ultimately enters into the nucleus [32]. Our results indicated maximum activation of JNK occurred around 12 h after exposure to sulindac and DCA. This appears to result in the translocation of cytochrome c into the cytoplasm and cleavage of PARP 18 h after initial treatment with sulindac and DCA (Figure S6A). As a positive control for these experiments we treated cells with 100 µM of etoposide, an apoptosis-inducing agent. Under sulindac and DCA combination treatment, enhanced nuclear fluorescence can be observed in a majority of cells that are actively undergoing apoptosis (Figure S6B).

Detailed analysis of whole cell immunofluorescence experimental data revealed that ~94% of cells not treated with either sulindac or DCA showed punctate, mitochondrial cytochrome c fluorescence with little diffuse staining in the cytoplasm or in the nuclei. In contrast, after sulindac treatment, 81% of cells showed diffuse, distinct cytoplasmic fluorescence, and very little nuclear fluorescence. After DCA treatment, ~83% of cells showed diffuse, distinct cytoplasmic fluorescence, and <5% of the cells showed strong nuclear fluorescence. However, when the cells were treated with both sulindac and DCA, ~72% of cells showed both nuclear and cytoplasmic fluorescence and ~11% of cells showed strong nuclear fluorescence. These results suggest that the released cytochrome c from the mitochondria may initiate the intrinsic apoptotic pathway functioning in the sulindac and DCA mediated cancer killing.

14.4 DISCUSSION

The present study is an extension of our previous work, which demonstrated that sulindac made cancer cells, but not normal cells, more sensitive to oxidative stress [7]. In these previous experiments sulindac was pre-incubated with the cells for 24–48 hours and then the sulindac was removed before the cells were exposed to either TBHP or H_2O_2 for 2 hours. It was evident from the previous experiments that sulindac pretreatment made the cancer cells much more sensitive to the oxidizing agent resulting in a large increase in ROS and loss of mitochondrial function [7].

It seemed reasonable, based on these results, that sulindac in combination with compounds that affected mitochondrial respiration would result in selective enhanced killing of cancer cells, but not normal cells. In the present study using A549 and SCC25 cancer cell lines, the combination of sulindac and DCA enhanced the killing of these cancer cell lines, but not normal lung or skin cells. Our results on the amounts of DCA needed in whole cells are consistent with what has been reported previously [28], [33], [34]. In our system, the IC50 for DCA with SCC25 cells is 23 mM and for A549 cells is 35 mM. The IC50 for normal keratinocytes is >50 mM and for normal lung cells (MRC5) is ~40 mM. The results also indicated that the cancer cell death that was observed involves ROS production, JNK activation, and mitochondria initiated apoptosis. With regard to a lack of effect on normal cells, it has been shown that sulindac protects normal lung cells against oxidative damage resulting from TBHP exposure [7] and we have also recently reported sulindac can protect cardiac cells against oxidative damage resulting from ischemia/reperfusion through a preconditioning mechanism [35].

To our knowledge there are now at least 8 compounds, including our studies with TBHP, H_2O_2 and DCA, that have shown enhanced and selective cancer killing in the presence of sulindac [7]–[9], [12]–[15]. Although their metabolic targets within the cell are known, and are different, it is quite likely that they all, directly or indirectly, cause cell death in the presence of sulindac through a mechanism that involves an alteration in mitochondrial respiration and ROS production [10], [12], [14]. It seems likely that when one finds a drug, that in combination with sulindac, selectively kills cancer cells, but not normal cells, the mechanism of killing involves oxidative stress leading to mitochondrial dysfunction. Altered respiration may be a common factor in these experiments using sulindac/drug combinations and the present results using DCA support this view. It is quite possible that the sulindac effect may be related to the observations made more than 50 years ago by Warburg, who noted that normal cells prefer respiration to obtain their energy, whereas cancer cells prefer glycolysis, due to a defect in the respiratory chain [16]. This basic difference in mitochondrial respiration between normal and cancer cells may make cancer cells more sensitive to oxidative stress [36], [37]. It seems that sulindac may amplify this fundamental difference in the biochemistry of normal and cancer cells. Although how sulindac sensitizes cancer cells to drugs that affect mitochondrial respiration is still not clear, but is under active investigation. Spitz and coworkers [38], in studies on glucose deprivation of cancer cells, have come to a similar conclusion regarding the differences in metabolism between normal and cancer cells. In line with these results, another recent study has shown that pharmacological inhibition of lactate dehydrogenase could result in selective cancer killing [39]. In the latter studies it was shown that the enhanced selective killing of cancer cells also involved ROS production, and the effect seen was attributed to an altered respiratory process in the cancer cells.

It should be pointed out that the combination of sulindac with an oxidizing agent or drugs that may affect mitochondrial function has already been tested clinically. Meyskens et al. 2008 showed that the combination of sulindac with DFMO had a significant effect on the recurrence of colon polyps and appearance of colon cancer in a 3-year clinical study [13]. We recently reported the use of sulindac, with H_2O_2, in a proof of concept clinical study for the topical treatment of actinic keratoses [15]. One of the disadvantages of using this combination was the need for two topical formulations since the compounds could not be stored for long periods without destruction of the sulindac by the H_2O_2. In addition, one cannot use H_2O_2 for treatment of internal tumors since it cannot be taken orally. However, the combination of sulindac and DCA could be delivered as a single formulation amenable for topical use, and the two compounds can be used orally. In fact, for several years, DCA has been used clinically to lower lactic acid levels in patients suffering from lactic acidosis [40]–[42]. DCA also has been used as an anti-cancer agent in vitro and in vivo using several different cancer cell lines indicating that mitochondrial metabolism in cancer cells could be a new therapeutic target [18], [20], [22]. Michelakis et al., (2010) have shown that treatment with DCA "remodels" mitochondrial metabolism in glioblastoma patients with reversible toxic effects. It should be noted that both sulindac and DCA are affordable, relatively non-toxic and can be taken orally. If the combination proves to be successful in vivo it will add a new dimension in cancer treatment as both the drugs target mitochondrial metabolism in multiple cancers [22].

In summary, our studies using the combination of sulindac and DCA suggest that sulindac selectively makes cancer cells more sensitive to agents that affect mitochondrial respiration resulting in oxidative stress and mitochondrial dysfunction. These results could be related to the respiration defect in cancer cells, originally observed by Warburg [16]. Studies aimed at understanding the fundamental differences between how cancer cells and normal cells respond to sulindac and agents that affect mitochondrial function are currently under investigation.

REFERENCES

1. Boolbol SK, Dannenberg AJ, Chadburn A, Martucci C, Guo XJ, et al. (1996) Cyclooxygenase-2 overexpression and tumor formation are blocked by sulindac in a murine model of familial adenomatous polyposis. Cancer Res 56: 2556–2560.
2. Taketo MM (1998) Cyclooxygenase-2 inhibitors in tumorigenesis (Part II). J Natl Cancer Inst 90: 1609–1620.
3. Taketo MM (1998) Cyclooxygenase-2 inhibitors in tumorigenesis (part I). J Natl Cancer Inst 90: 1529–1536.
4. Rao CV, Rivenson A, Simi B, Zang E, Kelloff G, et al. (1995) Chemoprevention of colon carcinogenesis by sulindac, a nonsteroidal anti-inflammatory agent. Cancer Res 55: 1464–1472.
5. Vogt T, McClelland M, Jung B, Popova S, Bogenrieder T, et al. (2001) Progression and NSAID-induced apoptosis in malignant melanomas are independent of cyclooxygenase II. Melanoma Res 11: 587–599.
6. Richter M, Weiss M, Weinberger I, Furstenberger G, Marian B (2001) Growth inhibition and induction of apoptosis in colorectal tumor cells by cyclooxygenase inhibitors. Carcinogenesis 22: 17–25.
7. Marchetti M, Resnick L, Gamliel E, Kesaraju S, Weissbach H, et al. (2009) Sulindac enhances the killing of cancer cells exposed to oxidative stress. PLoS One 4: e5804.
8. Soriano AF, Helfrich B, Chan DC, Heasley LE, Bunn PA Jr, et al. (1999) Synergistic effects of new chemopreventive agents and conventional cytotoxic agents against human lung cancer cell lines. Cancer Res 59: 6178–6184.
9. Jiang TT, Brown SL, Kim JH (2004) Combined effect of arsenic trioxide and sulindac sulfide in A549 human lung cancer cells in vitro. J Exp Clin Cancer Res 23: 259–262.
10. Jin HO, Yoon SI, Seo SK, Lee HC, Woo SH, et al. (2006) Synergistic induction of apoptosis by sulindac and arsenic trioxide in human lung cancer A549 cells via reactive oxygen species-dependent down-regulation of survivin. Biochem Pharmacol 72: 1228–1236.
11. Jin HO, Seo SK, Woo SH, Lee HC, Kim ES, et al. (2008) A combination of sulindac and arsenic trioxide synergistically induces apoptosis in human lung cancer H1299 cells via c-Jun NH2-terminal kinase-dependent Bcl-xL phosphorylation. Lung Cancer 61: 317–327.
12. Minami T, Adachi M, Kawamura R, Zhang Y, Shinomura Y, et al. (2005) Sulindac enhances the proteasome inhibitor bortezomib-mediated oxidative stress and anticancer activity. Clin Cancer Res 11: 5248–5256.
13. Meyskens FL Jr, McLaren CE, Pelot D, Fujikawa-Brooks S, Carpenter PM, et al. (2008) Difluoromethylornithine plus sulindac for the prevention of sporadic colorectal adenomas: a randomized placebo-controlled, double-blind trial. Cancer Prev Res (Phila Pa) 1: 32–38.
14. Seo SK, Jin HO, Lee HC, Woo SH, Kim ES, et al. (2008) Combined effects of sulindac and suberoylanilide hydroxamic acid on apoptosis induction in human lung cancer cells. Mol Pharmacol 73: 1005–1012.

15. Resnick L, Rabinovitz H, Binninger D, Marchetti M, Weissbach H (2009) Topical sulindac combined with hydrogen peroxide in the treatment of actinic keratoses. J Drugs Dermatol 8: 29–32.
16. Warburg O (1956) On the origin of cancer cells. Science 123: 309–314.
17. Whitehouse S, Cooper RH, Randle PJ (1974) Mechanism of activation of pyruvate dehydrogenase by dichloroacetate and other halogenated carboxylic acids. Biochem J 141: 761–774.
18. Bonnet S, Archer SL, Allalunis-Turner J, Haromy A, Beaulieu C, et al. (2007) A mitochondria-K+ channel axis is suppressed in cancer and its normalization promotes apoptosis and inhibits cancer growth. Cancer Cell 11: 37–51.
19. Stacpoole PW, Kurtz TL, Han Z, Langaee T (2008) Role of dichloroacetate in the treatment of genetic mitochondrial diseases. Adv Drug Deliv Rev 60: 1478–1487.
20. Wong JY, Huggins GS, Debidda M, Munshi NC, De Vivo I (2008) Dichloroacetate induces apoptosis in endometrial cancer cells. Gynecol Oncol 109: 394–402.
21. Cao W, Yacoub S, Shiverick KT, Namiki K, Sakai Y, et al. (2008) Dichloroacetate (DCA) sensitizes both wild-type and over expressing Bcl-2 prostate cancer cells in vitro to radiation. Prostate 68: 1223–1231.
22. Michelakis ED, Sutendra G, Dromparis P, Webster L, Haromy A, et al. (2010) Metabolic modulation of glioblastoma with dichloroacetate. Sci Transl Med 2: 31ra34.
23. Michelakis ED, Webster L, Mackey JR (2008) Dichloroacetate (DCA) as a potential metabolic-targeting therapy for cancer. Br J Cancer 99: 989–994.
24. Cory AH, Owen TC, Barltrop JA, Cory JG (1991) Use of an aqueous soluble tetrazolium/formazan assay for cell growth assays in culture. Cancer Commun 3: 207–212.
25. Staley K, Blaschke AJ, Chun J (1997) Apoptotic DNA fragmentation is detected by a semi-quantitative ligation-mediated PCR of blunt DNA ends. Cell Death Differ 4: 66–75.
26. Chou TC, Talalay P (1984) Quantitative analysis of dose-effect relationships: the combined effects of multiple drugs or enzyme inhibitors. Adv Enzyme Regul 22: 27–55.
27. Papandreou I, Goliasova T, Denko NC (2010) Anticancer drugs that target metabolism: Is dichloroacetate the new paradigm? Int J Cancer 128: 1001–1008.
28. Stockwin LH, Yu SX, Borgel S, Hancock C, Wolfe TL, et al. (2010) Sodium dichloroacetate selectively targets cells with defects in the mitochondrial ETC. Int J Cancer 127: 2510–2519.
29. Babbar N, Ignatenko NA, Casero RA Jr, Gerner EW (2003) Cyclooxygenase-independent induction of apoptosis by sulindac sulfone is mediated by polyamines in colon cancer. J Biol Chem 278: 47762–47775.
30. Selimovic D, Ahmad M, El-Khattouti A, Hannig M, Haikel Y, et al. (2011) Apoptosis-related protein-2 triggers melanoma cell death by a mechanism including both endoplasmic reticulum stress and mitochondrial dysregulation. Carcinogenesis 32: 1268–1278.
31. Zhang S, Lin Y, Kim YS, Hande MP, Liu ZG, et al. (2007) c-Jun N-terminal kinase mediates hydrogen peroxide-induced cell death via sustained poly(ADP-ribose) polymerase-1 activation. Cell Death Differ 14: 1001–1010.

32. Nur EKA, Gross SR, Pan Z, Balklava Z, Ma J, et al. (2004) Nuclear translocation of cytochrome c during apoptosis. J Biol Chem 279: 24911–24914.
33. Heshe D, Hoogestraat S, Brauckmann C, Karst U, Boos J, et al. (2011) Dichloro-acetate metabolically targeted therapy defeats cytotoxicity of standard anticancer drugs. Cancer Chemother Pharmacol 67: 647–655.
34. Madhok BM, Yeluri S, Perry SL, Hughes TA, Jayne DG (2010) Dichloroacetate induces apoptosis and cell-cycle arrest in colorectal cancer cells. Br J Cancer 102: 1746–1752.
35. Moench I, Prentice H, Rickaway Z, Weissbach H (2009) Sulindac confers high level ischemic protection to the heart through late preconditioning mechanisms. Proc Natl Acad Sci U S A 106: 19611–19616.
36. Deberardinis RJ, Sayed N, Ditsworth D, Thompson CB (2008) Brick by brick: me-tabolism and tumor cell growth. Curr Opin Genet Dev 18: 54–61.
37. Vander Heiden MG, Cantley LC, Thompson CB (2009) Understanding the Warburg effect: the metabolic requirements of cell proliferation. Science 324: 1029–1033.
38. Ahmad IM, Aykin-Burns N, Sim JE, Walsh SA, Higashikubo R, et al. (2005) Mi-tochondrial O2*- and H2O2 mediate glucose deprivation-induced stress in human cancer cells. J Biol Chem 280: 4254–4263.
39. Le A, Cooper CR, Gouw AM, Dinavahi R, Maitra A, et al. (2010) Inhibition of lac-tate dehydrogenase A induces oxidative stress and inhibits tumor progression. Proc Natl Acad Sci U S A 107: 2037–2042.
40. Stacpoole PW, Harman EM, Curry SH, Baumgartner TG, Misbin RI (1983) Treat-ment of lactic acidosis with dichloroacetate. N Engl J Med 309: 390–396.
41. Stacpoole PW, Barnes CL, Hurbanis MD, Cannon SL, Kerr DS (1997) Treatment of congenital lactic acidosis with dichloroacetate. Arch Dis Child 77: 535–541.
42. Stacpoole PW, Gilbert LR, Neiberger RE, Carney PR, Valenstein E, et al. (2008) Evaluation of long-term treatment of children with congenital lactic acidosis with dichloroacetate. Pediatrics 121: e1223–1228.

There are several supplemental files that are not available in this version of the article. To view this additional information, please use the citation information cited on the first page of this chapter.

AUTHOR NOTES

CHAPTER 1

Competing Interests
The authors declare that they have no competing interests.

Author Contributions
SSB helped design and perform the experiments and wrote the manuscript. YV helped design the studies and prepare the manuscript. AX, AO and VL provided technical support. FC prepared and provided normal and cancer associated fibroblasts. EMC supervised, directed and designed all studies and wrote the manuscript. All authors read and approved the final manuscript.

Acknowledgments
We thank Dr. Erik Sahai, Cancer Research UK London Research Institute, for input on the manuscript and for providing cancer associated fibroblasts. FC was supported by a Cancer Research UK grant CRUK_A5317. YV was supported by a Michael Smith Foundation for Health Research/ Crohns' and Colitis Foundation of Canada Trainee Award and is a recipient of a postdoctoral fellowship from the Canadian Institutes for Health Research (CIHR). EMC is supported by operating grants from the CIHR and the Canada Foundations for Innovation (CFI). He holds a CSL Behring Research Chair and a Tier 1 Canada Research Chair in Endothelial Cell Biology, is an adjunct Scientist with the Canadian Blood Services, and is a member of the University of British Columbia Life Sciences Institute.

CHAPTER 2

Competing Interests
The authors declare to have no competing interests.

Author Contributions

MA planned and performed experiments. AB performed experiments. RS provided essential contributions to fluorescent activated cell sorting. IR and KF designed and supervised the study. MA, KF and IR wrote the manuscript. All authors read and approved the final manuscript.

Acknowledgments

This work was supported by the 'Novartis-Stiftung für Therapeutische Forschung'. We also would like to thank Constance E. Brinckerhoff for providing the MMP-1-reporter construct.

CHAPTER 3

Competing Interests

The authors declare that they have no conflicts of interest.

Author Contributions

BR and CLA conceived and designed experiments. BR and SC performed experiments and analyzed data. KD contributed the SNapShot assay and acquired and interpreted the SNaPshot data. BR and CLA drafted the manuscript. All authors read and approved the final version of the manuscript.

Acknowledgments

This work was supported by the following: R01 grant CA80195 (CLA), ACS Clinical Research Professorship Grant CRP-07-234 (CLA), Breast Cancer Specialized Program of Research Excellence (SPORE) P50 CA98131, and Vanderbilt-Ingram Cancer Center Support Grant P30 CA68485; DOD Breast Cancer Research Program post-doctoral award BC087465 (BNR) and NCI K08 CA143153 (BNR). The breast cancer SNaPshot screen was performed in the Vanderbilt Innovative Translational Research Shared Resource supported by the Vanderbilt-Ingram Cancer Center and the TJ Martell Foundation. Technical assistance was provided by Donald Hucks, MS.

CHAPTER 4

Competing Interests
The authors declare that they have no competing interests.

Author Contributions
MAL suggested the idea and designed the all research process, took part in acquision of clinical data, analyzed & interpreted of all the data, finally drafted & revised the manuscript. JHP performed the experimental, analyzed the data and reviewed the manuscript. SYR was in charge of collecting all the clinical data, analyed and interpreted the data. STO & WKK supplied all the tissue specimen, collected and interpreted data, and reviewed & commented the manuscript. HNK interpreted the pathological findings, immunohistochemical staining, reviewed & commented the manuscript. All authors read and approved the final manuscript.

Acknowledgments
This research was supported by the financial support of the Catholic Medical Center Research Foundation made in the program year of 2010 and Seoul St. Mary's Clinical Medicine Research Program year of 2009–2011 through the Catholic University of Korea.

CHAPTER 5

Competing Interest
The author declares that they have no competing interest.

Author Contributions
AZ and SH participated in the design of this study, and they both carried out the study. XS collected important background information, together with LD, and performed the statistical analysis. XB and NW participated in the design and helped to write the manuscript. All authors read and approved the final manuscript.

Acknowledgments

This work was supported by grant from the Science & Technology Bureau of Changzhou (No.CJ20130029).

CHAPTER 6

Competing Interests

The authors declare that they have no competing interests.

Author Contributions

MQ, WB, and JW carried out the design of the experiments, performed most of experiments, and drafted the manuscript. TY and XH participated in the molecular biology experiments and statistical analysis. YL made the figures. XW was involved in financial support, the design of the experiments, data analysis, and final approval of the manuscript. All authors read and approved the final manuscript.

Acknowledgments

This work was supported by National Natural Science Foundation of China (No. 81072139, No.81172476) and the Young Scientific Research Project of Shanghai Municipal Health Bureau (No.20124Y045). We thank Dr. Yuyang Zhao for providing us with the plasmid PWP1/GFP/Neo-AR and its negative control PWP1/GFP/Neo (Department of Urology, Shanghai First People's Hospital Affiliated to Shanghai Jiao Tong University School of Medicine, Shanghai, China). We also thank Dr. Xin Luo, who implemented apparatus and reagent management in the laboratory.

CHAPTER 7

Competing Interests

The authors declare that they have no competing interests.

Author Contributions

The study was conceived by JL, DG, YZ and CK. Experiments were carried out by JL, ZZ and CK. Statistical analysis was carried out by JL. Manuscript was written by JL. All authors read and approved the final manuscript.

Acknowledgments

This research was funded by the National Natural Science Foundation of China (No. 81172438).

CHAPTER 8

Competing Interests

The authors declare that they have no competing interests.

Author Contributions

YJ and SYC drafted the manuscript. All the authors have read and approved the final manuscript.

Acknowledgments

This work was supported in part by grants from the National Nature Science Foundation of China (Grants 31201095, 81071903 and 81072069) and West China Hospital of Sichuan University. No institution was involved in data interpretation, writing the article, or the decision to submit the paper for publication. The authors are indebted to all reviewers for their kindly reviewing of the manuscript.

CHAPTER 9

Competing Interests

The authors declare that they have no competing interests.

Author Contributions

MTB conceived the study. LE, TH, and EL did genomic characterization of human PDA samples. MA provided analysis of cell line expression data. NT and HY performed RNAi studies. MK did protein expression analysis of cell lines. The PDA mouse model studies were done by PP-M, and DAL. Construction, scoring, and analysis of TMA were done by DA, TK, PR, and CP. MTB, RR and DDVH reviewed all patient-based data. PP-M, CP, and MTB wrote the paper. All authors read and approved the final manuscript.

Acknowledgments

We thank Drs Alistair G Rust, David J Adams, and David Tuveson for helpful discussion and critical review. This work was supported by the American Association for Cancer Research/SU2C Pancreatic Dream Team, NCI PO1 (5P01CA109552-04) 'Targets to Therapeutics in Pancreatic Cancer', and NCI 1R21CA137687 'High Definition Clonal Analyses of Archival Pancreatic Adenocarcinoma Samples'.

CHAPTER 10

Competing Interests

CAL receives consulting income from Arno Therapeutics, Inc. This interest has been reviewed and managed by the University of Minnesota in accordance with its Conflict of Interest policies. CRH declares that she has no competing interests.

Author Contributions

CRH and CAL together led the initial design and conception of the manuscript. CRH led the writing of the first and all subsequent drafts of the manuscript. CAL contributed significant written and editorial inputs to the manuscript at every stage. Both authors read and approved the final manuscript.

Author Information

CAL joined the University of Minnesota (Departments of Medicine and Pharmacology) faculty in 1999. Her research is focused on steroid hormone action in breast cancer progression. Her laboratory studies the role of cross-talk between growth factor-mediated signaling pathways and steroid hormone receptors, using the human progesterone receptor as a model receptor. CAL holds the Tickle Family Land Grant Endowed Chair of Breast Cancer Research at the University of Minnesota. She is the Director of The Cancer Biology Training Grant (T32) and the Cell Signaling Program Lead within the Masonic Cancer Center. CAL is Editor-in-Chief of the journal Hormones and Cancer (jointly held by The Endocrine Society and Springer). CRH is a senior post-doctoral fellow in the laboratory of CAL.

Acknowledgments

The authors would like to thank Dr. Andrea R. Daniel (Minnesota) for her critical review of this manuscript, and Michael Freeman (Minnesota) for editorial assistance.

CHAPTER 11

Competing Interests

The authors declare that they have no competing interests.

Author Contributions

GGM, JHT, SAM and JPI conceived and designed the experiments. GGM, JHT, TR, SP, MRE, JJ, NJR, HL, TT, AT and PR performed the experiments. GGM, YL, XYZ and SL analyzed the data. GGM, JHT, SAM and JPI wrote the paper. All authors read and approved the final manuscript.

Author Information

GGM and JHT are first co-authors. SAM and JPI are senior co-authors.

Acknowledgments

This work was supported in part by an MD Anderson Research Trust Fellow Award, funded by the George and Barbara Bush Endowment for Innovative Cancer Research (SAM) and DOD Breast Cancer Research Postdoctoral Fellowship (JHT). GGM was also funded by the following non-for-profit organizations: Fondation Nadine Midy, Foundation Nelia et Amadeo Barletta (FNAB) and the Association pour l'Aide à 'la Recherche et l'Enseignement en Cancérologie (AAREC).

CHAPTER 12

Competing Interests

The authors declare that they have no conflict of interest.

Author Contributions

CR, ZV: designed the paper, carried out analysis and interpretation of data and wrote the manuscript. IT: conducted statistical analysis. VT: conducted data analysis, carried out data interpretation. WG: critically revised the manuscript. All authors read and approved the final manuscript.

CHAPTER 13

Competing Interests

Xiulan Su holds a patent on ACBP-L (Chinese national patent: ZL96122236.0 and international patent: A61K35/28).

Author Contributions

Conceived and designed the experiments: XL S. Performed the experiments: C D, LY S, XM W and HW C. Analyzed the data: JL Z, LY S and Z C. Drafted the manuscript: LY S, JL Z, XL S and Z C. All authors read and approved the final manuscript.

Acknowledgments

This work was supported by the National Natural Science Foundation of China [No. 30860327, China], Major Project of the Affiliated Hospital of Inner Mongolia Medical College [No. ZD9809, China], and NIH/NIDCD intramural research projects [No.Z01-DC-000016] for Zhong Chen. We thank Dr. Ke Yang from Beijing University Medical Health Center for helpful discussions and providing cell lines. We also thank Cindy Clark (NIH library) for helpful editing the manuscript.

CHAPTER 14

Funding

Funding assistance from the National Institutes of Health to KA (grant 5K01CA95620) and HW (grant R15 CA122001) and from the State of Florida to HW (SURECAG grant R94007) to carry out this work is gratefully acknowledged. The funders had no role in study design, data collection and analysis, decision to publish, or preparation of the manuscript.

Competing Interests

The authors have declared that no competing interests exist.

Acknowledgments

The authors express their thanks to David Brunell for help in determining the combination indices and to Edna Gamliel for assistance with the cell culture.

Author Contributions

Conceived and designed the experiments: KA SK KDS HW. Performed the experiments: KA SK. Analyzed the data: KA SK HW. Contributed reagents/materials/analysis tools: KDS. Wrote the paper: KA SK HW.

INDEX

Milton Keynes UK
Ingram Content Group UK Ltd.
UKHW022057141024
449569UK00031B/1665